The First Century

A History of the 28th Infantry Division

Prepared under the direction
of
Colonel Uzal W. Ent

Robert Grant Crist
EDITOR

This book was produced and technical assistance provided by Stackpole Books, Harrisburg, PA 17105.

Library of Congress Cataloging in Publication Data
Main entry under title:

The First century.

Includes index.
1. Pennsylvania. National Guard. 28th Infantry Division—History. I. Ent, Uzal W., 1927-
II. Crist, Robert Grant.
UA423.5 28th.F57 355.3'7'0948 79-22275
ISBN 0-8117-0620-6

DEDICATION

To the men of the Pennsylvania National Guard whose century this was and who so ably performed the deeds recorded here. They set the example of foresight, wisdom, dedication, sacrifice and honorable service to their communities, Commonwealth and nation.

May we who now serve and those who will wear the uniform of the 28th Infantry Division in the future be inspired to preserve and perpetuate a noble trust.

Contents

List of Maps

Foreword

REMARKABLE AS IT may seem, no one has ever published a history of America's oldest Army division, the 28th Infantry, Pennsylvania National Guard. Only short monographs dealing with parts of the Division or with short periods of its history have seen print. This book, therefore, is a "first."

It is appropriate that the omission be corrected finally in the year of the Divisional centenary. That which follows is something less than what would have appeared had money for publication been abundant and more time available. Its authors are a small group of men pressed into duty early in 1979 who used spare time after business hours to make the book possible. It was a labor of love but one which love was not supposed to blind. With the triumphs and commendations are defeats and criticisms.

The director and editor both agreed that the story of the 28th should include within space limitations the origins of its constituent elements. Accordingly, the period covered is virtually all of the nearly three centuries of Penn's province.

This book, like so many others, permits a reader to infer that the more things seem to change the more they resemble something that has already transpired. Examples are the term "Army Advisors" and the acronym "ACSAT" (Active Component Support at Annual Training). Both are relatively new. The "Army Advisors," of course, are Regular Army officers and non-commissioned officers assigned to counsel and assist the commands of the Army National Guard. ACSAT personnel are small contingents of company or battalion size forces from the Regular Army which are detailed to assist during the annual field training periods. 1979 is the second year of ACSAT. However, the concepts behind the Advisors and ACSAT were conceived and tried with success almost 100 years ago in Pennsylvania. In similar fashion, the recruiting program currently underway duplicates in almost all particulars a promotional effort proposed by an officer of the Indiana National Guard in 1913.

In the pages that follow, Benjamin Franklin's casual amateurs become determined professionals in the art of artillery and farmers with fowling pieces evolve into men with BAR's and bazookas. What do not change are the long hours required for training, the prolonged separation from hearth and home, and the camaraderie that distinguishes the volunteer, citizen soldier.

If this book helps develop a reasoned pride in readers for the citizen-soldier, it will have been worth it all.

Acknowledgments

Many people contributed to the development of this book. At the risk of neglecting to mention everyone, I at least will make an attempt here to name and thank those who did most to make this publication possible.

DONORS

The P. H. Glatfelter Company of Spring Grove, Pennsylvania, generously donated one half of the paper used to publish this volume.

ITT Terryphone offices in Harrisburg spearheaded a drive for funds, resulting in donations of $4,400.00. In addition to a generous contribution by ITT, the following firms contributed to the fund:

Brockway Glass Co. Foundation
Florig Equipment Co., Conshohocken
Oliver B. Cannon & Sons, Inc., Philadelphia
National Guard Association of Pennsylvania
Northern Central Bank, Williamsport
Pennsylvania Power & Light Co.
Pittsburgh Forging Co., Pittsburgh
The York Water Co., York
Regimental Corporation, First Regt of Infantry, Philadelphia

Over 500 people and organizations joined with certain units of the Division in Sponsorship and Friendship programs, raising $47,000.00. Contributors to these two programs are rightfully listed at the beginning of the book.

CONTRIBUTING AUTHORS

A number of dedicated and hard-working students of history, some scholars of note in various fields, contributed chapters to this book. Each has a by-line for the chapter he contributed. I thank each of them.

CONTRIBUTORS OF MATERIAL AND PICTURES

Virtually every separate unit and battalion of the Pennsylvania Army National Guard contributed copies of their lineages and other historic material. These proved most valuable. Of particular value, however, was the manuscript of a history of the Division which the author, Mr. Raymond G. Carpenter, generously loaned to me as we worked on the present volume.

Pictures came from many sources. These included previous regimental and Division histories as well as pictures from the following sources: 1st Regiment Infantry Museum (103d Engr Bn), J. Craig Nannos, Director, First Troop, Philadelphia City Cavalry Museum, Henry D. Lavinos, Curator; Company of Military Historians, "Military Uniforms in America"; Second Penna. Regiment, Continental Line, Inc. (photos by Jack Swern); National Archives, Wash., D.C.; Military Photograph Archives, Pentagon; The J. Craig Nannos Collection. Archives, Penna. Dept. of Military Affairs.

PICTURE PROCESSING

Over 5,000 pictures were researched and examined in making the selection of about 200 which finally went into the book. Both the 28th Infantry Division Public Affairs Section and the 109th Public Affairs Detachment assisted in selecting and processing the pictures. However, special thanks are due to MAJ Robert M. Fisher, CPT Gary J. Milgate, MSG Marcus J. Fretz, SP6 Oliver E. Wagner of the 28th, CPT Frank P. Brennan, Jr and SSG Peter C. Wilcox of the 109th, Major Clinton Tennill, CPT J. Craig Nannos and CW3 Arthur H. Kramer, all of the Department of Military Affairs.

WATER COLORS

Mr. Donald Troianni of Stanford, Connecticut painted the truly outstanding water colors which are contained in the center of the book.

BOOK COVER

The keystone and soldiers depicted on the cover of this book were conceived by SGT William T. Wilfong III, HHC, 1-111 Infantry as the logo for the 28th Infantry Division's Centennial celebration. The idea was then artistically rendered by SSG Jay W. Bushong, II, HHC, 28th Infantry DISCOM. This logo was fittingly chosen to illustrate the cover of the book.

EDITOR

I gratefully acknowledge the excellent work and advice rendered by our editor, Mr. Robert Grant Crist. His expert touch is manifest throughout the book.

Many other people helped in this project. COL Richard E. Thorn, Mrs. Marguerite E. Foultz and CW2 Kenneth C. Kelsey all deserve recognition.

Readers should note that an annotated copy of the book's manuscript, indicating sources used, is on file in Headquarters, 28th Infantry Division.

One last note. Many people contributed much to the completion of this book. Because of constraints of time and money, the final book could not include many things which otherwise would have been included. I accept responsibility for the decisions which were made concerning these matters.

COL Uzal W. Ent

FRIENDSHIP ROSTER

John B. Allard
276th Army Band
John P. Athey
Stanley N. Ayers
Cornelius E. Batta
MAJ William H. Bepler
MAJ & Mrs. William M. Bozzelli
Tom Brotherwood
SSG & Mrs. Jerome F. Buchner
Harold J. Camish
CPT E. E. Campbell
Francis J. Carey
1SG & Mrs. John D. Carlantonio
G. Robert Carr
Russell T. Carvin
Glenn W. Clark, Jr.
James H. Clement
Jack Colbaugh
CH(MAJ) Calvin H. Cole
MAJ & Mrs. Kenneth M. Cox
William H. Dixon
SGT George W. Dobson
SGM George Duvak (Ret)
LTC & Mrs. John W. Ent
William A. & Charlotte R. Ent
James V. Ferguson
John P. Forr
W. Charles Gallagher
Peter A. Garafola
Dominic C. Guisto
Leonard Gold
In Memory of CW3 Marion O. Gorby, by
SFC & Mrs. Wayne L. Gorby & Family
H. W. Haas, Jr.
Leo E. Hafele
Ellsworth L. Hall
BG William E. Hall
Mr. & Mrs. Darryl R. Hamm
Leon H. Hancock
Betty R. Hann
Robert J. Hartlieb
BG Charles S. Hendricks
CW3 Clinton L. Herbster
William E. Hobbs
Ivan H. Hollabaugh
Mr. & Mrs. Homer A. Hummel
2LT Joyce R. Hunt
Mr. & Mrs. Frank J. Kanes
SGT Frederick Jo Keller, Jr.
William Kirk

Arnold Knicker
George T. Knouse
Raymond H. Lesmer
Alfred W. & Patricia Lysinger
MAJ Elwood F. Mapes (Ret)
Albert R. Marsico
CW3 Nicholas J. Mascuilli
George F. Mills
CSM John P. Moran
BG W. A. Morgan
COL William V. Munhall
John C. McIlvaine
Officers & Gentlemen of First Troop
Phila City Cavalry, Trp A, 1-104 Cav
Officers & Men of 1-111th Inf
Officers & Men of HHC, 2-111th Inf
Officers & Men of Co A, 2-111th Inf
Officers & Men of Co B, 2-111th Inf
Officers & Men of Co. C, 2-111th Inf
Officers & Men of CSC, 2-111th Inf
CSM Richard S. Orr
CPT & Mrs. James C. Pakala
2LT Barry D. Parsons
F. R. Plumb
Robert W. Pocklington
Lawrence Pugh
SFC & Mrs. John W. Richards
PSG Victor M Ricker III
SFC Ulysses Rodriguez
SFC William E. Santlarsci
SGT John R. Schott
Jean Schweig
MG Richard M. Scott
CW4 George H. Seely
LTC Robert L. Shoeman
Mary R. Silvis
Mr. Charles L. Smith
MAJ & Mrs. David E. Smith
PSG John C. Spangler
Frederick H. Stafford
LTC Bruno C. Terlizzi
1LT Charles D. Thomas
CPT Paul D. Trembeth
COL C. Uhler
John Wallace
CPT Raymond R. Wallace
BG Charles E. Widmer (Ret)
SSG William T. Wilfong III
Alfred C. Young
COL Christopher A. Young
Francis W. Zabicki
LTC Stephen J. Zubach

ROSTER OF SPONSORS

CPT William G. Aboud
BG Hall F. Achenbach
MAJ & Mrs. John V. Alfandri
COL William H. Allen (Ret)
MAJ & Mrs. Robert E. Altier
CW3 George W. Ames
Paul Andruzzi
MAJ Landry Kyper & Carol K. Appleby
COL Charles B. Armagost
COL Robert A. Armstrong
MAJ Heinrich N. Babb
BG James H. Bair
BG Cornelius O. Baker
CPT William M. Barnes
CPT & Mrs. Robert Barziloski
MAJ Charles E. Bearinger
CSM Dale J. Beaston
CPT Samuel F. Bell
SSG Dennis J. Benson
CSM Robert R. Berkey
MAJ Donald R. Berry
COL Jack C. Betson
CPT Luis E. Biamon
MAJ John E. Biddle, Jr.
BG Albert J. Blair (Ret)
Blakely Borough Council
CHAP Herbert R. Blouch
Blue Mountain Chapter AUSA
William W. Bodine, Jr.
CHAP Donald M. Bohn
CPT D. Guy Bolton
MG Fletcher C. Booker, Jr.
MAJ Terrill D. Boring
CSM Albert A. Borsello
Henry E. & Mary M. Bower, Jr.
LTC Allen N. Boyer
James J. Boyle
MAJ William F. Bozzelli
Carl E. Brandt
BG Albert G. Branyan
MAJ James R. Brehm
LTC George E. Breslin
MAJ & Mrs. Clarence A. Bricker
CHAP (MAJ) Earl M. Brooks
LTC Henry P. Brown
COL Paul K. Brown
LTC Thomas I. Brunton
CPT & Mrs. Joseph A. Burdenski
LTC Frank F. Burgess (Ret)
BG John F. Burk, Jr.

Robert F. Buchanan
MSG & Mrs. Robert L. Bucher
Mr. & Mrs. Charles Campbell
COL Elton M. Cannon MD
Raymond G. Carpenter
G. Robert Carr
LTC Robert D. Carroll
BG Robert M. Carroll
Paul E. Carson
LTC & Mrs. William D. Carter
Ralph Cassel
MAJ Paul J. Catalano
PSG & Mrs. Charles L. Certato
LTC William E. Chamberlain
SFC Herman W. Clemens
LTC Lawrence J. Cochran
SFC & Mrs. Allen R. Collins, Jr.
Mr. & Mrs. Thomas E. Collins, Jr.
Commander, Spt Co, 2-112th Inf
Commanding Officer, 56th Bde
Company A, 28th Avn Bn
Company A, (-Det 1) 2-112th Inf
Company B, (-Det 1) 2-112th Inf
Company B, 228th S&T Bn
CPT & Mrs. William G. Confer & Family
LTC & Mrs. E. J. Conway
MAJ & Mrs. Robert J. Conway
LTC John F. Costello, Jr.
CSM Frank Covello
CPT Robert P. Coyne
COL Kenneth R. Craig
SGM & Mrs. Robert H. Crossley
Lt Paul Cunningham
MAJ Richard A. Daddona, Jr.
MAJ Francis J. Damico
SGT & Mrs. George A. Davis
LTC Austin F. Deller
Daniel J. Delmastro
Det 1, Co B, 2-112th Inf
Mr. & Mrs. Albert DeWolf
CPT Daniel A. DiBruno
LTC Francis W. Ditzler
COL Armand DiVincenzo
Albert W. Dogan
Donald J. Donaghy
William A. Dougherty
COL Henry E. Dreher
MG William H. Duncan
LTC James L. Dunkelberger
MAJ Terry L. Eckert
CPT James W. Eckley III
CPT & Mrs. Paul F. Edgar, Jr.
Palmer D. Edmunds
Enlisted Personnel, HHC, 28th DISCOM
Lowell Kent Fesmire
1SG Norman Feust
MAJ Donald Ficco
COL George B. Flanagan (Ret)
MG Henry K. Fluck
Wilbur G. Fox
Harold J. Frady
Mr. & Mrs. Paul B. Frederick
David E. Fritchey
COL Earl Fuller, Jr. (Ret)
William J. & Suzanne Gallagher
Charles A. Garvey
BG Robert M. Gaynor, DSC
SGM Robert W. Geiger, Sr. (Ret)
Louis R. Gentile
MSG James Gillespie
BG William Buchanan Gold, Jr.
CPT & Mrs. Thomas E. Gordon
Stanley Louis Gorski, II
BG Merrill W. Goss
CPT Joseph A. Gottwald, Jr.
Peter J. Graybash
COL Donald W. Griffin (Ret)
MSG James Franklin Griffith
COL Robert A. Guenthner
SSG & Mrs. Raymond C. Hackney
MAJ William F. Haenn
SGM Edward H. Hall
BG Laurence L. Hamacher (Ret) and Francis
CW4 Joseph M. Harrison
MG Joseph A. Healey
LTC Charles Heisterkamp III, MD
2LT Thomas C. Helm
COL Francis C. Hennesy
Raymond F. Henry
LTC Joseph E. Herkness
CPT & Mrs. Richard S. Hess
MAJ William O. Hickok
BG Daniel W. Hildreth III
COL & Mrs. Alfred G. Hill (Ret)
BG Edward A. Hitchin
LTC & Mrs. Joseph M. Hodgson
LTC Raymond H. Hoffman (Ret) &
Anna R. Hoffman
SGM Fred A. Houdenshield, Jr.
CPT Norman K. Howard

COL Edward M. Huber
COL William Huber
LTC Charles G. Huch
F. M. Huffman
William W. Huffman
Hq Btry, 28th DivArty
COL Frank T. Huray (Ret)
George H. Imhof
MAJ Robert L. Irvin
MAJ David E. Irving
George Iyoob, Jr.
LTC John Jackson (Ret)
CPT Richard C. Jacobs, Jr.
COL Vernon E. James
CPT George W. S. Johnson, Jr.
COL Francis E. Jones, Jr.
LTC Theodore M. Kaczmarczyk
Francis J. Kafka
MG Nicholas P. Kafkalas
Fred W. Kaiser
CPT Harold K. Kaler
BG Moe Katz (Ret)
COL Lawrence A. Keller
MAJ Nicholas R. Keller
SGM & Mrs. Kenneth C. Kelsey
MAJ George E. Kemp
CPT Joel D. Kenderdine
CPT George R. Kenny
PSG Kenneth P. Keogh
CPT Alexander Kerr
MAJ & Mrs. Allen L. Kifer
Harry Kimmel
Knelly-Podany Post #616, American Legion
LTC Thomas A. Knobloch
MAJ Tommy Klutts, Jr.
Frederick L. Koch, Sr.
Gerald L. Kochel
CSM Michael F. Komarny
COL Frank G. Koronkiewicz
LTC Raymond M. Kowalski
1LT Nancy E. Kozak
Charles G. & Helen S. Kraus
LTC & Mrs. William F. Kuba
COL Albert G. Kuhn
LTC Gethin J. Kurtz
Mr. & Mrs. Chester J. Kuzminski
SGM Carl S. Kwaczek
CSM Salvatore M. Laganelli
CPT Marcel F. Lamour
Leopold Larkins, Jr.
Robert E. Laskowski
BG Harold J. Lavell
LTC George R. Law
SFC Michael D. Leonard
COL Brooke Montgomery Lessig
LTC William M. Lickfield
MAJ Charles Edward Liken
BG W. Armin Linn (Ret)
LTC John T. Loftus
George W. Lynch
MAJ James F. Mahoney
SFC T. J. Manchak, Jr.
MSG William H. Markley
MAJ John L. Martin
Mr. R. G. Martin
SGT Vazquez O. Martinez
28th Div Materiel Management Center
Robert May
1SG & Mrs. Charles L. Mayer (Ret)
Men of Company E, 28th Avn Bn
Men of the Mess, Hq Btry, 28th DivArty
SP4 Russell D. Menkins
COL Richard D. Merion
Clarence H. Miller
Dennis Paul Miller
Paul H. Miller
BG Rodman D. Miller (Ret)
Sen. William J. Moore
Dr. (LTC) & Mrs. Edward R. Morasco
Mr. Thomas Morrow
Mount Pleasant Vol. Fire Dept.
LTC Ralph F. Mueller
W. Stanley Mumford
COL George J. Mumma
CPT J. H. Myers
MAJ James W. MacVay
CPT Charles D. McCall
John F. McCann
CPT Maurice J. McCarthy III
Frank H. McClelland
MAJ & Mrs. Charles F. McClintock, Jr.
BG John E. McDonald
MAJ & Mrs. James J. McGill
SSG Michael T. McGinnis
BG Leo T. McMahon (Ret)
LTC & Mrs. Robert R. McQue, Sr.
CPT J. Craig Nannos
National Guard Assoc. of the US
NCO Club, HHC, 1-111th Inf
Carl W. Nelson

LTC John A. Newns
COL Steve C. Nicholas
Rep. & Mrs. Fred C. Noye
Old Forge Veterans Club
2LT Thomas E. Oldham
MAJ David E. Pace
COL & Mrs. Albert G. Palmateer, Jr.
MAJ Edward W. Parks
MAJ Andrew B. Pastorek, Jr.
Pennsylvania American Legion
MSGT & Mrs. Arthur E. Peoples
COL & Mrs. Percival E. Pereira
BG & Mrs. Robert E. Peters
Eli P. Plaskow
COL Peter Platten
Albert Donald Powell
WO1 & Mrs. John E. Prendergast
CPT Albert G. Price III
COL Joseph J. Prusaitis, Jr.
1SG Eugene E. Raudenbush
LTC Elton D. Reep
CPT & Mrs. Joseph D. Reeves
COL David E. Reiber, MC
LTG John Remetta
Guilleomo A. Reyes
LTC & Mrs. Edward A. Reynolds
MAJ & Mrs. John L. Rhodes, Jr.
MG George J. Richards (Ret)
LTC Robert J. Richards (Ret)
Clarence N. Richwine
Mr. & Mrs. James H. Roden
COL David E. Rodgers
BG Daniel Rogers
LTC Stanley W. Roots, Jr.
BG & Mrs. J. C. Rosborough
Andrew L. Ruffatt
BG John C. Ruggaber
COL Ralph J. Ruscello
MAJ & Mrs. Paul M. Russo
MAJ Albert Rutherford
COL Edward A. Sahli
BG & Mrs. Gerald T. Sajer
LTC James B. Samuels
Colonel William P. Sandy
MAJ Jay Sarajian
LT Michael P. Sariano & Son Sean
LTC James J. Scally
Samuel C. & Esther M. Schaffer, Jr.
CPT William J. Schaming
Frank Scherle
PSG Robert R. Schleifer
Bernard E. Schmick
1SG James T. Schmidt
George W. Schuler
SP5 William Schurter
BG Thomas W. Scott, Jr. (Ret)
Mr. & Mrs. William A. Seeds
MAJ Luke L. Shade
MAJ Miles J. Shaffer, Jr.
COL Thomas Sharpe
PSG Gary F. Shoemaker
LTC Ernest M. Shott
Mark G. Shultz
HQ, 28th Signal Bn
CW2 & Mrs. Orlando A. Signorini
Slovak American Beneficial & Sportsman Club
1LT Lawrence E. Smarr
COL Duane R. Smith
SSG Stephen M. Smith
BG & Mrs. Frank H. Smoker, Jr.
Francis G. Smysor
MAJ Leroy D. Snelbecker
CPT Burton H. Snyder
MSG & Mrs. Nevlin K. Snyder
LTG Richard Snyder (Ret)
1SG & Mrs. Richard N. Snyder
CPT Dealvia Spafford
CPT Philip W. Spence
MAJ Nelson W. Spengler, Jr.
LTC Gilbert R. Steele, Jr.
COL Severino Stefanon
LTC & Mrs. John F. Stever
MAJ Walter L. Stewart, Jr.
LTG Daniel B. Strickler
MAJ & Mrs. James B. Stodart, Jr.
SFC Brian K. & Gail E. Stoner
MAJ & Mrs. James T. Sullivan
Anthony G. Swano
Staff SGT Peter Szpak & Mrs. Cnythia L. Szpak
CPT Thomas L. Taylor
228th S&T Bn
COL Edward S. Terpien
CW4 George Robert Thomas
COL Tyssul G. Thomas (Ret)
LTC Ronald L. Thompson
SP4 Rosa L. Thompson
COL Richard E. Thorn
MG Frank Townend
MSG John A. Triszczuk
LTC William Troxell

BG Hans W. Tschudin
CSM Charles B. Tuttle, Sr.
LTC Peter Uram
US Army Readiness Group, Indiantown Gap
Mr. & Mrs. James A. Vail
COL Sam Valella (Ret)
1LT Gary F. Valinski
Arthur Varney
Clyde O. Vaughn
Mr. & Mrs. Robert S. Voshelle
CPT James L. Walsh
Mr. & Mrs. Roy Waltz (Ina)
SP4 Irvin D. Warner
Charles A. Waychoff
LTC William K. Weaver, Jr.
Earl F. Webb
BG Walter D. Weikel
BG & Mrs. John J. Weinzettle
WO1 Richard E. Wenrich
LTC Herbert C. F. Werner
CPT Michael D. Werner
Proctor Wetherill
MSG William L. Wheeler
LTC Howard K. White
MG Aylwyn P. Williams (Ret)
Gary L. Williams
Thomas M. Williams
SSG Russell S. Williamson, Jr.
BG & Mrs. Paul K. Winner (Ret)
1LT John Roy Wise II
Wayne A. Witherow
SSG Robert L. Witty
CPT Norman L. Wolfe
CW4 James J. Wolford
Wolo Off Assoc. 228th S&T Bn
MAJ Edward P. Woods
MSG Earl H. Woomer
Timothy J. Wright
LTC Jacob Yablonsky
MAJ Donald L. Zechman
MAJ Paul L. Zeigler
CPT Harold L. Zendt
CW4 Shuman W. Zimmerman
COL & Mrs. Harold Zuber (Ret)

CHAPTER 1

The Colonial Militia

Russell F. Weigley, Ph.D.

IF THE FOUNDER of Pennsylvania had been able to control the Commonwealth's destiny, there would have been no military history of Pennsylvania and no Pennsylvania National Guard division. William Penn extended his Holy Experiment not least as an experiment in applying the principle of nonviolence cherished by the Society of Friends. Penn was determined to save Pennsylvania from war through Christian dealing with all of the colony's neighbors. His vision of his Commonwealth was appropriately depicted when the Quaker painter Edward Hicks repeatedly portrayed Penn in his treaty-making with the Indians as standing in a setting suggesting Isaiah's "Peaceable Kingdom" of "the wolf [who] also shall lie down with the lamb." Props were lions, cows and other friendly beasts all congregating together in peace. Yet the history of colonial Pennsylvania eventually departed far from the vision of the Peaceable Kingdom. By the late colonial era, Quaker Pennsylvania, like the other British colonies in North America, possessed a military organization out of which would grow the later National Guard. Moreover, it was employing the organization in war.

There was from the beginning an ambivalence about Penn's pacifism. In his capacity as proprietor of a province, as distinguished from his role as private individual with a conscience, Penn was well aware that in an imperfect world governments have to deal forcefully with their neighbors and that indeed all government is founded ultimately on force. He knew that his government would have to recognize these facts, his private convictions notwithstanding. Certainly King Charles II, who had lived through the turbulence of Cromwell's commonwealth, when he granted Penn the proprietary charter for Pennsylvania was thoroughly aware of the role of force in politics. The King fully intended that his colony of Pennsylvania should be protected by force whenever necessary. The Pennsylvania charter of March 4, 1681, conferred upon the Proprietor authority "to do all and every other Thing which unto the Charge and Office of a Captain-General of an Army belongeth, or hath accustomed to belong," which customarily included power to command troops, commission military officers, and define the duties of the military force.

In accordance with English tradition and with the various frames of government that Penn from time to time issued for Pennsylvania, the executive authority could command armies, but the legislature had to appropriate funds to finance them. While William Penn was willing enough to have non-Quakers as Deputy Governors meet Pennsylvania's military obligations in his behalf, in part so he could assure himself of retaining his colony, other members of the Society of Friends felt less pressure to compromise the principle of nonviolence. Such other members of the Society of Friends dominated the early Councils and Assemblies, the legislative bodies, and they would not readily involve themselves in any support of military action whatever.

Penn's proprietorship included the Three Lower Counties, the present state of Delaware. The settlements there, largely non-Quaker, felt especially exposed to possible French maritime raiding and asked the provincial government to create a militia. Every other English colony in America had a compulsory-service militia, requiring every able-bodied freeman of appropriate age to bear arms and to undergo military training. The Pennsylvania government refused the petition.

The provincial government also continued to refuse repeated orders from the Crown to grant military appropriations. When the Crown endorsed New York's request that Pennsylvania supply eighty armed men or a sum of money equivalent to their expenses for the defense of Albany, and the Quaker colony still did nothing, the Lords of Trade and Plantations recommended to the King that the Pennsylvania charter be vacated. The Crown responded by appointing a soldier, Colonel Benjamin Fletcher, as military commander of all provinces from Massachusetts Bay to Maryland and governor of a unified New York and Pennsylvania. Fletcher arrived at Philadelphia in April 1693 and vigorously set out to disallow all antimilitary laws and to form a new assembly that would cooperate in imperial defense. The most he could accomplish was to pry from the Pennsylvania Assembly an act of November 7, 1696 to grant £300, as Markham told Fletcher, "for ye supply & relief of those Indians of the five Nations yt are in friendship with ye English with Necessaries of food & raymt."

The War of the Spanish Succession erupted in 1702, a conflict known in the American colonies as Queen Anne's War. It precipitated the organization of the first Pennsylvania military force, the first forerunner of the National Guard, but in circumstances that could only aggravate Quaker distrust of military preparations and hinder subsequent efforts toward military defense.

Queen Anne's War took on an especially dangerous cast for Pennsylvania with the news early in 1702 that New France had negotiated a treaty with the Iroquois. The Iroquois, the powerful Five Nations, were suzerains over most of the Indians of eastern Pennsylvania. The Five Nations' military strength was such that as Logan wrote to Penn, who was back in England: "If we lose the Iroquois, we are gone by land." In November 1702, Deputy Governor Andrew Hamilton asked the Assembly to establish a militia. Knowing the sentiments of the Assembly, Hamilton also tried to recruit a militia company on his own authority, sending drummers through Philadelphia and plying prospective recruits with rum. By now there was a growing Church of England population and influence in Philadelphia, but the Anglicans refrained from cooperating with Hamilton's militia plan in hopes of using the military issue to bring down the Quaker proprietorship. Nothing still visible by the light of history came of the drum-beating and drinking.

Hamilton died in 1703 with the defense question at the usual impasse. While William Penn's enemies were toasting the likely revocation of the Pennsylvania charter over the military issue—literally in the midst of such a tavern scene—the next deputy governor, John Evans, arrived in Philadelphia on February 2, 1704, with specific authorization from Penn to raise a militia. The next time that New York issued its habitual annual call for defense assistance Evans pressed the Pennsylvania Provincial Assembly hard for military action. When it refused, Evans, like his predecessor, acted on his own authority—but with clearer sanction from Penn—to recruit his own militia, called the Governor's Guard. Evans soon claimed to have ten companies. He had been accompanied to Pennsylvania by the Proprietor's son, also named William; and Billy Penn, who was in his late twenties, renewed his family's on-again, off-again martial tradition by drilling the Governor's Guard on Society Hill in Philadelphia. Pennsylvania had its first military force.

The force did not amount to much. The enemies of the Penns scoffed that it mustered not Evans's claimed ten companies but only some forty bedraggled men, and this report was surely much closer to the truth than Evans's.

The Assembly complained that Evans's "Militia were drawn up in a military posture to awe the Electors at Philadelphia, in the year 1704," but Evans was painfully aware of the true capacity of the Governor's Guard to awe nobody. On May 14, 1704, a courier raced into Philadelphia bearing a letter addressed from Governor John Seymour of Maryland to Evans, with the exciting news that a French fleet lay off the mid-Atlantic coast. Evans quickly convened the Council and secured the agreement of its non-Quaker members to call Pennsylvania to arms. On each of the next two days an additional similar report of French naval activity arrived, May 16 bringing the sheriff of New Castle County crying that six French ships were already in the Delaware and that they had bombarded Lewes into ruins. Evans scurried through an excited, even panic-stricken Philadelphia, urging all able-bodied men to muster on Society Hill. A number of Quakers were suspicious of the whole business, however, thinking that if the French were so close there ought to be yet more reports and refugees from the south. James Logan suggested that he and Evans go down the Delaware together to reconnoiter. Evans refused. Logan went by himself and soon confirmed from a shallop sailing upstream that nothing out of the ordinary had occurred in the Three Lower Counties. Evans had manufactured a war scare to raise a militia. The effect of his absurd method was to discredit a militia more than ever before.

The escapade helped hasten Evans's deserved recall. Through the rest of Queen Anne's War Pennsylvania's military activity reverted to the routine of calls for assistance from the Crown and other colonies, met by the Quaker Assembly with noncooperation or, when the Crown pressed diligently, to a grudging vote for minimal appropriations of funds in terms carefully eschewing any direct avowal of warlike purposes. In 1709 the Crown asked Pennsylvania for 150 men as part of a 1,500-man force from this colony, New Jersey, New York, and Connecticut. The Assembly offered £500. In response to similar calls in 1711 the Assembly voted £2,000 "*for the Queen's use*," but the war ended in 1713 before the money was collected.

War returned to the British Empire and its American colonies in 1739, when one Captain Jenkins displayed in England with grisly relish an ear that he said had been cut off his head by Spanish coast guards. A long-standing English resentment of Spain's restrictions upon foreign commerce with Spanish America thereupon erupted into the "War of Jenkins' Ear" against Spain. This conflict merged in 1744 into a renewal of the struggle for imperial supremacy between Great Britain and France, called King George's War, the American phase of the War of the Austrian Succession. By this time, Pennsylvania had become a well-populated colony, reaching far up the Schuylkill, west through Lancaster, and beyond the Susquehanna to Shippensburg. The population of 100,000 by 1740 was diverse as well as large and fast-growing, German and Scotch-Irish as well as English and Welsh. The Friends had become a minority, and, while the German sects were also opponents of violence, the majority of Pennsylvanians no longer felt bound by religious scruples against bearing arms. William Penn's heirs were only nominally Quaker or had turned Anglican and were not pacifists. The Friends, nevertheless, retained disproportionate social and economic influence as well as control of the Assembly. The stage was set for a complex struggle to resolve the issues of defense and military organization in Pennsylvania.

In 1740 the British government called on the American colonies for 3,500 soldiers to serve as an American regiment in the British establishment for operations against the Spanish in the Caribbean. The plan was for the colonial assemblies to feed the troops and transport them to the theater of war, where British command would take over. A quota of eight companies was assigned to Pennsylvania. Governor George Thomas, who was the more sympathetic to this design because he was a planter from Antigua, set out to enlist the companies on his executive authority. He met an unexpectedly hearty response, seven hundred men under arms. Too many of them for the satisfaction of the province's property holders, however, were indentured servants recruited with the promise of freedom from their indentures. By resorting to such an expedient, Thomas doubly offended an Assembly disdainful of military enterprises. The legislators retaliated by voting a peculiar defense money bill, providing that no funds should be released until all the indentured servants in the governor's companies were returned to their owners, and until recruiting officers gave their guarantees to enroll no more servants. The Assembly went on to prosecute the recruiting officers for depriving masters of their property without due process of law.

Infuriated, Governor Thomas advised the British government that Quakers must be excluded from office, or Pennsylvania would remain indefensible. Hearing of this opinion, the Assembly denounced the Governor as an enemy of the people. James Logan, by now an elder statesman, tried to persuade the Friends' Yearly Meeting that if they insisted on strict adherence to the principle of nonviolence they should voluntarily abandon participation in politics: because government is necessarily founded on force, "Friends as such in the strictness of their principles ought in no manner to engage in it." The Governor's activities had too much heated the political atmosphere for Logan's plea to enjoy even the remote chance of acceptance that it might have had in quieter times.

Indeed, in September 1740 Governor Thomas's eight companies had embarked at Philadelphia and New Castle for Jamaica, there to rendezvous with the rest of the Caribbean expedition. From March 9 to April 11, 1741, the British and colonial force laid siege to Cartagena, Spain's principal seaport on the northern coast of South America. Unhappily, the siege failed, as so did a bloody assault. The besiegers, including the Pennsylvanians, suffered heavily from yellow fever and other scourges of the tropics. The failure owed much to mismanagement of almost every conceivable sort: tactical, strategic, logistical, and hygienic—notwithstanding the circumstance, a captain from Virginia, Lawrence Washington, chose afterward to name his Potomac River estate in honor of the naval commander of the Cartagena expedition, Admiral Edward Vernon. Clearly, there was nothing in this affair to dispose Pennsylvanians more favorably toward waging war.

Yet James Logan's efforts to convince the Quakers they must face up to the problem of war may not have been altogether without effect. The pressure of a discontented Crown presently brought the Assembly to vote £3,000 for the King's use, as well as to end the furor over the governor's military companies by simply compensating masters for the loss of their servants to the military. After King George's War broke out, the Assembly voted £4,000 for beef, pork, flour, wheat, and "other Grain." The printer and rising political leader Benjamin Franklin persuaded Governor Thomas that "other Grain" could mean grains of gunpowder, and Thomas used the money to buy munitions for the colonial expedition against the fortress of Louisbourg, a far more fortunate venture than the one to Cartagena. In 1746 the Assembly actually voted to equip the four Pennsylvania companies of 100 men each which had been requested by the Crown as part of two colonial corps. Governor Thomas recruited the companies during the summer, and they left Philadelphia on September 4 to help defend Albany. Their service, however, was short; they were discharged October 31.

At the same time the Assembly remained adamantly opposed to any general militia law that might provide an ongoing military force and a permanent defense for the province. This refusal of more than temporary compromises of the Quaker conscience gnawed at the fears of such men as Logan and Franklin, who thought that amidst imperial and Indian wars, Pennsylvania's own boundaries could not go unscathed forever. In July 1747, fears approached abrupt realization. A new report of hostile vessels in the Delaware was no mere repetition of John Evans's hoax. A French privateering sloop, manned largely by Spaniards, captured a ship from Antigua, murdering its captain, and landed marauders on two plantations just below New Castle.

The Assembly met in August and, as usual, took no action, contending that the Delaware River was long and the danger already past. Three more French privateers promptly showed up in Delaware Bay, however, and rumor soon had it that at least six French ships from the West Indies intended to rendezvous during the coming year to plunder Philadelphia itself. Leading non-Quaker citizens of the city agreed that Franklin ought to employ his literary talents to compose a plea for self-defense. On November 17, 1747, there appeared in consequence the pamphlet *Plain Truth; or, Serious Considerations of the Present State of the City of Philadelphia and Province of*

Pennsylvania. By a Tradesman of Philadelphia. The anonymous tradesman warned that the danger was not only from the sea but from the Indian frontier as well, and he argued that "the sentiments of true Pennsylvanians, of fellow-countrymen," could not rest content when either seaboard towns or western farm settlements were threatened. All Pennsylvanians should unite in the common defense. Because the Quaker leadership of the province would not show the way to defense, and the great merchants apparently would not finance projects that the Quakers did not support, the "middling people" of Pennsylvania must take the initiative themselves, "the farmers, shopkeepers, and tradesmen." " 'Tis computed that we have at least (exclusive of the Quakers) 60,000 fighting men, acquainted with firearms, many of them hunters and marksmen, hardy and bold."

On November 21 Franklin met with about a hundred Philadelphia tradesmen, many of whom were ready to sign his draft of articles of military "Association" at once. Franklin preferred to wait until he received also, two days later, the approval of "a great meeting of the principal gentlemen, merchants and others." The next night the mass signing-up of volunteers commenced, according to one of the more modest estimates gaining 300 signatures in three days. Franklin printed the "Form of Association" in his *Pennsylvania Gazette* for December 3, to spread the military activity throughout the province. Companies of fifty to 100 volunteer Associators were to elect their own captains, lieutenants, and ensigns, one of each of these commissioned ranks to a company. The companies of a county were to form one or more regiments, the company officers electing a colonel, a lieutenant colonel, and a major. By December 6, some 600 Philadelphia Associators met at the State House on Chestnut Street, marched to the Philadelphia Court House on High (Market) Street at Second, and divided themselves into companies according to wards and townships. Eventually as many as 10,000 Associators may have signed the muster rolls throughout the province.

Pennsylvania thus gained its first military force of armed and organized citizens in formidable numbers. But the Association was not quite an official military force representing the government of the province. Through Richard Peters, the secretary of the province, Franklin kept the Penns in England apprised of his military plans and activities, and the Proprietors were not expected to object. No governor was present on the scene, Thomas having succumbed to poor health and gone home. Anthony Palmer, president of the provincial Council, informed the Association on December 7 that their activities were "not disapprov'd" and that he would commission the officers they elected. Notwithstanding the extralegal nature of the Association, there is considerable merit in the judgment of the writers of the standard history of Philadelphia that: "The epoch of William Penn's empire ceases with Nov. 21, 1747," the date of Franklin's meeting with the tradesmen to commence founding the Association.

Because it was the defense of the Delaware that posed the immediate need for the Association, the military leaders promptly looked to the planting of fortified batteries along the river. Franklin; William Allen, who would soon be Chief Justice of the province; Thomas Hopkinson, first president of the American Philosophical Society; Edward Shippen and Charles Willing of prominent mercantile families; and others organized a lottery to raise funds to buy cannons. The Associators were able to purchase thirty-nine guns from Boston and to order more from London. To tide Pennsylvania over, Franklin and several other commissioners called on Governor George Clinton of New York early in 1748 to request a loan of guns. "He at first refus'd us peremptorily," according to Franklin; "but at dinner with his council, where there was great drinking of Madeira wine, as the custom of that place then was, he softened by degrees, and said he would lend us six. After a few more bumpers he advanc'd to ten; and at length he very good-naturedly conceded eighteen. They were fine cannon, eighteen-pounders, with their carriages... ."

In April the Association mounted its first battery at the foot of Philadelphia's Society Hill, which was then still a real hill. By June a larger battery called "the Association Battery" rose below Old Swedes Church, where the Philadelphia Navy Yard would be in the nineteenth century. By August this battery had twenty-seven guns. In November 1750 thirteen more were added to make a total by that time of more than fifty; the latest were a gift from the Penns, in tangible expression of their approval of self-defense.

Meanwhile, those Associators who adhered to their terms of enlistment trained and drilled weekly. But no hostile ships or soldiers ever tested the Pennsylvania Association of 1747-1748. French privateers appeared again off Cape Henlopen, and a Spanish brigantine approached New Castle in the early summer of 1748, to be driven off by guns mounted there. By that time, however, Great Britain had made peace with France and Spain through the Treaty of Aix-la-Chapelle of April 19, 1748. The news of peace reached Philadelphia on August 24, and military activities for the most part ceased.

The Treaty of Aix-la-Chapelle could not offer more than a brief truce in the colonial rivalry of France and Great Britain for predominance in North America. In the middle of the eighteenth century both powers were pushing this rivalry toward a showdown, and contending ambitions converged particularly upon a strategic junction of rivers claimed by Pennsylvania: the Forks of the Ohio. The struggle for the site of Pittsburgh was destined to assure the termination of the nonviolent era of William Penn and to turn his province finally to an acknowledgment of force as the foundation of government and of arms as the ultimate reliance for the security of the state in a tumultuous world.

The Proprietors' contribution of guns to the Association Battery for the last war had not yet reached Philadelphia when the complaints of French agents in the territory drained by the Ohio River set in motion the events leading to a new, and climactic, war. English fur traders and land buyers, said the agents of Louis XV, were suborning the loyalties of the Indians along the Ohio and threatening France's claims to the river that linked her outposts on the St. Lawrence and the Mississippi. The Marquis de Duquesne, Governor General of New France, dispatched an expedition that in 1753 built Fort Presqu'isle at present-day Erie and Forts Le Boeuf and Machault where the towns of Waterford and Franklin now stand, as the beginning of a chain of fortifications between Canada and the Ohio River.

The Pennsylvania Assembly, still dominated by the Quakers, refused to heed the urgings of Thomas Penn, son of the founder, that it take steps to defend the claims of the Proprietors to lands beyond the Appalachian Mountains. The most the Assembly would do was to support with gifts a conference at Carlisle called by the province's principal Indian agent, George Croghan, to try to hold the western Indians to friendship with the English. Virginia, however, also claimed the area of the Forks of the Ohio, and this colony responded more energetically to the French threat. In December 1753 Major George Washington of the Virginia militia carried to Forts Machault and Le Boeuf Governor Robert Dinwiddie's demand for French withdrawal. When the French refused, Dinwiddie promptly sent a militia company to build a Virginia fort at the Forks. A force of 500 French with Indian allies arrived from Fort Le Boeuf in April to drive the Virginians from their uncompleted stockade. On the site the French built a more formidable structure called Fort Duquesne. Dinwiddie dispatched militia to try to recapture the Forks, but on the 4th of July Washington, now a lieutenant colonel, had to surrender the stockade he had erected at the Great Meadows along the way and abandon the effort, because he had collided with superior French strength.

Thus without the help of Pennsylvania the energies of Virginia did not suffice to turn back French encroachment. London thereupon decided that the British Empire in America required British regular soldiers. In 1755 the government dispatched to Virginia Major General Edward Braddock with the 44th and 48th Regiments of Foot. Braddock was to recruit these regiments to war strength in America, command all British military forces in North America, and with militia supporting his regulars, eject the French from the Forks and from any other footholds inside British claims.

Early in 1755, the Pennsylvania Assembly went so far as to vote £20,000 for the King's use in the new war, to be raised by issuing paper money redeemable in ten years. The Governor of Pennsylvania, now Robert Hunter Morris, attempted to reduce the redemption period to five years, for fear that excessive reliance on paper money might anger the British authorities and thereby threaten another forfeiture of the Pennsylvania proprietorship, which he was pledged to the Penns to protect. As usual, there were movements under way to transform Pennsylvania into a royal colony. The effect of the Governor's action was to touch off a new squabble between the Proprietors and their Governor on the one hand and the Assembly on the other, which again practically nullified any official military action by Pennsylvania. The Friends hoped to exploit military emergencies to give them the leverage to maintain and even enhance their political power against the Proprietors and Governors. As the most recent survey of Pennsylvania history observes of the comfortable, well-to-do Quaker leadership of the middle eighteenth century: "It has often been stated that the Quakers stayed aloof from the French and Indian War because of their pacifism, but more likely the preservation of their political power was their main interest."

The Quakers were given to understand that General Braddock nourished a strong prejudice against them, and the Assembly was not above sending the ubiquitous Franklin to meet with the general at Frederick, Maryland, and attempt to disabuse him of his prejudice. Franklin discovered that Braddock was hard put to gather an adequate quantity of wagons to support his proposed expedition against Fort Duquesne—he had wrested only twenty-five from Maryland—and the Philadelphian was able to win a certain share of favor for Pennsylvania by contracting for 150 wagons with four horses each and 259 pack horses in Lancaster, York, and Cumberland Counties. At Franklin's urging the Pennsylvania Assembly curried further favor by sending off to every British officer with Braddock a horse laden with such items as a good Gloucester cheese, two well-cured hams, two gallons of Jamaica spirits, and two dozen bottles of old Madeira.

Such was the extent of Pennsylvania's contribution to the Braddock expedition. It was not enough to ward off the famous disaster near the crossing of the Monongahela on July 9, 1755, following which the Indians all along the Pennsylvania frontier felt the wind blowing in France's favor and rose up to reclaim old lands now held by the Pennsylvania whites. The Delawares and Shawnees sent two especially troublesome war parties raiding eastward in the early autumn, one ranging from Bedford toward Carlisle, the other crossing the Susquehanna to harry settlements along the Blue Mountains almost as far as the Tulpehocken area of northwestern Berks County.

On July 20, soon after he heard the news of Braddock's defeat, Governor Morris informed the Assembly that as an encouragement to Pennsylvania to help her sister colonies expel the French from the Ohio, the Proprietors were willing to offer 600,000 to 700,000 acres of good, well located land, valued at about £90,000 Pennsylvania currency, for distribution among those who would take up arms, 200 acres to go to every private soldier, larger amounts to officers. The Assembly responded with a bill to appropriate £50,000—to be raised by a tax on all estates, including proprietary lands; the Assembly knew the Governor was forbidden by the Penns to approve the taxation of proprietary lands. The defense issue seemingly could not be disentangled from the struggle for control of Pennsylvania politics. Though Indian raids within a hundred miles of Philadelphia provoked even some of the Mennonites to arm themselves, the Pennsylvania metropolis and its immediate environs remained safe and calm, and the Assembly persisted in refusing to act.

Morris thereupon encouraged a new Association on the pattern of 1747, and men throughout the province, especially in the counties immediately under the Indian threat, signed the muster rolls for new volunteer companies. The Governor or his agents again commissioned officers chosen by election, or sometimes they issued direct commissions.

Before the new Association could do much to safeguard the frontier, the back country at last, and understandably, lost patience with the Assembly's self-righteous professions of conscientious scruple against arms. A new Assembly elected in October 1755 and meeting on November 3 offered little sign of abandoning old attitudes. From among the usually stolid Pennsylvania Germans a mob then rose up in late November to gather at the Womelsdorf home of Conrad Weiser, another provincial Indian agent and himself a German, and to demand that Weiser lead them to a confrontation with the Assembly. On November 25, the "Dutch mob" marched into the Assembly hall of the State House to demand defense legislation forthwith.

Franklin had allied himself with the Assembly leadership in various assaults upon proprietary power, but now he warned the Quakers that time had finally run out for their old habits. They must either follow all the other colonies by authorizing a provincial militia or lose the political power for which they had battled so long and so deviously. At the same time, however, as an inveterate conciliator and mender of political fences, Franklin also offered a militia bill so scrupulously protective of the right of conscientious objectors not to give military service that the product would be not a militia based on military service as a general obligation of citizenship as in the other colonies, but in practice a volunteer force. Franklin's militia would differ from the Association in little except authorization by the Assembly. Under the German farmers' anger and Franklin's cajolements, the Quakers split. While a group of "principled" Friends led by Israel Pemberton abstained, enough other Friends followed Isaac Norris in joining the Assemblymen from the West to pass a militia law.

The Militia Act of November 25, 1755, was followed quickly by a Supply Act of November 27, appropriating £60,000 for various "heavy charges for the King's use," thus still referring to military measures by circumlocution. Nevertheless, Pennsylvania at length possessed as close an approximation of the militia forces of the other British colonies as it was destined to get during the colonial era. In this sense the militia force founded upon the Militia Act of 1755 is the most direct colonial ancestor of the Pennsylvania National Guard and the 28th Division.

But the Quaker influence still otherwise limited defense to thin gruel. Not only was militia service voluntary. In addition, the men were not to be taken more than three days' march beyond the settlements and not to be kept more than three weeks in garrison. The officers continued to be elected. The Assembly voted no equivalent of the British Mutiny Act to regulate discipline, and the obedience and order of the militia were uncertain.

Nor could mere legislation subdue Indian raids. During the night before passage of the Militia Act a war party of Shawnees had attacked a settlement of white and Indian Moravians at Gnaddenhuetten, on the site of present Lehighton, burning the town and murdering everybody who did not manage to run away into the forest. The Shawnees continued to prey upon the Moravian settlements above Bethlehem through most of December. Governor Morris proposed a chain of defenses along the line of the Blue Mountains from Easton to Mercersburg, with sizable stockades able to hold a militia company and many families to be erected at twenty-mile intervals and smaller blockhouses for temporary shelter to rise at five-mile intervals.

The Supply Act provided for a committee of seven to disburse the defense funds, and Franklin chaired the committee. On December 18 Franklin set out with James Hamilton, a former Governor and a member of the provincial Council and friend of the current Governor, and Joseph Fox, the Quaker chairman of the Assembly committee on accounts, to travel the Blue Mountain line and hasten the completion and manning of the defenses. The rotund, unmilitary Franklin received the unofficial title of "general" from some of the frightened settlers and refugees whom he visited; he labored conscientiously enough that in part at least he merited the honorific. But the committee system of managing funds was an expression of the persistent distrust between Governor and Assembly, and a series of committees were destined to go on disbursing defense funds as long as the province voted such funds, without the continuity of a permanent administrative agency, and to the detriment of sound management and prompt payment of accounts.

The first companies formed to garrison the frontier posts mustered from thirty to seventy-five men, with the smaller ones carrying a captain and a lieutenant as officers and the larger ones adding an ensign. By the spring of 1756 companies were for the most part reorganized to a uniform size of three officers and fifty men. A force to serve through a year's campaign now evolved with the enlistment of men at six dollars a month plus arms, blanket, and rations: a step away from a short-service militia toward a more stable provincial military force. On April 15, moreover, the Assembly passed a law modeled on the then current version of the British Mutiny Act, so the troops could be subjected to military discipline. With 1,400 officers and men on the rolls the province formed its companies into three battalions and eventually into the two-battalion Pennsylvania Regiment and the single-battalion Augusta Regiment, named for the fort at the junction of the North and West Branches of the Susquehanna. Colonel William Clapham of the Augusta Regiment and Lieutenant Colonels Conrad Weiser and John Armstrong of the 1st and 2nd Battalions, respectively, of the Pennsylvania Regiment commanded, under the Governor as commander in chief. In accordance with the military custom of the times, the regimental and battalion officers each retained command of a company, with the second officer of that company designated captain lieutenant and exercising the day-to-day, working command.

By February 1756 the chain of forts had been substantially completed along the Blue Mountain line and extended up the Susquehanna to Fort Augusta as well as up the Juniata to Fort Granville, near where Lewistown now stands. The 1st Battalion of the Pennsylvania Regiment garrisoned the posts east of the Susquehanna, the 2nd Battalion most of those west of the Susquehanna, while the Augusta Regiment headquartered at the fort whose name it bore.

With the Assembly still hesitant to vote military measures unless almost bludgeoned by the populace and still feuding over other issues with the Governor and the Proprietors, Pennsylvania would not and could not accept a call from Major General William Shirley, Governor of Massachusetts Bay and Braddock's successor as Commander in Chief in North America, to join with neighboring colonies for a new expedition against Fort Duquesne in 1756. The neighboring colonies did not respond favorably either. Governor Morris and several prominent non-Quaker citizens of Pennsylvania had appealed to Parliament, however, to put a final end to Quaker obstructionism. After considerable deliberation, the British government proposed to sponsor a bill in Parliament that would compel all provincial officeholders and legislators to swear to an oath—thus forcing every practicing Quaker to depart from active politics. The Society of Friends in England reacted by offering to persuade the Pennsylvania Friends to resign from the Assembly, if Parliament would refrain from erecting such an obstacle that would bar Quakers

The Pennsylvania Regiment 1756-1758 (Courtesy The Company of Military Historians).

from office even in peaceful times. The bargain was struck, and in consequence the elections of the autumn of 1756 returned only eight pacifist Quakers out of thirty-six members of the Pennsylvania Assembly. This shift of power permitted passage of somewhat stronger military laws than the Militia Act of 1755, a circumstance fortunate for defense because on July 7, 1756, the King had disallowed the 1755 law on the ground that it was "rather calculated to exempt Persons from Militia Service than to encourage and prompt them." Since the royal veto was not known in Pennsylvania until October 15, it had little practical effect; by that time the Militia Act of 1755 was due to expire in fifteen days anyway, and the political climate was changing.

Meanwhile the mere passive defense of the Pennsylvania cordon of stockades had proven as unsuccessful as passive defenses usually are in accomplishing more than a blunting of hostile attacks. Lieutenant Colonel Armstrong, substantial citizen of Carlisle and man of influence through the western part of the province, proposed in the summer to lead a raid against the Delaware Indian town of Kittanning on the Allegheny River, the source of many of the raiding parties. Soon after, Armstrong learned that his brother, Lieutenant Edward Armstrong, had been killed while resisting an attack on Fort Granville led by a Delaware known as Captain Jacobs, who operated from Kittanning.

The Governor approved Armstrong's plan. On July 30, the colonel led about 300 volunteers from all the companies of his 2nd Battalion out of Fort Shirley in Cumberland County (present Shirleysburg, Huntingdon County). Following the Frankstown Traders' Path to the west, Armstrong neared Kittanning on September 7 and sent forward a scouting party. Guided over the final mile or so less by the scouts than by the beating of the drums in the town, he attacked at dawn the next morning, achieved surprise, killed about forty Indians including Captain Jacobs, drove away the rest, pursued as far as he felt he safely could (there were reports of French and Indian reinforcements approaching), freed a number of prisoners, burned the town, and destroyed an impressive stock of ammunition. The Delawares moved their principal village west to

The Royal American Regiment (Courtesy The Company of Military Historians).

Logstown on the Ohio, where Ambridge stands today. They could thus be shielded by Fort Duquesne, but their ability to harass the Pennsylvania frontier was proportionately reduced. John Armstrong's raid on Kittanning was the first notable offensive military achievement of the province of Pennsylvania. Unhappily the moral effect was much undercut by news that on August 14 the French had captured Oswego on Lake Ontario along with several regiments of colonial troops.

By November, the Indians thus once more encouraged were not only harrying western Cumberland County, killing and capturing settlers and militiamen in the area where Colonel Armstrong had built an outpost called Fort Loudoun, but had again followed the war path as far east as Berks County, to the vicinity of Fort Lebanon. The new Commander-in-Chief in North America, Lord Loudoun, for whom Armstrong had named the fort, felt obliged to bolster the Pennsylvania defenses with a battalion of regulars, part of the 62nd (later the 60th) Regiment, the Royal Americans.

The Indian raids eased off during the winter of 1756-1757, largely because the French withdrew all but a skeleton garrison from Fort Duquesne. In the summer, however, with the French back in strength at the Forks to encourage marauding, neither the impact of the Kittanning expedition nor the presence of regular soldiers prevented the Indians from scourging the frontier again all the way to the Tulpehocken settlements. Pennsylvania enlisted enough men to retain its three battalions in its forts, but the new Governor, Lieutenant Colonel William Denny, a regular army officer, felt he lacked the resources for anything better than a return to a cordon defense. Several local counterattacks were largely the work of Colonel Armstrong in the area of Cumberland County, abetted by Royal Americans based at Carlisle.

In fact, British forces were on the defensive almost everywhere throughout America. Their loss of Fort William Henry in July was as severe a reversal and as much an encouragement to the French and Indians as the fall of Oswego the preceding year. Before the end of 1757, however, William Pitt became George II's First Minister, and Pitt moved with prompt energy to turn the course of the war. By April 1758, Brigadier General John Forbes had arrived in Philadelphia with over a thousand men of Montgomery's Highland Regiment. With these troops plus some 350 of the 1st Battalion of the Royal Americans under Lieutenant Colonel Henry Bouquet, nearly 350 of the Independent Companies of South Carolina (also regulars), forty artillerymen of the Royal Regiment of Artillery and a train of guns, and, it was hoped, some 5,000 militia from various provinces, Forbes was to mount an expedition against Fort Duquesne.

To contribute Pennsylvania's proper share to this venture, Governor Denny proposed to raise the military force of the province to 2,700 men. Pennsylvania reorganized its troops into two enlarged battalions, of which the 2nd Battalion of the Pennsylvania Regiment and the Augusta Regiment formed the foundations. The reformed Pennsylvania Assembly proved cooperative enough to appropriate £100,000 for the King's use, making the desired force of 2,700 possible.

The vanguard of Forbes's regulars, commanded by Colonel Bouquet as Forbes's deputy, rendezvoused at Carlisle on May 24 with the Pennsylvania battalions, commanded by Lieutenant Colonels John Armstrong and James Burd. Forbes had decided to march west to the Forks not by Braddock's route but along the Raystown Path from Carlisle to Raystown (now Bedford) and from there on a road he would build. Unlike Braddock, who had tried to leap to the Forks in a continuous march, Forbes would carefully build stockaded supply depots along the route, so that in case of a reverse he need not retreat all the way back to the main settlements. Bouquet believed that the Pennsylvania troops should lead the way out of Carlisle because they knew the country. A Swiss veteran of European wars, Bouquet found the Pennsylvanians shockingly undisciplined and ill-equipped, and efforts to arm them better contributed to delaying the departure from Carlisle. On June 24, however, Bouquet led the advance of the Pennsylvanians and some Virginians into Raystown, where he began building Fort Bedford. The Virginians refused to labor upon the fort without extra pay, but the Pennsylvanians worked, and Bouquet rewarded them with a gill of rum for every man every day.

By September 3 Colonel Burd was in the van and had reached Loyalhanna, where he commenced building an entrenched camp that later became Fort Ligonier. But Burd had nearly outrun his supplies and was soon down to one day's provisions, so that Forbes appealed to Governor Denny for Pennsylvania to furnish additional wagons and teams beyond the more than 100 already assembled, in June. On August 20 the now-compliant Pennsylvania Assembly voted inducements to citizens who would offer four good horses and a wagon and would agree to carry a load of at least 1,400 pounds, in addition to forage for the horses, from Lancaster to Raystown. By October Forbes himself was in Raystown and commenting adversely on the rascality of Pennsylvanians, but his quartermaster general was soon praising the great trains of wagons on their way from Lancaster.

The expedition had suffered a setback in September when Major James Grant of the regulars with an advance force of regulars and provincials, including about 100 Pennsylvanians, allowed himself to be enveloped on the last high hill before Fort Duquesne. The Pennsylvanians, scenting danger, hastily departed before the completion of the envelopment and the opening of Grant's main battle. At least they could serve another day.

Forbes's slow depot-building march nevertheless had a quality of inexorability. The French at Fort Duquesne were confident enough through the summer and into the autumn, but they gradually came to realize that they lacked the means to halt permanently the patiently determined Scots general and his unprecedentedly powerful and well supplied British column. The Indian allies of the French came to the same realization, and they grew receptive to Forbes's diplomatic efforts to persuade them to shift their allegiance. In late November, John Armstrong with his Pennsylvania battalion and Colonel George Washington with his comparable force of Virginians took the van and led the army from Chestnut Ridge across Bushy Run to Turtle Creek. After dark on November 24 most of Forbes's army gathered on the hills around Turtle Creek. Then the sound of a great explosion echoed from the direction of the Forks. The French had despaired; the next evening the Forbes expedition marched into the abandoned ruins of Fort Duquesne.

Forbes was unwell, and he returned to Philadelphia to die there in the spring of 1759. Brigadier General John Stanwix succeeded him and began building the new Fort Pitt at the Forks of the Ohio. Though the retreat of the French to Canada deprived the Indian raids of their mainspring, enough marauding bands still prowled the forests that Pennsylvania kept its enlarged force of 2,700, or at least that large a table of organization strength, in the field for the next two years.

As part of the price of wooing the Delawares and the Shawnees away from their French allegiance while Forbes pushed westward, George Croghan and Conrad Weiser had agreed in 1758 to the Treaty of Easton, whereby Pennsylvania retroceded to the Indians a considerable tract in the western part of the province drained by the Ohio River and its tributaries. Soon, however, the relative peace brought to the frontier by the French withdrawal, and enforced by regular and Pennsylvania troops, tempted increasing numbers of white settlers to take up land in this very area. The Treaty of Paris of 1763 removed France permanently from North America, but the new white settlements upon territory the Indians regarded as theirs both by rights and as compensation for their raising the British standard helped touch off yet another war of the Pennsylvania frontier, the 1763 uprising associated with the name Pontiac, an Ottawa.

Burying their own differences in fear and resentment of the Anglo-Americans, the Indians of the Northwest swept up small garrisons throughout the trans-Appalachian country, including Forts Presqu'isle, Le Boeuf, and Venango—the former French Port Machault—in Pennsylvania. The hostiles laid siege to Forts Pitt, Ligonier, and Bedford, guarded by small British garrisons and crowded with refugees, while even Fort Augusta seemed precariously held. By June war parties again prowled as far east as the Cumberland Valley.

The provincial army that Pennsylvania kept on active duty through the latter seasons of campaigning against the French had been demobilized in 1761. That year the Assembly authorized an active force of only 300 to garrison the far western posts until November and thirty men for Fort Augusta, which retained its garrison continuously for several years thereafter. When Pontiac's uprising set the frontier aflame, the committee for the disbursement of military funds acted on its own initiative because the Assembly was not sitting, and on June 11, 1763, it ordered the enlistment of 100 men to reinforce Augusta. The Assembly promptly returned to session and on July 6 voted to restore the active militia to a total of sixteen fifty-man companies. Several companies of Associators once more took up arms.

Some soldiers of the Pennsylvania army joined the few companies of the Black Watch and Montgomery's Highlanders which were led west from Carlisle by Colonel Bouquet on July 19. The total force amounted to only about 500 men. Arriving at Bedford on July 25, Bouquet there picked up thirty additional men of the region as guides. But in the desperate battle of Bushy Run fought on August 5-6, where for a time it appeared Bouquet might suffer a repetition of Braddock's disaster, it was the disciplined maneuverability of the Highland companies that permitted first the feigned retreat and then the bayonet charges that turned the tide against the uprising.

Bouquet relieved Fort Pitt, but his small force could not alone complete or assure the pacification of the frontier. Continuing Indian raids precipitated the retaliatory attacks of the "Paxton Boys" against reservation Indians in December 1763. The Pennsylvania Assembly therefore voted a provincial army of over 1,000 for 1764, of which 250 would garrison forts and the rest campaign with the British. Sporadic raids persisted to mar the new year, but Bouquet also campaigned westward to overawe the Indians in their own towns. By December 5, 1764, Governor John Penn, grandson of the founder, felt able to declare warfare at an end in Pennsylvania. The provincial army was again demobilized, except for a garrison retained at Fort Augusta until June 13, 1765.

A posse comitatus of about 500 men raised from militia companies helped assert Pennsylvania's claims in the Wyoming Valley against Connecticut during the Second Yankee-Pennamite "War" of 1773-1775. During 1774 the Pennsylvania Assembly supported a force of some 100 volunteer rangers in Westmoreland County for several months to guard Pennsylvania's interests against both Indians and Virginia in the war waged by Governor Lord Dunmore of the latter colony upon Indians of the Ohio River area. But the history of active campaigning for the colonial Pennsylvania militia, the pre-Revolutionary forerunners of the Commonwealth's National Guard, is concentrated in the years 1755-1764, from the yielding of the Assembly's Quakers to frontier demands for protection after Braddock's defeat, to the suppression of Pontiac's uprising.

Until those few years of intense military activity the military history of Pennsylvania had been markedly different from that of any other of the British colonies in North America, with their militia forces based on compulsory service by nearly all able-bodied males of appropriate age. Derived from medieval England, the principle of a universal obligation to military service had for a time taken on new life in America, while it was expiring in the mother country. From the French and Indian War onward, however, the history of Pennsylvania's citizen-soldiers merges into the mainstream of the history of such soldiers throughout what would become the United States. By that time, population had swelled enough, and the frontier had receded enough from the centers of population, in all the colonies, that the compulsory service of all militiamen was rarely invoked, even in a single area of a colony. In all the colonies the men who served in active campaigns had come to be, for the most part, those who volunteered to fill the quotas assigned to their counties by the provincial military authorities. Throughout the British colonies, including Pennsylvania, the volunteer militiaman had emerged as the prototype of the volunteer National Guardsman of later days.

CHAPTER 2

Revolution to Whiskey Rebellion

John B. B. Trussell, Colonel, USA Retired

THE APPROACH OF the Revolutionary War found Pennsylvania with a precedent for military organization but, alone among the British colonies in America, with no military units actually in being. The province had never had a militia in the strict sense of the term (that is, an organization in which enrollment was mandatory, with short-term active duty, at least on a selective basis, being compulsory). Instead, at times of crisis it had employed a system of voluntary "military *associations*," differing from conventional militia in the one important respect that membership was not required by law.

The growth of controversy and tensions leading to the convening of the First Continental Congress also brought steps toward the creation of some sort of military capability when, on June 4, 1774, a meeting at Hanover Township in Lancaster (now Dauphin) County adopted a number of resolutions. One of these was to organize on a military basis to protect the people's rights. There is no conclusive evidence that any organization was actually formed, but the fact that the resolution prescribed a flag for such a unit implies, although it admittedly does not prove, that the unit in fact came into being. Fully documented, by contrast, is the creation on November 17, 1774, of the Philadelphia Troop of Light Horse (later redesignated as the First Troop Philadelphia City Cavalry), consisting of twenty-eight men. By December a company had been organized in York County, with at least one more in existence in the same county by mid-February, 1775.

The trend indicated by these and corresponding actions throughout Pennsylvania was given increased impetus by news of actual hostilities in Massachusetts in April 1775. For example, on May 22 the Northampton County Committee of Safety passed resolutions, one of which urged all free men of the county to arm themselves and muster for military training, and in Westmoreland County, a battalion was formed on May 24, with John Proctor, a prominent local figure, elected as its colonel.

The Provincial Assembly on June 30 formally legitimized these and the similar organizations that had come spontaneously into being throughout the province: it passed a resolution approving them and recommending that all counties that had not yet done so provide weapons, ammunition, and equipment for their use. On August 19, a set of standard rules—the "Articles of Association"—for their administration and discipline was approved. Within the next few months, officers had been

Photograph of the Original Color of the First Troop, Philadelphia City Cavalry.

commissioned for a total of fifty-three authorized Associator battalions.

Notwithstanding the response suggested by the number of these units, Pennsylvanians were by no means unanimous in supporting military efforts against the British government. Some were pacifists on religious grounds; many sought to remain neutral; and many others remained at least covertly loyal to the Crown. The county Committees of Safety represented the more radical political faction, and it was to them that local control and administration of the Associator units was assigned. In many cases they did not hesitate to use this authority to intimidate and even to suppress such political opposition as developed. The numerical strength of the Associator units, moreover, represented a substantial—and *organized*—political power base, enabling its members to demand a number of reforms in recognition of their commitment to serve. Among these was a more equitable geographical apportionment of representation in the Assembly, which was still largely dominated by the southeastern counties—the same counties where pacifist and loyalist sentiment tended to be concentrated. In brief, the Associators thereby were a factor in bringing about the changes embodied in the radical new state Constitution put into effect in September 1776. They also used their power to insist on steps to penalize men who refused to enlist in local companies. As early as November 1775 the Assembly was prevailed upon to impose a fine of two pounds, ten shillings on these so-called "Non-Associators." In May 1776 all Non-Associators were required to surrender any weapons in their possession. To avoid such penalties many men whose zeal for the Revolution was at best tepid began to enlist, with the eventual result that the Association became only nominally voluntary, being diluted with half-hearted "patriots" to the

Associators of the City and Liberties of Philadelphia, 1775. The 111th Infantry and 103d Engineers trace their lineages to these Associators. (Courtesy The Company of Military Historians).

detriment and eventual breakdown of the system. Before that happened, however, the Associators did furnish the basis for a substantial military contribution—providing Pennsylvania's portion of what was called the "Flying Camp."

That organization came into being after Congress, in June 1776 asked Pennsylvania, New Jersey, Maryland, and Delaware to provide a total of 10,000 militia for a special, temporary pool of unit reinforcements, to be enlisted until the following November 30, to augment the army under George Washington which was in the vicinity of New York City awaiting an expected British descent. Pennsylvania's quota was 6,000 men, 4,500 of whom were to be Associators. The gap would be filled by Colonel Samuel Miles's State Regiment of Riflemen and Colonel Samuel Atlee's State Battalion of Musketry, two special organizations (these were so-called "State Regulars," not part of the Associator structure) enlisted for long-tour, full-time active duty, but in the service of the State rather than of the Continental Congress.

To determine the procedures for raising the 4,500 Associators requested, delegations consisting of two officers and two privates from each Associator battalion convened at Lancaster on July 4. In an assertion of their independence of the Assembly these delegates on their own initiative assigned quotas to the various counties so that the total could be met, authorized these men to serve outside of Pennsylvania, and elected a "first brigadier" (Daniel Roberdeau) and a "second brigadier" (James Ewing) to command the Pennsylvania contingent.

The ease with which the volunteers were obtained varied from county to county. York County, for example, sent no less than five battalions (upwards of forty companies) to Perth Amboy, New Jersey, to join the Flying Camp, but only fourteen of these companies were mustered into that organization, the others returning home six weeks after they had left. Other counties, on the other hand, failed to meet their full quotas, and the total never reached the 4,500-man goal. Miles's and Atlee's organizations refused to serve under Roberdeau and Ewing, as they had not been represented at the Lancaster conventions; consequently, they were attached to a Continental brigade stationed on Long Island. All the same, some fifteen Associator battalions served at one time or another in the Flying Camp, the largest number on duty at any single time being ten.

For example, as of the end of July five Philadelphia battalions and one Chester County battalion were assigned to the Flying Camp in New Jersey, and a four-company battalion of riflemen from Northampton County arrived late in the month, bringing the total to 2,557 men. During August, the Northampton County battalion moved to Long Island, being joined there before the August 27 battle by a battalion from Berks County, one from Lancaster County, elements of a Chester County battalion, and one York County company. The five Philadelphia battalions in New Jersey, however, went home, being replaced by a second Chester County battalion, two from York, and one each from Bucks, Cumberland, and Lancaster counties. Even without the Associators on Long Island, the Pennsylvania strength with the Flying Camp now totalled 2,880. At the end of the following month, what with normal losses (not to mention the men killed, wounded, and captured at the Battle of Long Island—from which, for example, only eighteen men of the York County company returned to duty), it had dropped to about 2,700, but the arrival of replacements brought it to almost 4,000 by October 31.

While the bulk of the Associators served at Fort Lee, New Jersey, some were posted to guard the passes north of White Plains, New York. In addition, elements of one of the York County battalions were part of a raiding force which carried out a successful hit-and-run attack against British troops on Staten Island on the night of October 15. Substantial elements of at least three Associator battalions were lost after a courageous struggle when Fort Washington, New York was captured on November 16. After other erosion had occurred during the army's retreat through New Jersey, the Pennsylvania contingent of the Flying Camp was down to 2,082 by November 30, the date marking the end of the enlistments. Even so, 1,173 of Pennsylvania's Flying Camp veterans, representing five different battalions, were still with Washington on December 31.

These troops had taken part in the December 26 attack on Trenton, New Jersey. In addition, some 3,000 Associators served in various aspects of that operation. As an inducement, on December 12 the Assembly had offered bonuses varying from five to ten dollars (depending on the speed of response) to encourage Associators to volunteer for six weeks of active duty. Some of these men guarded the banks of the Delaware between Trenton and Philadelphia; the remainder, from Philadelphia and from Lancaster, Cumberland, and Chester counties, were scheduled to take part in the assault itself. Part of these made the crossing with Washington, the remainder being assigned to units scheduled to cross downstream but were blocked by ice in the river.

After the Hessians were defeated, numbers of the Associators went home. Although some Associator units replaced them—for example, a Northampton County battalion joined the army at the first of the year—the total Associator strength on active duty for the Princeton campaign (January 2-3, 1777) dropped to something under 1,700.

Princeton represented the Association's swansong. For reasons previously mentioned, the system had become inadequate to the need. A genuine militia law was clearly necessary; further, with the adoption of a new state constitution in September, 1776, and the establishment of control by the radical political faction, such a law had become politically attainable. In fact, one was enacted on March 17, 1777.

In outline, it provided that all white males between the ages of eighteen and fifty-three who were capable of bearing arms would be enrolled, with only ministers of the gospel, masters and faculties of colleges, purchased servants, and certain officials being exempt. On the basis of population distribution, each geographical area inhabited by fifty to eighty eligible men was to form an infantry company. (Existing artillery companies—the term "battery" was not used in an organizational sense—and cavalry troops, in which membership was voluntary, were not affected.) The infantry companies of each county were to be organized into battalions at the rate of eight companies per battalion. All officers were to be elected by the men for three-year terms, but initially only freeholders were eligible for election (this restriction was lifted on June 19, 1777). An annual schedule of ten days of training by company and two of training by battalion was prescribed. The men of each company were to be assigned by lot to eight "classes" of equal size.

Active duty was fixed at a maximum of sixty days, but to avoid depopulating any area of its physically fit manpower, militiamen were to be called up as individuals, by classes, rather than by units. Once mustered, the members of the class in question would be organized into provisional companies ("marching companies") for the duration of the tour, and all such companies of a county would be grouped into one or more provisional battalions. Classes were to be called up in numerical sequence, and no class could be ordered to a second tour until all other classes of that county had been called for duty. Eventually small fines were established for missing drill days, and a substantial fine (forty pounds) for failing to march with a class called for active service. It was permissible, however, to send a substitute on active duty, provided the substitute himself had not been called up.

Aside from the virtue of always leaving some men at home for local protection and to keep the

Color Guard, Second Pennsylvania Regiment (Courtesy 2d Pennsylvanian Regiment, Inc.).

economy functioning, this system had the merit of sharing the burden of military service equitably. On the other hand, it guaranteed that the members of a marching company would never have trained together as a body, and would not in most cases even be acquainted with each other. From the standpoint of purely military effectiveness, therefore, it had inherent drawbacks.

While the Pennsylvania militia law brought the enrollment of an estimated 60,000 men, it is doubtful that more than a fraction of the training companies adhered fully to the training schedule. Because some counties bore a disproportionate burden of the active-duty calls, it has been said that "tens of thousands" of the men enrolled never saw active service. Nevertheless, provisional militia battalions served to augment the Continental Army during the Brandywine, Germantown, and Whitemarsh campaigns of 1777; detachments maintained a screen between British-occupied Philadelphia and the Continental Army's encampment at Valley Forge during the winter of 1777-1778; individual companies in frontier counties performed valuable service against the Indians; and numbers of others made less spectacular but genuinely useful contributions by carrying out such functions as guarding prisoners of war. As a forerunner of what was to be a frequent nineteenth-century militia role, there was even an instance of militia being called on to put down civil disturbance when, in October 1779, the Philadelphia Light Horse, reinforced by artillery, dispersed a mob that was trying to burn the houses of unpopular political figures.

To examine some of these activities in more detail, at the onset of Sir William Howe's move against Philadelphia in September 1777, an entire division of two brigades of Pennsylvania militia—almost 3,000 men—was called up to reinforce Washington. At Brandywine, on September 11, this force was posted to guard a ford downstream from what became the southern limits of the battlefield. When the Continentals retreated, the militia withdrew without having been engaged. At the Battle of Germantown on October 3, the Pennsylvania militia division formed the right assault column, assigned the mission of turning the enemy left flank. Once again, it was so far on the edges of the main action that it was not actually involved in combat and advanced no closer than light artillery range. However, its fieldpieces bombarded the enemy troops in front of them, keeping the enemy deployed there from being free to reinforce their hard-pressed comrades at the center of the line, and the militiamen were the last Americans to leave the field when the Army withdrew.

The story was somewhat different in early December, when the British made what proved to be an abortive move against the Americans drawn up along an east-west line of hills at Whitemarsh. Stationed on the American right flank, the Pennsylvania militia met the enemy's opening probe on December 5. On Washington's orders, one Pennsylvania unit of 600 men bravely rushed forward to block the attack. Met by disciplined volleys, they took cover behind trees and fences. In the process of rallying them to resume the assault, Brigadier General James Irvine was shot out of the saddle and captured. By that time, some forty or fifty of the militiamen were down, killed or wounded, and the attack was abandoned. Another Pennsylvania militia detachment, under Brigadier General James Potter, was also beaten off when it tried to flank the advancing British. Even so, General Howe decided not to attempt to follow up his success and halted his men in place.

On December 7, shifting their main effort to the right, the British made another tentative probe. While they were moving into position Potter again led his militia against them but, as before, was beaten off. A determined British assault never developed, however, and on December 8 Howe marched his army back to Philadelphia.

Successive militia call-ups had taken place piecemeal from as early as April 1777, so there was considerable turnover as units completed their sixty-day tours. Consequently, while some of the militia battalions at Germantown had also been at Brandywine, none of those were still on duty for the engagements at Whitemarsh. On January 5, 1778, shortly after the Continentals moved into Valley Forge, the last Pennsylvania militiamen in active service were released. Within a matter of days, however, the first of a series of new calls brought some 500 to active duty, to serve under Brigadier General John Lacey in patrolling the area north of Philadelphia defined by an arc between the Schuylkill and Delaware rivers.

Lacey did everything he could, but he never had enough manpower to be more than sporadically effective in obstructing enemy foraging. Moreover, early on the morning of May 1, a strong British raiding party struck his camp near the Crooked Billet Tavern. Because of a light rain, Lacey's outposts were under shelter instead of guarding the approaches. The raiders advanced undetected, one element blocking the Pennsylvanians' line of retreat while another drove in from the front. Although Lacey was awakened suddenly, he rallied most of his men and led them in cutting a way through the blocking force, but his command took heavy losses. When the affair

Pennsylvania Artillery Crew with 4 Pound Cannon (Courtesy, 2d Pennsylvania Regiment, Inc.) ***Photo by Jack Swern.***

The Pennsylvania Line (Courtesy, 2d Pennsylvania Regiment, Inc.) *Photo by Jack Swern.*

was reported to Washington, he ordered courts-martial for the officers who had been in charge of the outposts, one being acquitted but the other cashiered.

When the British evacuated Philadelphia and moved to New York in June 1778, the requirement for militia service in the state's more settled areas ended for the time being, but along the frontiers the story was different.

Because of Indian raiding from the fall of 1777 throughout all of 1778 Northumberland County rotated its militia on active service so that some of its units were continuously on duty. In Westmoreland County, while the response to individual Indian raids was usually undertaken spontaneously without benefit of a formal militia call, militia companies were frequently ordered to active-duty tours to garrison local forts and blockhouses or for particular operations.

Not all of these had good results. In February 1778, for example, the Continental officer commanding the Western Department, Brigadier General Edward Hand, led some 500 militia from Westmoreland County to seize British supplies at an Indian town at the site of what is now Cleveland, Ohio. Delayed by bad weather, the force got only as far as an Indian village located at modern New Castle. There, frustrated at finding the village deserted except for an old man and several woman. They then went about ten miles farther, fied orders, killing the man and at least one women. They then went about ten miles farther, until they came across four Indian women and a young boy, and killed all except one woman. This episode, which came to be called "the Squaw Campaign," was widely condemned.

The Tory and Indian attack on the Wyoming Valley in early July 1778, precipitating the "Great Runaway" when settlers along both branches of the Susquehanna fled their homes in panic, brought a flurry of militia calls, with Northampton County units on duty guarding their frontier and companies from Cumberland, Lancaster, and Berks counties being sent to reinforce the Northumberland County militia and the handful of Continentals at Muncy. Some of these militiamen were still on duty to take part in Colonel Thomas Hartley's offensive expedition to the north that Fall, and helped inflict a stinging defeat on Indians who attacked the force at Wyalusing on September 29.

Before the war was out, Pennsylvania militia in the southeastern part of the State was again called on for service. That Fall, it was decided to move British and Hessians, who had been held at Boston since the surrender of General John Burgoyne at Saratoga in October 1777, south to Virginia. At York, the column of some 4,500 prisoners was turned over to Pennsylvania militia, who guarded them on the final leg of their transfer to Charlottesville.

Farther to the north, assignment of additional Continentals to the Muncy-Sunbury region eased the demands for militia on that front during the first part of 1779. Late in the Spring, however, these troops were transferred to take part in the Sullivan expedition assembling at what is now Wilkes-Barre for a punitive march against the Indians in upstate New York. That left Northumberland County again vulnerable, so once more militia units from other counties had to be called in until the Sullivan operation was over and Continental troops could be brought back to Sunbury.

In the west, jurisdictional disputes developed between the county authorities, seeking to retain their militia for local defense, and the commander of the Continental Army's Western Department, who wanted militia to augment his troops for larger strategic plans. A major conflict arose when Colonel Archibald Lochry (as Westmoreland County Lieutenant, the chief administrator of the local militia structure) tried in 1779 to withdraw two companies from the garrisons of forts under Continental command in order to guard threatened areas closer to their homes. Colonel Daniel Brodhead, now the Department Commander, objected; when Lochry insisted, Brodhead refused to authorize the rations the companies would need to make the move. Both Lochry and Brodhead protested to the State's Supreme Executive Council.

Because the militia constituted the power base of the party in power and the county lieutenants tended to be, in effect, the local political "bosses," the militia viewpoint in such quarrels was usually the one upheld. Indeed, the Council sometimes went so far as to advise militia commanders that they were under no obligation to "accept the insults" of Continental officers by obeying their orders. Thus, when leaders of the militia detachments accompanying Brodhead on a punitive expedition in the Summer of 1779 disagreed with one of his decisions, they did not hesitate to withdraw their men before the operation was completed.

The expiration of the 1777 Militia Law brought a new Act on March 20, 1780. Reflecting only minor innovations, it prescribed a training schedule which now included brigade as well as company and battalion drill days but reduced the annual total required, and established new fines (increasing by rank) for missing these drills. New three-year commissions were issued, many incumbent officers asking to be replaced. To the confusion of future historians the battalions in each county, although remaining otherwise unaltered, were renumbered according to the relative seniority of their commanders. Pay for active duty, previously fixed at Continental Army rates, was raised to correspond to prevailing rates for common labor. The categories of exemptions were increased, but the President of the Council was now authorized to call up as many as four classes from any county at a time.

Shortly after this law came into force action was begun to draw on the militia for individual reinforcement of Pennsylvania's Continental regiments. The first step was to require each militia company to furnish one of its members for seven months' active duty. Later in the year a new "levy" was imposed, this time employing a somewhat different method to obtain men to serve not for seven months but for the duration of the war. For that purpose, all Pennsylvania taxpayers were divided into 2,700 classes, each class equal in the total of its taxable property; every class had to provide a recruit or pay a fine. In 1781, to obtain recruits for eighteen months' service, this system was employed again.

While a considerable number of these "new levies" did reach Continental regiments, their total never approximated 2,700, not to mention 5,400. Of those who were brought on duty, moreover, many did not join the Continentals but became members of the special militia units,

13th Pennsylvania Regiment, Continental Line (The Pennsylvania Regiment), 1776-1778 (Courtesy The Company of Military Historians).

5th Pennsylvania Regiment, Continental Line, 1777-1783 (Courtesy The Company of Military Historians)

formed for relatively long tours of frontier service, which were known as "rangers" or "ranging companies."

These were the outgrowth of the need for forces in being to react to the Indian threat without the delay of a militia call-up. They might be enlisted for any period from four months to the duration of the war, sometimes for defensive duty in their home regions and sometimes for a particular campaign. In the Spring of 1781, for example, Northumberland County raised two such companies, one for seven months and one "for the duration." A record also exists of a ranging company being authorized as late as December of the same year.

An example of ranging companies formed for a specific operation is provided by the authorization, in March, 1781, for Colonel Archibald Lochry to raise two companies in Westmoreland County to accompany George Rogers Clark, a Virginia militia general, on a campaign against Detroit. Lochry's force did not start down the Ohio River until July 25, several days after Clark and the main body had left. Still trying to catch up, by August 24 Lochry's party was well into the Ohio region when it was ambushed by Indians, forty-two of the rangers (including Lochry) being killed and the rest captured.

Periodic requirements for militia in all parts of the State had continued throughout the war. In August, 1780, reports of British troop movements in New Jersey brought out militia from Philadelphia. When Cornwallis moved into Virginia in May, 1781, the prisoners of war being held there were sent to join others already at Reading, Lancaster, and York, where additional militia companies were ordered to duty to guard them. In October, when fear that the British in New York might move to aid Cornwallis, besieged at Yorktown, Berks County militia was called up, marching to Newtown in Bucks County. A few weeks later, a Northampton County company was on duty to bury the dead killed in an Indian raid.

Even after peace with Britain was concluded, conflict with the Indians seemed certain to continue. Therefore, when the 1780 Militia Act expired in 1783, it was immediately renewed, although some minor changes indicate that somewhat less sense of urgency was felt—annual training days were cut to six, the fine for failing to march when called up was substantially lowered, and active duty pay was reduced to Continental Army levels. As it happened, Pennsylvania itself remained largely peaceful despite the Indian troubles just beyond its borders in Ohio. Therefore, after Congress in 1792 passed a national Militia Act which imposed less stringent requirements for militia service, Pennsylvania's implementing legislation, passed in 1793, was modified accordingly: the age ceiling was lowered from fifty-three to forty-five, and militiamen were required to provide their own arms (except for cannon). Organizationally the Pennsylvania militia was formed into no less than seventy-two infantry regiments in twenty-two brigades, further grouped into nine divisions. With these were seven artillery companies and eight cavalry troops. The class system was retained but was so phrased as to provide a basis only for asking men to volunteer, not for ordering them to active duty.

Governor Thomas Mifflin had no difficulty in obtaining volunteers for his initial use of militia under the new law, easily raising a detachment in the spring of 1794 to protect surveyors laying out towns in the newly acquired Erie Triangle. President Washington, however, feared that Pennsylvania's actions would cause the tribes of that area to join the hostile Indians confronting General Anthony Wayne and his Regulars in the Old Northwest. He demanded the withdrawal of the militia and surveyors; and Mifflin, whose clashes with Washington dated from early in the Revolution, grudgingly complied.

In July of the same year a new crisis, "the Whiskey Rebellion," came to a head.

The trouble stemmed from Federal attempts in western Pennsylvania to tax whiskey, which the western Pennsylvanians contended was discriminatory. Officials trying to collect the tax were shot at, some were tarred and feathered, and some had their houses burned. President Washington was alarmed by the reports. The Regular Army was fully preoccupied with Indian problems, but before he could call militia into Federal service, the law required that he must have a ruling from a Federal judge that the judicial system in the affected area had collapsed. Accordingly, Washington asked United States Supreme Court Justice James Wilson, of Philadelphia, to consider the matter. However, Wilson replied that without further verification, the reports did not support such a ruling. On August 2, therefore, Washington turned to Mifflin and asked him to call out the Pennsylvania militia on his own authority. Aside from the political and personal differences between the two men, Mifflin knew that to use the militia to suppress the protests would be unpopular throughout the State and initially refused Washington's request. When he finally agreed, it was only on condition that Washington would give him time to convene the General Assembly and to send a commission to try to settle the problem, and that any militia call-ups would be initiated and publicly justified by Washington himself.

Unknown to anyone in Philadelphia the situation in western Pennsylvania had grown even more threatening. Local leaders in the region, acting without authority, issued a call for all the militia in the area to assemble under arms on August 1 at Braddock's Field—not to restore order, but to constitute a force for further defiance. Once gathered, these men had to be restrained from marching on Pittsburgh to burn the town, and they refused to return to their homes.

On August 4 Justice Wilson issued the ruling Washington needed. Three days later, after word of the Braddock's Field assembly had reached Philadelphia, a presidential proclamation was issued calling on the protestors to disperse by September 1. Simultaneously a warning order went out to New Jersey, Maryland, Virginia, and Pennsylvania to fill quotas totalling almost 13,000 militia, who were to be ready to march when called. Pennsylvania's share was 5,196 men. On September 9 Mifflin called on Philadelphia and eleven counties to provide these troops.

Immediately he ran into opposition. Only Dauphin and Montgomery counties were willing to comply. In a rapid tour of the reluctant areas, however, Mifflin exhorted the militiamen to accept their duty, to such effect that by September 19 he had persuaded some 4,400 to start marching toward the designated assembly point at Carlisle. Joined there by the New Jersey contingent, they left early in October, made rendezvous with the Virginia and Maryland forces, and on October 19 arrived at Bedford. In the face of such overwhelming force the "rebellion" collapsed. No violence occurred, and the militiamen were sent home on November 17.

In an organizational sense, by the time of the Whiskey Rebellion the Pennsylvania militia had achieved a blend of features of both its previous approaches to military needs. The concept of purely voluntary "Associations" had been compelled by circumstances to give way to a system based on mandatory enrollment and mandatory service for those called; this had then been superseded by a structure which retained compulsory enrollment but, in practice, depended on volunteers from the enrolled militia for active duty. In essence the outlines had been established of a method of raising forces which would continue in effect until the Civil War.

CHAPTER 3

1800-1859

Albert J. Habersberger, Major, Hq, 28th DISCOM

WITH THE ADVENT of the Nineteenth Century the organizations that made up the Pennsylvania Militia provided more of a social focal point for the people of the various localities of the Commonwealth than a meaningful military purpose. The units in the years 1800-1850 reflected different themes that provided communities with geographic, ethnic, racial, class and (in some cases) common trade identity. Very little time was devoted to actual military training. The units seemed to devote themselves to providing community services in the form of social outings and grand shows of splendor through parades and ceremonies.

However, this general lack of preparedness did not seem to detract from the Pennsylvania Militia's willingness to participate in the defense of the Commonwealth and the country. They were called upon many times during the period to fight and die for the preservation of the freedom and independence of the infant nation. The first half of the nineteenth century would see Pennsylvania Militiamen participate in both the second war with England, more commonly known as the War of 1812, and the Mexican War. In addition to these major wars fought in defense of the nation, citizen soldiers helped to insure the peace and stability of the Commonwealth during many occasions of civil disturbance.

The Militia of the Commonwealth of Pennsylvania was composed of both volunteer units and of the enrolled militia at this time. Men between the ages of eighteen and forty-five were compelled by law to be members of enrolled militia units. The image of the volunteer militiaman that we have institutionalized and has created much folklore skirts reality, which was that conscription was forced on most of the military men of the period.

The time was one of great pomp, ceremony, and showy events. Units had colorful names which in some cases identified the area where the unit was raised, its military purpose or the ethnic heritage of the members of the unit. Typical unit names were: Northampton Rifles, First Troop Philadelphia City Cavalry, Washington Blues, Cadwalader Grays, Hibernian Greens, Steuben Fusileers, Irish Volunteers, Duquesne Grays, and the Wyoming Yagers. Members of these units somehow were able to surpass the commonality of the unit names with the ostentation of their uniforms. It was a time of brilliant color, plumes, epaulettes, swords and sashes. The behavior of the units in like manner manifested itself in much ado and grandeur. Each man had to provide his own weapon and uniform. The practice led to great differences in degrees of uniformity from unit to unit. The units with wealthier members had more colorful, and in some cases more patterned, uniforms. Officers were the biggest offenders of the quest for uniformity. They designed their own colorful uniforms in fashions that generally did not conform to the uniforms of others.

Units of the Pennsylvania Militia were required to muster twice a year for a day's training. These muster days primarily consisted of marching, drilling, and taking the opportunity to display the distinctive uniforms. Those who did not have weapons paraded about with whatever they could improvise as a suitable weapon. The extemporization was necessary because many men resisted buying and using their own uniform and weapon. Drills usually ended with a gala social gathering which generally involved the entire community.

The number of citizens involved in military service 1800-1850 was considerable. Over ten percent of the population of the United States was involved in some form of military service. Rank was awarded freely, and there appeared to be a general officer in practically every community in the nation. Just prior to the War of 1812 the Commonwealth of Pennsylvania had no less than sixteen divisions, each commanded by a major general. For the most part junior officers were elected and senior officers commissioned or appointed to their positions. These positions were held for generally four to seven years.

At the beginning of hostilities of the War of 1812 the Pennsylvania militia was not a well trained, disciplined fighting unit prepared for duty as integrated members of a formal army. The State had tried to form some cohesiveness into its military structure under the provisions of the Military Act of 1802. However the militia was in a disorganized condition, lacking both modern weaponry and trained soldiers.

There were, of course, no enemy activities during the War of 1812 in Pennsylvania. The war proved to be short and might be summarized by depicting the major events in the time sequence that they occurred. The initial activity of the war took place on the Canadian border, with Detroit and Chicago falling into enemy hands. The United States was repulsed in the attempt to invade Canada through Niagara. Then Commodore Oliver H. Perry's naval victory over the British near Put-in Bay secured Lake Erie and Detroit. In 1814 a British force came up the Chesapeake Bay and captured Washington. The British invasion force was repulsed at Baltimore, and peace came with the treaty of Ghent in 1814.

With the outbreak of hostilities, the Federal government drafted a quota of fourteen thousand men from Pennsylvania. Governor Simon Snyder immediately endorsed the war and ordered the adjutant general to activate the required militia units. At this time the State's militia was divided into two active divisions. The First Division, commanded by Major General Isaac Worrell, was established in Philadelphia with units from the eastern part of the State. The Second Division was formed from units in the north and western parts of the State and was commanded by Major General Adamson Tannehill. The Second Division mustered two thousand men each at Meadville and Pittsburgh. For the war, portions of these contingents provided defensive forces in Ohio and provided blocking forces against British eastward advances. In the early part of 1813 a thousand Pennsylvania militiamen were moved to Erie in order to protect the war ships that were being built there for Perry's fleet.

At Erie an artillery battery, which traces its lineage to the present day 109th Field Artillery, distinguished itself by participating in the sea battle for Lake Erie. The battery, under the command of Captain Samuel Thomas, originated in Kingston, Luzerne County. Perry's fleet was divided, with some ships still remaining in the harbor in Erie. The Commodore had to move into the protected harbor to unite his fleet. The battery commanded by Captain Thomas provided Perry with the cover fire necessary to execute his move. Captain Thomas's battery participated in a cannonade battle with the British squadron in order to support the Admiral's necessary link-up. Further, four of Captain Thomas's men volunteered to serve as cannoneers on Perry's fleet, a valuable gesture, for the fleet was short of seamen cannoneers. The four men participated in the great naval battle for Lake Erie.

With the battle for Lake Erie won, Captain Thomas's unit was attached to a regiment which invaded Canada. They captured the town of Malden and Sandwich and crossed the Detroit River to attack the British and Indian forces which were occupying the fallen community of Detroit. Captain Thomas's men were in the lead units of the assault on Detroit. When the British fled the outpost, the main elements of the American army pursued, with the exception of a contingent of fourteen men who were left under the command of Captain Thomas. The remainder of the battery continued to serve with the main elements and participated in the battle of Thames, with Lieutenant Ziba Hoyt commanding the troops.

At the same time Pennsylvania Militiamen were serving at Erie and in Ohio, elements of the First Division were ordered to defend Philadelphia. They initially established themselves, one thousand strong, at Shell Pot, a location north of Wilmington, Delaware. Additional troops under the command of General Thomas Cadwalader were ordered by General Snyder to Camp Dupont, also near Wilmington, to protect the Delaware and Elk rivers. A new camp called Camp Gainer was established near Marcus Hook, Pennsylvania, and was posted with many Pennsylvania units, including the First Troop Philadelphia City Cavalry.

Additionally many troops from the central counties of Pennsylvania were activated by General Snyder after the fall of Washington when a British march on Baltimore was anticipated. Some Pennsylvania troops participated in battles in Maryland as augmentation and attachments to Maryland units. Several examples were Captain Michael H. Spangler and his 100 York volunteers, who were attached to a Baltimore troop, and the units from Hanover of Captains John Bair and Frederick Matzgers, which joined Maryland forces in the battle of North Point.

After the Treaty of Ghent signaled the end of the War of 1812, the nation entered the "Thirty Years' Peace." For the most part the words describe accurately the time for Pennsylvania militiamen. While Regular Army units and other states' militia were engaged in battles with the Indians in the Seminole Wars of 1817, 1832-1841, and the Black Hawk War of 1832, the Pennsylvania militia activity was of considerably less magnitude. However, all was not peaceful in Pennsylvania. Militiamen were involved in serious civil disturbances at home. The state of preparedness of the militia did not improve greatly during this period. There was an effort on the part of the Commonwealth to improve the State-owned stores of weapons and ammunition. Additionally there was an effort to create greater uniformity in the military dress of the militiamen.

In September of 1824 Lafayette made an official visit to Philadelphia. The First City Troop and the First County Troop provided the military escort for Lafayette during his stay in the Commonwealth. It was on this visit to the United States that Lafayette referred to the militia as the "Guarde Nationale," recalling the name of the Parisian militia that he had commanded in 1790. This was one of the early influences in the development of the designation "National Guard" for the United States militia forces.

A major political conflict occurred in Harrisburg in 1838 that brought national focus on the Pennsylvania political scene and the Pennsylvania militia. In the election of 1838 the incumbent Governor Joseph Ritner, of the Anti-Masonic Party, was defeated by the Democrat, David Rittenhouse Porter. Disputes arose concerning the validity of the election in the case of many legislative seats. It was apparent that the leaders of the legislature, including abolitionist Thaddeus Stevens, had been unseated. It was a period when emotional matters played a major role in the election process. For example, issues raised by Anti-Masonry Party candidates were incendiary and personal, as were those shouted by the slavery and anti-slavery partisans. Governor-elect David R. Porter denounced anti-slavery agitation and called for action to quell it. Into this scene was called the Pennsylvania militia's First Division, under the command of Major General Robert Patterson. A potentially riotous situation had developed in Harrisburg. In effect, 1,000 militia men arriving in Harrisburg found the outgoing Governor trying to use them as political pawns. Governor Ritner attempted to influence General Patterson's troops to bring pressure on a political solution. General Patterson refused to participate in the political controversy although announcing his intention to protect public property. As a consequence of the General's adamancy, the Governor relieved these troops. The confrontation subsequently became known as the "Buckshot War," in reference to the type of ammunition the militia were ordered to employ.

The early part of the 19th Century was an era in which anti-Catholic sentiment was prevalent in many parts of the United States. Pennsylvania was no exception, and Philadelphia was the scene of many bloody and destructive religious riots. Roman Catholic churches and church properties were destroyed, and many people killed and injured. The Pennsylvania militia played an important role in containing these riots and protecting lives and property. The most serious of the riots in Pennsylvania, occurring during 1844, were called the "Native American Riots." These anti-Catholic riots were blamed on the influence of the Native American Party, which had been formed in 1835 for the purpose of opposing Roman Catholic growth and influence in the country. The group opposed immigration and the granting of citizenship rights to certain ethnic groups, particularly Irish Catholics.

The riots that took place in the Spring and Summer of 1844 were mostly in Kensington, Philadelphia. Mobs sacked and burned Roman Catholic churches and came into armed conflict with militia men. Men of the Montgomery Hibernian Greens were set upon by the mobs, and one soldier was killed. The City Guards, under the Command of Captain Joseph Hill, fought a gun battle with the rioters, who were not only firing small arms but employing cannon in their attack on Captain Hill's men. The City Guards killed and wounded many men in the melee.

The period of time between conclusion of the War of 1812 and the beginning of the War with Mexico proved to be controversial for the Pennsylvania Militia. In addition to the Buckshot War and the Native American Riots, the militia found itself involved in numerous race riots, tradesmen's riots, and strikes. Service in these domestic disturbances demonstrated the need for maintaining a strong well-disciplined militia in order to "keep house" in a State that proved to be extremely volatile.

The War with Mexico was a conflict that came as no surprise to most Americans. Mexico had warned that admission of the Republic of Texas into the United States would provoke a war. Mexico was most distressed at American expansionist activities particularly on the Texas border. The United States Government moved troops commanded by Brigadier General Zachary Taylor to an area which in the opinion of Mexico was too close to the Rio Grande River, across from the Mexican town of Matamoros. General Taylor's men were attacked by the Mexican military under the command of General Mariano Arista on April 24, 1846. Pennsylvania Militia units played a major role in the eventual victory over Mexico. President James K. Polk levied two regiments from the Commonwealth of Pennsylvania. Governor Francis R. Shunk had no trouble meeting this levy, because ninety Pennsylvania companies had already volunteered for service. These two regiments were called the First Pennsylvania Volunteers and the Second Pennsylvania Volunteers. The two regiments were organized by the following companies:

Washington Grays of Philadelphia. An officer of this famous unit in the uniform worn prior to the Civil War. When the State formed a single Division, the Washington Grays became Company G, 1st Regiment Infantry.

First Pennsylvania

Co. A, Pittsburgh Blues
Co. B, Washington Rifles
Co. C, Monroe Guards
Co. D, City Guards

Co. E, Washington Light Infantry
Co. F, Philadelphia Light Guards
Co. G, Jefferson Guards
Co. H, Cadwalader Grays
Co. I, Wyoming Artillerists
Co. K, Duquesne Greys

Second Pennsylvania

Co. A, Reading Artillerists
Co. B, American Highlanders
Co. C, Columbia Guards
Co. D, Philadelphia Rangers
Co. E, Westmoreland Guards
Co. F, Stockton Artillerists
Co. G, Cameron Guards
Co. H, Fayette County Volunteers
Co. I, Hibernia Greens
Co. K, Captain William Small's Company

Two additional companies were assigned to the Second Pennsylvania regiment; these were primarily composed of volunteers from the State of New Jersey. The Commanders of the Regiments were Colonel F. W. Wynkoop and Colonel William B. Roberts, respectively. In December of 1846, and January of 1847, these regiments arrived in New Orleans in preparation for movement to the war zone. They camped at the old battle grounds of the War of 1812, about six miles south of the city of New Orleans. During January and February the troops were transported to the Island of Labos in order to stage an invasion of Southern Mexico. Many soldiers became sick because of unsanitary conditions on the twenty-day journey. The respite on the Island of Labos gave many of them the opportunity to recuperate. The troops made preparation for the attack on the city of Vera Cruz and the much feared castle of San Juan D'Ulua. The regiments were assigned to the Division which was commanded by General Robert Patterson, a Pennsylvanian, who had

Dress uniform of the First Troop, Philadelphia City Cavalary.

1st Regiment Pennsylvania Artillery. A Segeant of this regiment in the uniform worn between 1855 and 1861. This regiment became the 1st Regiment Gray Reserves and then the 118th and 119th Pennsylvania Volunteer Infantry Regiments during the Civil War.

led the militia during the farcical Buckshot War in Harrisburg.

On March 2, 1847, the troops embarked for Vera Cruz and landed three miles south of the city on March 9th. The battle lasted twenty days, and on March 29th the siege ended with the city's surrender. The units then rested until their march on the City of Mexico. On April 18th they were involved in the battle of Cerro Gordo, under the commander, General Winfield Scott. He had to undertake the task of attacking Mexico's General Santa Anna's 11,000 well emplaced troops with only 8,500 men. The fight was fierce with about 4,000 Mexican soldiers either being killed, wounded, or captured. The Pennsylvania units were reportedly in the frontal attack of the battle. General Santa Anna barely escaped capture.

The Army then marched on to Puebla, leaving some units from Pennsylvania behind at Jalapa to garrison the town and protect the lines of communication. Pennsylvania units, including the Wyoming Artillerists, who were designated Company I, First Pennsylvania Volunteers, were involved in a minor battle at the pass of La Hoya. This was fought to defend a wagontrain from Vera Cruz against guerrilla attack.

When General Scott marched on the City of Mexico, Pennsylvania units took part in all of the major battles, although some units were left behind to garrison Puebla. These units eventually engaged Santa Anna in his last push to salvage something from complete defeat. The units that were in the assault on the City of Mexico fought in the battles of Contreras, Churubusco, Molino del Rey, San Páscual, Belen Gate and Chapultepec. Many of these battles proved costly, with the loss of many Pennsylvanians. After the victory at Chapultepec the American flag was raised by Captain Samuel Montgomery and Captain E. C. Williams; the same flag was raised over the City of Mexico by Captain Williams on September 13, 1847, when it fell to the Americans. However, Santa Anna escaped the city with a sizable force, which later proved to be the nemesis of the Pennsylvania units remaining at Puebla.

Six companies of the First Pennsylvania were garrisoned at Puebla, as well as one company of dragoons and a detachment of artillery—in all, 600 soldiers, six pieces of light artillery and one iron gun. They did not receive any real trouble from the inhabitants of the city of 8,000 during their occupation. Unfortunately the companies did figure in Santa Anna's plan to attack the United States forces along its supply and communications route in an attempt to salvage an already lost war. Santa Anna for a month laid siege to the troops at Puebla without ever really allowing his troops to become decisively engaged. The city was relieved in mid-October. Santa Anna did not engage the relief column but made his escape. The retreat proved to be the end of any real hostilities. The treaty of Guadalupe Hidalgo was signed on February 2, 1848.

Regimental Drum, 92d Regiment, Pennsylvania Militia, circa 1810-1840. (J.C. Nannos collection)

The Pennsylvania units did not return home until June, July, and August of 1848. However, the war proved to be costly for the Pennsylvania Militiamen who participated. Only about one half of those who had originally answered the call returned home. The Pennsylvania Militiamen, forebearers of the National Guard, had proved their value as fighting men.

CHAPTER 4

The Civil War

James A. Killian, Former Chaplain 213th Area Spt Group

THE NATION WAS at war. It was one that few wanted, but help was needed. Help could come from only one source, the civilian population. The Regular Army establishment at the time of the beginning of the war numbered only 16,435 officers and men, hardly enough to man forts, depots, and strategic locations. The recourse to civilian supplements was not a precedent-setting move by any means. It had been the citizen soldier, called for a brief term of duty, who bolstered the regular forces in previous wars or, in some cases, provided the entire striking force.

It must be noted that the vast numbers of Pennsylvania volunteers in the War Between the States included men serving in units which were specially formed to fill the manpower needs of this particular war and that many of these faded from the rolls as soon as the final peace was settled. These units and their men served well, bringing much deserved praise from those who led them, served with them and shared the hardships of long marches and counter-marches, and desperately fought battles and skirmishes. However, it is not these emergency units but the group of Pennsylvania volunteers whose lineage often began in pre-Revolutionary days and has extended to the present with which this chapter is concerned.

Jim Dan Hill in *The Minuteman in Peace and War* clearly points out that the writers who have for so long emphasized the unprepared attitude of the Union have failed to recognize that the neglect was only at the Federal level. Each State, in actuality, had its own armed forces, some of which were remarkably well organized, trained and equipped. Local unrest on the part of ethnic groups, opposition to the rising railroad expansion, and industrial tensions led to the organization and use of militia units in virtually every State. New York had more men enrolled, uniformed, armed and trained than the Regular Army force of the United States. Pennsylvania in December of 1860 counted nearly 19,000 officers and men in 339 infantry, eighty-four artillery, and fifty-three cavalry units, a force also larger than the entire Regular Army. A sizable force was ready to respond to the call of President Lincoln when hostilities commenced.

It was not just the common soldier who was the volunteer. The highest level of command in the Union army included men who must be accorded the same designation. It was not a war of the professional leader matched with the volunteer private. Most of the "Regular Officers" of the Union Army were, in reality, civilians. Most of them, like

Grant, Sherman, Slocum, Burnside and McClellan, had West Point educations, but had returned to civilian life. Sherman and Grant held temporary, active commissions in the United States Volunteers until mid-point in the war, when they were finally awarded commissions in the Regular Army.

President Lincoln's call for 75,000 men was directed to the governors and adjutants-general, with a specific quota of men and units given to each State. Pennsylvania was at first asked to raise sixteen regiments, most of which were to be infantry. Each was to have ten companies, conforming to Regular Army tables of organization, with the usual regimental staff added to raise the strength to 780. April 1861 saw Harrisburg, Wilkes-Barre and Philadelphia as centers for the State units to take the oath of service for a three-month term of federal duty. Units from Philadelphia, Reading, Allentown, Pottsville, York and Pittsburgh found themselves loading equipment, clothing and personal articles into available transport for shipment to troop concentration areas. For some of the Pennsylvania guardsmen this meant the Capital of the nation, which was in danger of imminent capture by Confederate forces massing in the neighboring state of Virginia. The Pottsville unit, then Companies B and H, 25th Pennsylvania Volunteer Infantry, with the Ringgold Artillery of Reading, the Allen Rifles of Allentown, and Logan Guards of Lewistown moved from Harrisburg to Washington in response to the urgent pleas from a Congress already planning a withdrawal from its endangered capital. The danger of these "First Defenders" came before the Confederate gray was ever seen, or the sound of a rebel rifle was heard.

Baltimore, a crucial rail junction center, was a center of men of Confederate sympathies who made an organized effort to prevent military aid from passing through. On April 18th the Pennsylvania volunteers, uniformed but unarmed, made their way through the city and were exposed to the howls, rocks and other missiles of a mob. In this movement the first blood actually shed in the war was that of Nicholas Biddle, a black servant of a captain of the Washington Artillerists of Pottsville, who was struck on the head, causing a scar which he carried proudly to the day of his death.

Pennsylvania Volunteer Infantryman of 1861.

How effective were these volunteers? A rumor arose that over 5,000 Pennsylvania men were arriving for the defense of Washington. Reputedly the Confederate leaders decided that the odds were too great to attempt an immediate attack. A consequence was the resolution passed by the national House of Representatives in July 1861 which paid tribute to the "530 soldiers from Pennsylvania who passed through the mob of Baltimore and reached Washington on the eighteenth of April last for the defense of the national Capitol."

A glorious beginning, especially in the terms of cost: a national capital saved, not a shot fired, and no casualties, save for the head wound of Nicholas Biddle. Sadly, this was the end of easy glory for the men of Pennsylvania. Some of the same units, including those of Pottsville, were to fall to the ill luck of poor leadership, and feel the onus of defeat

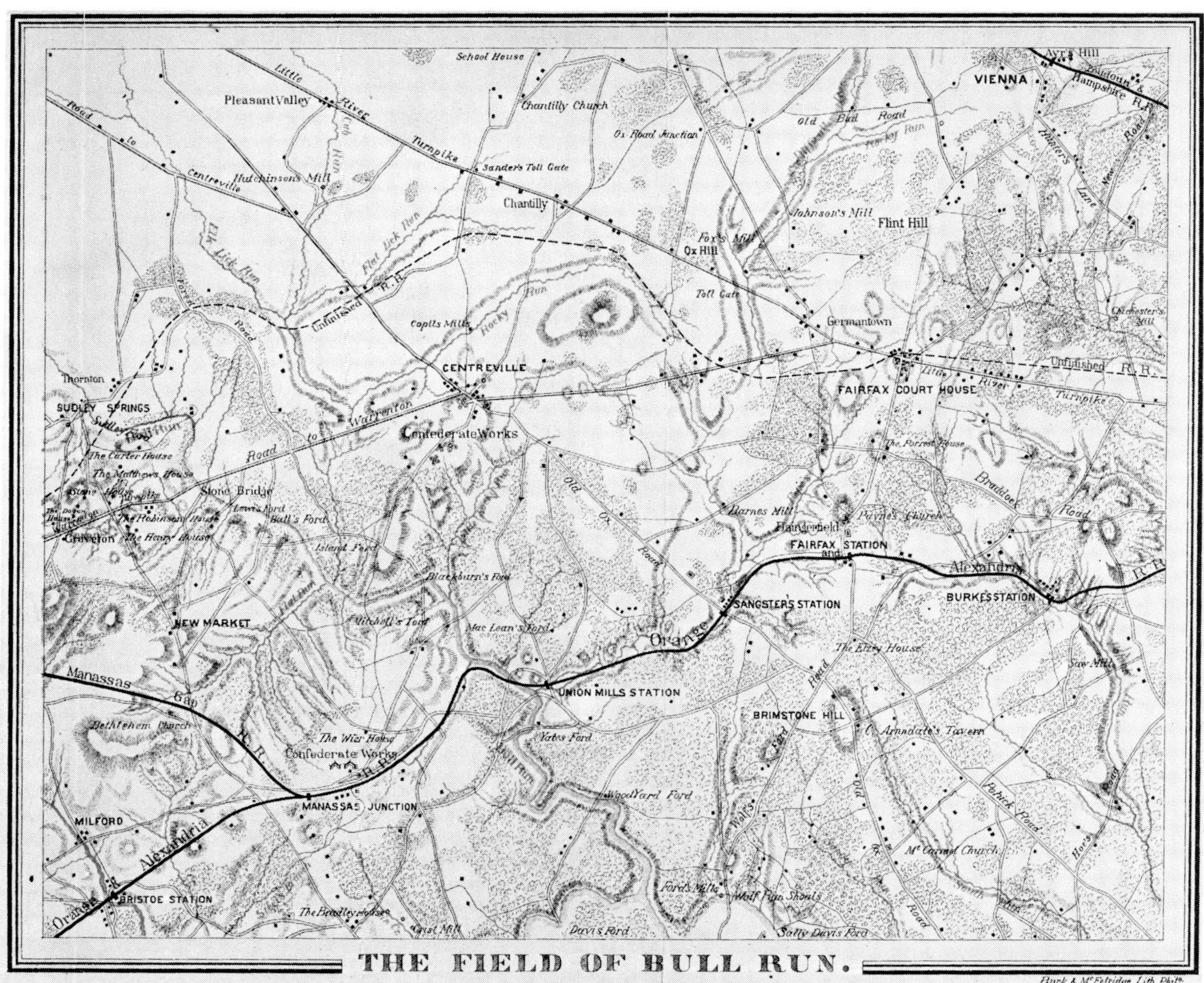

The Field of Bull Run

that same summer at Bull Run. Major General Irvin McDowell, only recently a major but now the army commander, completely lost control of his superior forces and watched them make a head-long flight back to the safety of Washington.

The problem of the three-month enlistment soon arose. By late July many Pennsylvania units were mustered out of Federal service. Lincoln issued call for more troops, but for some men the single three-month tour of duty sufficed. For others, however, a brief stay at home was merely a respite between enlistments. New units were formed, with new names, and sometimes, new commanders. Enough old faces were present in most new units to make the men feel at home, and to be ready once more for duty. With new uniforms, new weapons to replace some of the very old ones, and the promise of better planning in the way of camp design, ration supply and training, the men headed south to take up the cudgel of a war Lincoln said was to preserve the Union.

A look at some of those newly mustered-in units, now for three-year or duration of war terms, may help explain what happened between August 1861 and the major battles of 1862-65. The present 109th Field Artillery, then the 143rd Pennsylvania Volunteer Infantry of Wilkes-Barre, elected as its commander, Edmund L. Dana, who was a Mexican War veteran, a major general of militia, and a man with a reputation as an outstanding citizen. The 143rd was in training at Camp Luzerne until November 7, 1861, when it was sent

Raw Recruits at the first Battle of Bull Run.

Camp Logan - The first camp of the Pennsylvania Gray Reserves, 1861 (later the 118th and 119th Pennsylvania Volunteer Infantry Regiments.)

to the defense of Washington, duty which involved little fighting, heavy fatigue duty, and four times as many desertions as in any other time of the war. Six inches of snow in January helped to make things miserable for the men, and constant drill and review made for a period of time in which the letters home spoke of little but discontent and homesickness.

Elements of the 48th Pennsylvania Volunteer Infantry from Pottsville, Tamaqua, Minersville, Schuylkill Haven, and Cressona re-entered federal service in September 1861 and traveled to Camp Hamilton near the capital, where it underwent vigorous training in squad, company and regimental drill.

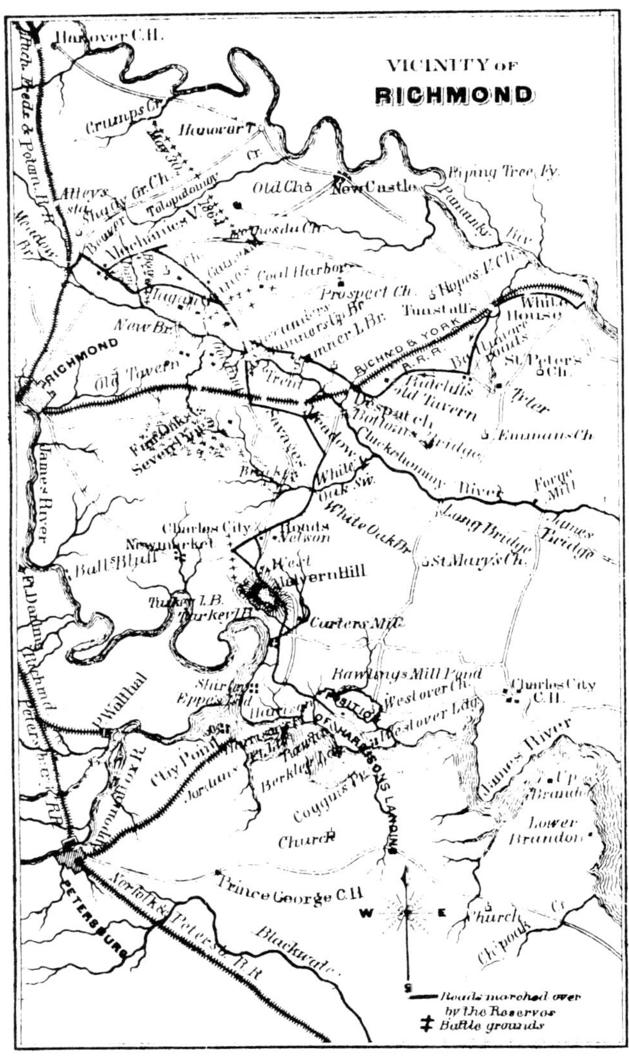

Vicinity of Richmond.

In 1862 action began with George B. McClellan, a Pennsylvanian who had gained prominence as a commander of the Ohio Volunteers in the West Virginia campaign, as the new commander of the Army of the Potomac. His decision to attack Richmond from the Virginia Peninsula resulted in a battle called the Seven Days (June 25- July 1), which saw him retreat from near Richmond. Troop C, 1st Philadelphia Squadron, was active in the Peninsula campaign.

After the disastrous second Battle of Bull Run, in which the ill-starred General McDowell and Major General John Pope lost several opportunities for victory, the men of Pennsylvania found themselves with McClellan at the town of Sharpsburg in Maryland. McClellan obtained copies of Lee's orders and had the opportunity to administer a beating to the army of Robert E. Lee. His slowness of movement, however, resulted in one more setback for the men in blue. The Battle of Antietam resulted in vast numbers of casualties, including many Pennsylvanians. Units representing eight still-existing units served at the battle of Antietam. In the West Wood, where it was said fell the heaviest fire of the war, poor staff work contributed to a slaughter of Union soldiers who had been massed so thickly that maneuver and cover was impossible. In this battle phase were four Pennsylvania regiments, termed the Philadelphia Brigade. They found themselves in reserve, and then, surprisingly struck with a flank attack which threatened to win the battle for the South.

The men from Reading and adjacent areas, as part of the 128th Pennsylvania Volunteer Infantry, (Co. E), fought with the forces of Major General Joseph Hooker in the East Wood and found themselves up against the veteran Confederate forces of John B. Hood and John Gordon. All in all it was a day of heroics, and of catastrophic bumbling by the high command. The Battle of Antietam, called a victory by a government starved for a victory, was at best no more than a stalemate and a portent of more vicious, horribly bloody battles yet to be fought.

On November 7, McClellan, the "Young Napoleon," was out, and in his place was installed the man with no confidence in his own ability, remembered chiefly for his whisker style, Ambrose E. Burnside. A mauled army has but two

Antietam

choices: to quit or to seek battle yet another time, in hope of an elusive victory. After some regrouping and some filling in of the decimated regiments and companies, the still mighty Army of the Potomac marched on to Fredericksburg. In the march were many who were embittered over the replacement of McClellan, who if not a brilliant combat leader, was loved for his apparent, and probably true, concern for the men in his command. Still, Washington had spoken, and Burnside, who delayed the takeover of command as long as possible, was given his chance.

Fredericksburg was a town that the army had been near in an earlier time. Now there were over 130,000 men on the march, ranging over a twenty mile column and requiring tremendous amounts of food and forage. Looting, once done on an individual basis in the dead of the night, now became regular routine. Bruce Catton tells of a Pennsylvania recruit in General Humphrey's division who recalled that as many as two hundred men from one regiment were arrested on that march with stolen goods in their possession. With only overnight confinement imposed on them, the men were convinced that "it was their duty to forage upon the inhabitants of the enemy's country." Thus did the Pennsylvanians, men of New England, New York, and the Midwest move on to the Rappahannock and Fredericksburg, a move frowned upon by Henry W. Halleck, general in chief, but approved by Lincoln.

The crossing of the Rappahannock was accomplished by boats and pontoon bridges, both of which were far away from the proposed crossing,

A pontoon bridge similar to those built under Confederate fire at Fredericksburg in 1862.

ordered not by military wire, but by a letter which took many days to be delivered. General Herman Haupt, chief of railway transportation, set himself to the task of ensuring that men and goods could move to the front for the attack and for relief and re-supply. In one of his ingenious plans he used Schuylkill County coal barges mounted on trains for non-stop supply runs from Washington south to the troops.

Bridges running right into Fredericksburg were being built, but the construction was held up by a force of 1,500 Mississippians who hid in basements and behind rude barriers until the engineers showed themselves on the bridges when they opened an intense fire that drove the engi-

Fredericksburg

neers to safety. Even what was considered the most thunderous cannonade ever heard, by massed Union guns which levelled the town, failed to drive out the sharpshooters. A boat crossing by New Englanders and fierce fighting in the town gave Fredericksburg to Burnside, but Lee still sat on a series of hills outside the town in entrenchments that he had taken time and great care to construct.

Burnside launched a frontal assault. The present 103rd Engineer Battalion, known then as the 119th Pennsylvania Volunteer Infantry, Gray's Reserves, in the Sixth Corps, commanded by Major General William Franklin, crossed the river, and under Major General W. F. Smith, was ordered to make a reconnaissance in force. They came under fire by heavy artillery but maintained their ground, though for the first time under fire. Franklin's forces, had they been properly relieved, might possibly have turned the tide and won the battle for Burnside. Total Union losses were 12,321 killed, wounded and missing, against Confederate losses of 5,309.

In the battle at Fredericksburg also were the antecedents of Troop C, 104th Cavalry, 108th Combat Support Hospital, 2nd Battalion, 112th Infantry, and the 1068th Military Police Company. No longer were these volunteers the raw rookies they had been shortly before. Recognition was not won in this war by fancy uniforms, sleek horses, and shiny weapons.

After the Fredericksburg campaign "Fighting" Joe Hooker relieved Burnside and popularized the distinctive insignia first used by the men of Major General Philip Kearney. The Sixth Corps, including the 119th Pennsylvania Volunteer Infantry, wore the cross on their caps.

With Winter there sprang up around Fredericksburg whole towns of log and tent hutments, each mirroring the design ideas of the individual builders but arranged in rows near bare areas reserved for drill and recreation. The moment of rest, however, was interrupted January 20, 1863, for the 119th Pennsylvania Volunteer Infantry, among others, who were ordered out for Burnside's second campaign, the infamous "Mud March."

Rain made necessary double and triple teams of horses hitched to artillery and boats, sometimes to no avail. There are records of 150 men assisting

Typical Civil War Camps

Going Into Camp

General View of Camps, Warrenton Junction

Winter Camp after a Snowstorm

A Footsore Straggler

In Heavy Marching Order

The Flankers

Drilling the Awkward Squad

The Company Cook

Stockaded "A" Tents

Oats for Six

Vignettes of Camp Life in the Civil War

Burnside's Famous Mud March

horses and still failing to move heavy equipment.

The "Mud March" signalled the end for Burnside, and the Army of the Potomac greeted yet another commander, Major General Joseph E. "Fighting Joe" Hooker. Noted by Secretary of the Treasury Salmon P. Chase "as a frank, manly, brave and energetic soldier, of somewhat less breadth of intellect than I had expected, however, though not of less quickness, clearness and activity," Hooker had been extreme in his ambitions to be commander-in-chief. The time had come to show his mettle, and his troops were ready. They had survived the march through the mud which claimed as many casualties as some battles; they survived a winter called by some their own Valley Forge. Now they were ready to follow a man who had a reputation for fighting.

Hooker set about getting his army into shape, weeding out the profiteers who had sold the rations allotted for his men. He abolished the practice of each man cooking for himself and set up company kitchens. Hospitals were renovated, and the sick were actually cared for. Malingerers were quickly sent back to their units. With the high command, as well, Hooker did his work, replacing incumbents with old friends. Some good appointments were made.

Hooker would not sit still, and the men marched again toward the Rapidan and Chancellorsville. The 72nd, 90th, 119th, 128th, and 143rd, all Pennsylvania Volunteer Regiments which are still represented in the present Pennsylvania Guard, were all part of the army which at last moved with confidence under a confident leader. The 119th in the Sixth Corps, under John Sedgwick, and the 143rd in Abner Doubleday's Division of the First Corps took up their separate positions, ready to help "Fighting Joe." In addition, the 48th Pennsylvania Volunteer Infantry, with members of the Washington Artillery and National Light Infantry from Pottsville, were part of the reserve which did not see action.

The Battle of Chancellorsville began and was decided only when Hooker lost his confidence (perhaps as a result of almost losing his life in an artillery barrage which struck his headquarters.) His will to drive and to win faded. Eventually, after losing several opportunities and many men, he ordered a withdrawal to winter camp at Falmouth. Almost the only advantage to the Union of

Camp Lafayette at Chadd's Ford. Companies A and B, 1st Regiment Artillery, Philadelphia Home Guard.

Pennsylvania Keystone Battery officers, Centerville, Virginia, June 16, 1863.

Pennsylvania Keystone Battery in battery front, Centerville, Virginia, June 16, 1863.

the battle was the blow to the Confederacy when General Thomas J. "Stonewall" Jackson was accidentally shot by his own men.

There was little time to speculate on what might have been, for Lee was on the move. Hooker stirred himself and tracked the Confederates so as to keep himself between them and the nation's capital. Pennsylvania units endured long, tiring marches in the days that followed, but men were tougher by 1862. They were also more cynical, perhaps less inclined to waste energy on unnecessary activities, and more inclined to be careful with their equipment and skilled in living off the land.

The game of "Musical Chairs for Commanders" was played again on June 28. A Pennsylvanian, George G. Meade, was named commander of the Army. A good man, he was combat tested, and liked by his men, even though not well known by the rest of the army which had now become his. It was fortunate that he was a man of action, for his army in following Lee, was approaching the little town of Gettysburg.

The honor list for Pennsylvanians in the great battle that ensued is not as long as those in some previous and some later battles. The high command had siphoned off some Pennsylvanians to garrisons, to Vicksburg, and to the deep South. However, for the men of Pittsburgh in the Hampton Battery, the men from the coal regions in the 143rd Pennsylvania Volunteer Infantry, the men of the Philadelphia area in the Philadelphia

Gettysburg and Vicinity.

City Squadron, Troop A, the 119th Pennsylvania Volunteer Infantry, the 90th Pennsylvania Volunteer Infantry, the 72nd Pennsylvania Volunteer Infantry, and others, there was a battle to be fought.

One of the most promising Pennsylvanians, Major General John F. Reynolds, commanded the First Corps, of which the 143rd Pennsylvania Volunteer Infantry was a part. It was Reynolds and his men, with others, who fought some of the early skirmishes of this great battle, and it was here that the nation lost a promising soldier when Reynolds died in combat. The 143rd produced its own hero at Gettysburg, Sergeant Ben Crippen. He was the colorbearer who held the colors high in the orderly withdrawal through the town after Confederate numbers grew to overwhelming proportions. An English observer with the Confederate army wrote in his dispatch:

> A Yankee colorbearer floated his standard in the field, and the regiment fought around it, and when at last it was compelled to retreat, the colorbearer retreated last of all, turning now and then to shake his fist in the face of the advancing Confederates. He was shot. General Hill [A.P.] was sorry when he met his fate.

Colonel Dana went on to state in his diary that Major Conyngham rallied the forces and saved the battle flag, which is now displayed in the Capitol rotunda in Harrisburg. Pennsylvania units fought as well with the Sixth Corps which had made a strenuous twenty-four hour hike to reach Gettysburg. Other histories, with more space and time, tell the battle story in detail.

July Fourth was a day of independence for those valiant men in blue and gray. The fighting had ended; Lee had been stopped, and it was time to head south and resume the war in the Confederacy. Men of the Sixth Corps were ready to pursue, but there were few to accompany them, and Lee's retreat was cautious and well-covered. With Gettysburg over, the men in Meade's and Lee's armies had to be cared for, with treatment for the wounded, burial for the dead, and re-grouping and re-organizing for those fit to fight again.

July Third, the end of this eastern battle, was also the end of a great western battle, Vicksburg, which permitted the opening of the Mississippi for the North. Pennsylvania was represented in this battle as well, with a unit from the present 111th Infantry from Philadelphia participating in that long siege and eventual victory.

The Blue and Gray on the Field at Gettysburg

In the Devil's Den

The new year came with no notable battles except for those who were involved. It was time for one more change in command for the Army of the Potomac. On March 9 Ulysses Simpson Grant, hero of the western campaign, officially took over the reins. Meade, graciously offering to step aside completely, was asked to remain by his old friend.

Grant proceeded to re-organize the army into more efficient fighting units, merging some of the old regiments into others and abolishing two of the existing five corps, a move begun by Meade, but attributed to Grant. The proud veterans bitterly resented this and found it difficult to adjust to new leaders and new comrades. However, the new, red-bearded commanding general inspired confidence in the real fighting men, who looked to him for an end to the ceaseless marching and countermarching that had so far produced no decisive victory.

They crossed the Rapidan again and headed south. On the path was a five by twelve mile patch of second-growth brush and forest that they had seen the year before—the Wilderness. It had to be traversed as they made their way toward Richmond. Unfortunately, Lee, with his almost mystical prescience, knew that they were in that terrible wasteland and moved to stop Grant. The Battle of the Wilderness began on May 5 with the usual skirmishing, followed by frantic moves to reinforce and by commands to advance and advance again. The problem created by the terrain was that no central command, brigade or even regimental command could function in the thickness of scrub trees and brush. The 143rd Pennsylvania Volunteer Infantry, one of the units reassigned from the old First Corps to the Fifth Corps, was in the thick of the fighting, led by Major General G. K. Warren, the engineer whose quick thinking had saved the Round Tops at Gettysburg. The 119th and 96th Regiments of the Sixth Corps, the 72nd Regiment of the Second Corps, and other Pennsylvania units shared in the bloody, desperate fighting of the Wilderness. The 119th Regiment, part of the Corps commanded by General John Sedgwick, was assigned to the Second Brigade of the First Division, under Colonel

Emory Upton—a pale, slight, thoroughly combative man who believed the best way to lead was by setting the example of being in the heat of the battle and not in the relative safety of the rear. (Upton would prove his leadership conclusively at Spotsylvania). But the Wilderness was not the place to prove leadership on a level higher than squadron or platoon. With communication practically impossible, only the sheer courage, pluck or foolhardiness of the common solider would truly decide the issue in the Wilderness. Finally, on May 7 the two armies found this particular battle was over and their relative locations virtually unchanged. Lee had ordered a withdrawal, and only an occasional skirmisher's spat told the men that there was a war yet to be won.

The men of the Pottsville area found themselves on the eve of that battle scattered among the 25th and 48th Pennsylvania Volunteer Infantry Regiments, plus one company in the 2nd Regiment of Pennsylvania Volunteer Emergency Militia, and ready for more fighting. The men of Allentown, Philadelphia, Wilkes-Barre and Scranton, and Western Pennsylvania were ready, too, and they soon got all they ever imagined. Spotsylvania was a portent of things to come, for the trench, hastily dug and later improved, became a significant way to multiply the effectiveness of his troops. The men of Lee got there first, but just barely. They were the ones who had to be dislodged, driven out, sent on their way in defeat.

The men of the 48th Pennsylvania Volunteer Infantry from the Pottsville area were part of the Ninth Corps, commanded by their former army commander, Ambrose Burnside. Their own commander was Lieutenant Colonel Henry Pleasants. They, along with other Keystone Staters, were to be involved in one more battle which would seem to be bloodier than the last, and, as history would seem to prove, not as bloody as the next. It was here at Spotsylvania that Emory Upton, with his mixed bag of Vermonters and Pennsylvanians, would try an assault on the Confederate lines which, while failing, would inspire a much larger attempt by several corps. The attack did succeed, although with the usual staff foulups. With Wright on the west, Burnside on the east, and Hancock in the center, every face of the attack had Pennsylva-

The Final Charge at Gettysburg

In The Wilderness

nia guardsmen present and fighting. The combat was totally personal, with clubbed muskets and cannon firing cannister at point-blank range. The Bloody Angle of the Battlefield offered the view the following morning of corpses piled four and five deep, with bodies so mutilated by repeated hits that they were unrecognizable. The 119th Pennsylvania Volunteer Infantry lost its commander, Major Henry Truefitt, and his successor, Captain Charles Warner, both in the fight in the Bloody Angle. Of 400 men on duty before May 5, the 119th lost 215.

The next step was one of continuous sparring with the enemy, as Grant dictated a sliding or flanking movement to the east, always in contact with Lee's army, but always moving and forcing Lee to leave his entrenchments and to follow, keeping between Grant and Richmond. The Wilderness and Spotsylvania, with the attendant side battles and skirmishes, cost the Union army over 33,000 casualties, enough losses to cause a commander to order a retreat for re-grouping, re-equipping, and re-thinking of future strategy. This was not the solution for U. S. Grant. It was to be a war of attrition, with expected large losses, bad for both, but worse for the smaller army from the South which lacked the means for replacement of men, equipment, horses, and all the other paraphernalia of war.

Finally the armies faced each other at another obscure crossroad, Cold Harbor, which was neither a harbor, nor, in that time of year, very cold. Richmond was now only about ten miles away, but Lee was there and the maneuver area narrowed. Cold Harbor was to be one more very costly battle. For once, the men in blue arrived first, in the person of Philip Sheridan's cavalry troopers, part of whom were the veterans of the 17th Cavalry, a regiment from Pennsylvania. Sheridan's men held on, while Horatio Wright and his Sixth Corps hurried to their aid. They attacked, and gained a trench line held by Lee. The 119th Regiment again felt the crush and tragedy of close combat, losing their acting adjutant, Lieutenant George Humes, among others.

The great assault on the Confederate lines was postponed for two days, owing to the great amounts of marching and countermarching beyond the lines to get into position, the extreme fatigue of the troops and situations such as that of General William F. "Baldy" Smith, whose 16,000 men brought from the Bermuda Hundred had little ammunition and were almost halved by straggling. The attack began against an army which had an extra two days to prepare and improve trenches and rifle pits. The attack took not much more than fifteen minutes to fall apart and to leave whole companies without a living soul. Not for months, under Grant, had there been such total failure, and yet, the men did not run, as, perhaps some with "sense" might have done. They dug and dug deep, making exit lanes so they could move back if necessary, to safety. Finally on June 12, Grant's army slipped out of its holes and trenches and moved in a long semicircle to a new position across the James River. For once, staff work seemed to come together, and the Confederate riflemen the next morning found only empty rifle pits and the usual trash left by an army. The objective was Petersburg. In that army moving up to Petersburg were the National Light Infantry and Washington Artillerists. They had come a long way since 1861 and now would have a chance to use their particular talents in this long, bloody siege. Delays after initial victories in isolated assaults gave Lee the time needed to bring his veterans into the trenches prepared by P. T. Beauregard's outnumbered defenders. The result was, as Catton has said, Cold Harbor all over again—no gains, only terrible losses. Grant, after seeing the futility of assault, ordered a recourse to trench warfare. Trenches, broad ditches, sandbags, logs, and the very useful abatis—brush and branches sticking out toward the enemy to impede or retard movement—all became part of the Union position. Those who see France in World War I as the beginning of trench warfare would do well to look at Petersburg and its miles of trenches, zig-zags, and bombproof shelters.

At Petersburg the Philadelphia Brigade was broken up and its survivors parceled out to other units. A company in the 24th Michigan, which had to earn its respect from other veterans, mustered a sergeant and a private and no others. Still there was an army, and the 48th Pennsylvania Volunteer Infantry found itself with a task it could probably do better than any other. Lieutenant Colonel Pleasants of the 48th passed on a remark made by one of his men to his division commander, who sent him to Burnside. Burnside approved the idea of digging under the trenches to a point where a charge could be used to blow a hole in Confederate lines. Tools and other needed equipment, promised but not delivered, were improvised. A former mine boss, Harry Reese, was picked, and the miners began their task. Within three weeks they had finished a tunnel 510 feet long, twenty feet underground, with a seventy-five-foot shaft crossing the "T". Four tons of powder were placed in the shaft and fuses set. June 29 became the time when Pleasants and his coal miners would either be vindicated, or made the butt of the both armies who had heard about a "mine," but knew very little of the facts. The charge was blown, a tremendous crater was made, and the way was opened to Petersburg, if only the generals had co-operated. Sadly for the men of the 48th who had done everything expected of them, and more, the attack was a disaster; no ground was gained and held.

Other units of Pennsylvania were busy at this time. The 119th Regiment went on a combined cavalry-infantry raid which did substantial damage to a necessary railroad which was supplying Petersburg, putting it out of action for several weeks. A footnote to the work of the 48th Regiment is the fact that before they began their mine they were instrumental in capturing the flag of a Tennessee regiment and re-capturing the flag of the 7th New York Artillery. The men involved in this deed were Sergeant Patrick Monaghan of Minersville and Private Robert Reid of Pottsville. Both men received the Medal of Honor for the feat.

The Shenandoah Valley, a "breadbasket" area for the Confederacy, claimed the attention of Grant, who wanted some of Lee's forces diverted from Petersburg and who understood the military necessity of blocking this route to the north. Consequently, he sent Sheridan into the valley with several corps, including the Sixth. First, this corps, under Wright, made a fast march to the defense of Washington and became the only unit actually to fight in the presence of the President, who had come out to watch the battle with Jubal Early's troops. The fray ended quickly enough, and orders were sent ordering the Sixth Corps to the Shenandoah Valley, where a new commander, William T. Sheridan, awaited them. Sheridan was a driving man who demanded much of his men,

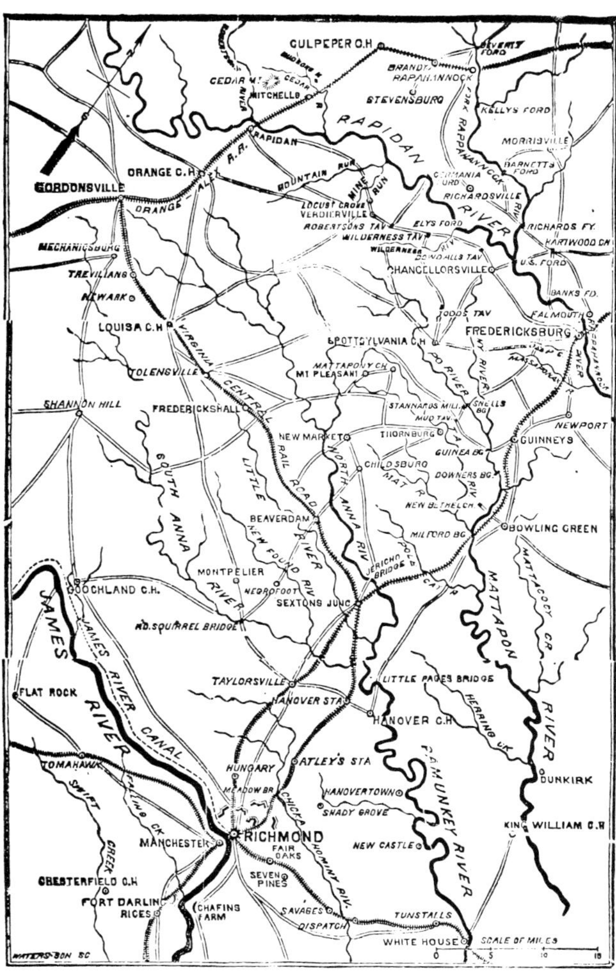

Map of Virginia, showing Fair Oaks, The Wilderness, etc.

and who was willing to give at least as much of himself. The men of the Allen Rifles, the National Light Infantry, and of the 47th and the 96th Pennsylvania Volunteer Infantry were among those men of Sheridan. Weeks were spent in rest, training, outfitting and re-organization. September 19 saw Wilson's cavalry and masses of infantry win over Early's forces. This victory at Winchester, coupled with one at Fisher's Hill and Sherman's capture of Atlanta, sealed the fate of the Peace Party back home. A brief, almost successful, drive by Early was slowed, then stopped, by the Sixth Corps. Sheridan, who was away, thundered back, rallying troops as he came, finally putting to rout the valiant, but outnumbered Confederates. The 199th and the 87th Pennsylvania Volunteer Infantry shared in the victorious Shenandoah campaign.

The 143rd Pennsylvania Volunteer Infantry was a part of a little known episode that should be noted. Moving out on December 10, 1864, to do more damage to the Weldon railroad which had been visited earlier by Wilson and the 119th Regiment, they liberated and consumed a supply of potent apple jack, ostensibly to help ward off the "miseries" of hard work and a freezing rain. A unit historian reported:

> Under these circumstances there was much disturbance caused by apple jack in which our regiment, probably being remote from the supply, took little active part. As a result, most likely, of this abstinence, we were placed on picket duty repulsing a charge by Rebel Cavalry and becoming rear guard next day.

PVT Lew McMakin, 1st Troop, Philadelphia City Cavalry, Troop No. 674.

Much more blood would be shed, but the tide was clearly against the South, and some of the units were already home. Men from the 72nd Pennsylvania Volunteer Infantry from the Philadelphia area were mustered out after their three-year enlistments expired in August 1864. (Individually, some surely re-enlisted, for assignment to other units.) This regiment had fought in all the major battles of the eastern campaign from Manassas through Petersburg. For the men of the Reading area in the 128th Pennsylvania Volunteer Infantry discharge came with mustering out in May 1863, after service at Antietam, Chancellorsville, and the general Virginia campaign of 1862. In the 90th Pennsylvania Volunteer Infantry, part of the First Corps at Fredericksburg and Gettysburg and of the Fifth in the Wilderness, the end of the war as a unit came in November 1864. For others, discharge came in the middle or closing months of 1865. The 143rd Regiment closed its Civil War career with garrison duty on Hart's Island in New York Harbor.

The 119th Pennsylvania Volunteer Infantry, possibly the busiest, most engaged regiments of the National Guard ancestors, as part of the Sixth Corps, was back at Petersburg after the Shenandoah campaign as part of the final assault. Finally it was one of the commands which harassed Lee right down to the meeting at Appomattox which sealed the doom of the Army of Northern Virginia. The Regiment was quickly moved to North Carolina, as insurance against a further stand by Joseph Johnston's Confederates. It was mustered out in June 1865.

Thus it was with the men of Pennsylvania, a portion, only, of a vast army of the Union, but one whose contributions were many and honorable. They were mostly volunteers, re-enlisting at times when they could have gone home. They were good fighters, led for the most part by intelligent, courageous men, whose major fault was the misfortune, on occasion, to be led on a higher level by incompetents, or "dalliers" who caused blood to be shed and lives to be lost by the failure to follow up on gains made by lower level decisions. The Keystone State, the only state of the North to know a major battle, had done its bit, with men and material. Its factories had excelled in production, and its towns, rural hamlets, and farms had provided their best—fathers, brothers, uncles, nephews, old and young, who did their best for Pennsylvania, America, and "Father Abraham."

CHAPTER 5

1866 to 1891

Uzal W. Ent, Colonel, Hq, 28th Inf Division

OVER 360,000 Pennsylvania men, organized into 210 regiments and numerous smaller formations, served in the Civil War. By 1866 this vast host had been discharged and returned to civilian life.

Reorganizing the Pennsylvania militia after the war was a slow process. The Commonwealth was divided into twenty, later twenty-one, geographic zones, called "divisions." These divisions had no standard organization but were composed of company districts. Not all men could serve in their home areas, for an Act of Legislature in 1864 had disenfranchised all who either had been guilty of desertion during the Civil War or who had failed to report when drafted. Lists of these men were secured and the offenders stripped of all rights of citizenship.

As reorganization continued in 1867, a problem was encountered because the State system required a minimum of eighty-three enlisted men for each new volunteer company, whereas, in the First Division (Philadelphia), only thirty-two enlisted men were required. Further, local Philadelphia law required everyone subject to military duty to pay a fine of two dollars per year if he did not belong to a military company. Funds derived from these fines were used to support existing militia units. The Adjutant General suggested a law standardizing the authorized strength of companies at fifty. He also suggested that each company be required to fill its roster to eighty-three men in time of war. He recommended that the fine be extended throughout the State but be reduced to one dollar in sparsely populated areas. Finally, he suggested that State law follow Philadelphia practice by paying each member of the militia up to twenty dollars per year for their service, using fine monies for the purpose.

By 1867, organization tables not withstanding, only seven of the twenty geographic divisions had any militia troops in them. There still was no uniformity of organization in these divisions. Men paid for their own uniforms and equipment, except for rifles, although they even had to pay freight charges for delivery of these weapons.

In July 1869, the First Regiment of Philadelphia established a precedent in Pennsylvania by holding an eight day summer camp at Cape May, New Jersey. In 1870 the name "militia" was dropped, and it became by law in Pennsylvania the "National Guard of Pennsylvania."

In 1870 the Adjutant General reported a provisional brigade of "colored" troops in Philadelphia, consisting of the 11th, 12th and 13th Regiments. The report also listed ten other separate companies of black troops which had been organized throughout the Commonwealth. Although the officers were white, these "colored" troops numbered almost 2,000 black Guardsmen.

By the end of 1871, there were nineteen regiments, three separate battalions and 175 separate companies of infantry, six of artillery and six of cavalry in the National Guard of Pennsylvania.

Prior to the formation of the Pennsylvania State Police in 1905, the Governor had no police or military force with which to restore order or enforce the law, except for the militia or National Guard. At the county level, most sheriffs' offices consisted of the sheriff and a few deputies. If the law enforcement problem became too great, the sheriff had to attempt to organize a posse. In order to secure further assistance he had to ask the Governor for the Guard. This is the reason why the National Guard had such a prominent part in the civil disorders which plagued Pennsylvania in the last quarter of the 19th Century.

In late 1870 Scranton area coal mining companies threatened a large pay cut, leading to a limited strike by their employees in December. The walkout became general in early 1871. By April many miners were destitute and ready to return to work. Rioting broke out between this faction and other miners who wanted to hold out. Major General Edwin S. Osborne's Ninth Division was sent to Scranton on appeal from the mayor of the city.

After alerting his division for duty General Osborne proceeded to Scranton. Upon arrival, about two o'clock on April 7, and discovering considerable confusion and disorganization, he telegraphed his commanders to report to him with their troops without delay. By 9 that evening, Guardsmen were posted throughout the city, and order was restored.

On the following afternoon, after consulting with the Mayor and local businessmen, the General withdrew most of the troops from the city. That night, Lieutenant Miles Wenner and Corporal William T. Care, both of the Hazleton Zouaves, were killed in an accident. (The record does not show the nature of the accident.) The General immediately ordered Companies C and E of the 15th Regiment back into the city. These two companies were reinforced by three more companies. In all, from April 7 to May 24, 1871, some 400 men were on duty within Scranton, and another 600 men were encamped at various points nearby.

The Guard provided security for meetings of men who wanted to return to work. On one occasion some forty men decided to return to work in one of the mines. A military escort was provided, to be continued on a daily basis. Angry men, women and children flocked to the scene and had to be held at bay with the bayonet.

On Wednesday, May 17, an unusually large crowd gathered to threaten and curse the escorted miners as they went slowly and silently through the mob, on the way home. Suddenly, stones were thrown from the crowd, striking several miners. A shot rang out, killing two of the mob. The crowd instantly fled, leaving the dead bodies where they fell. That shot served to halt the strike and restore order. On Monday, the twenty-second, work was resumed throughout the coal field, and the Guard was relieved on May 24.

General Osborne in his report mentioned that a miner was tried and acquitted for firing the fatal shot. William H. Zierdt, in his *Narrative History of the 109th Field Artillery* states that "... a squad fired into the rioters, killing two of them, ..." The question as to who fired was never resolved.

The strike of 1871 was only the first of a whole series of labor disorders which disrupted life in Pennsylvania during the late 1800's. Major strikes and riots occurred in 1872, 1874, 1875, 1877, 1892 and 1897. The strikes between 1874 and 1877 came as the result of a severe national economic depression in 1873. The worst disorders came in 1877 and 1892. Terrible and bloody as some of these upheavals were, the Guard gained much valuable experience in rapidly deploying, transporting, providing logistical support for, and commanding and controlling fairly large bodies of troops. The experience proved invaluable to the officers and men of the Pennsylvania Guard during the Spanish American War.

In July 1872 mobs attacked a number of saw mills near Williamsport, forcing the owners to close down. On the twenty-third, two local Guard units were ordered to duty in order to protect the mills and their workers. Major General Jesse Merrill was placed in command.

Five more companies were deployed to Williamsport. One company guarded the local jail, while other Guardsmen escorted civil authorities who were apprehending ringleaders of the mob. More troops were needed, and two companies from Shamokin joined the seven already at Williamsport. Guard posts were established in order to protect a line of saw mills some five miles long. Under this protection, some mill owners renewed work. Encouraged, others also reopened. By August 1, order was restored and the Guard released from duty.

A State law was passed in 1873, providing that a small amount of money be given to each company which passed its annual inspection and was otherwise qualified. As a result, $40,000 was divided among deserving units.

Adjutant General James W. Latta first recommended a reduction in the number of divisions in the National Guard of Pennsylvania. In his report of 1873, he pointed out that "twenty-one divisions, with that number of [generals] are too many for a force" which could never exceed 20,000 men. "The good of the service ... demands that the number of divisions be reduced," he concluded. General Latta also reported that the multitude of separate companies was being organized into regiments and battalions.

General Latta's recommendation for the reduction of the number of divisions was approved, when Governor John F. Hartranft on June 30 directed that the Commonwealth be divided geographically into ten divisions. All but eight infantry, four artillery and nine cavalry companies were organized into regimental or battalion formations.

On March 25, 1874, men from the shops of the Erie Railroad went on strike, in a wage dispute, seizing the Susquehanna Depot the following day and stopping all train traffic. Three days later Major General E. S. Osborne's Third Division, composed of the 9th Regiment, Wyoming Artillery, M'Clellan Rifles, 3rd Infantry Company and the Telford Zouaves—all from the Scranton-Wilkes Barre area—, were sent to Susquehanna. Upon arrival General Osborne requested the 1st Regiment of Philadelphia be sent to reinforce him. The 1st started to Susquehanna on the twenty-ninth, but because the train carrying the regiment could not get through the heavy snow, the troops marched the last four miles through snowdrifts, arriving at 12:10 P.M. the next day.

The Burgess of Susquehanna, W. J. Falkenburg, telegraphed the Governor, protesting the Guard's intervention. The Governor replied that, while he sympathized with the workers in not having been paid, they had illegally seized railroad property. Since National Guard troops had been sent to help the sheriff, the Governor continued, the burgess should contact General Osborne.

Under the General's orders railroad company officials cleared the railroad tracks, and passenger service was restored by the afternoon of March 29. After he talked to a delegation of strikers, all train traffic was resumed by noon the next day. The Guard was relieved of duty on the last day of the month.

The Coal Exchange attempted to reduce wages in late 1874, precipitating what became known as the "Long Strike" in the Schuylkill coal fields. The strike started January 1, 1875. Hazleton went out later that month, followed by the miners around Shamokin in early April. Except for workers in the Wilkes-Barre area, most other miners in the northern coal fields accepted the Coal Exchanges pay reduction.

Although no mining or transportation of coal took place in the entire area encompassed by Hazleton, Shamokin, Tremont, Pottsville and Shenandoah, matters remained relatively calm. In mid-March the Philadelphia and Reading Railroad Company tried to bring men into Shamokin from Berks County in order to operate their road. Striking miners fired on the train carrying the workers, ending the railroad company's ideas of running their line.

On March 31, Governor Hartranft ordered Major General Joshua K. Sigfried, commanding the Fourth Division, to duty in Schuylkill County. The Governor also issued a proclamation that day, aimed at the strikers. The combination of the proclamation, impending activation of the Guard and the energetic efforts of the Sheriff, Frank J. Werner, seemed to calm the situation.

Meanwhile, in late March the miners of the Lehigh region learned that the operators and coal transportation firms planned to combine. Believing this action would break the strike, miners and workers destroyed mining company property and intimidated other workers. Sheriff W. P. Kirkendall of Luzerne County appealed to the Governor for help. On April 4 the Governor authorized

Major General Edwin S. Osborne of the Third Division to assemble such troops "as you deem necessary... ." The General called in three separate infantry companies, an artillery battery and both the 1st and 9th Regiments.

The entire force was assembled at Hazleton April 7. Companies A and I, 9th Regiment, were sent to Buck Mountain. The next morning five companies of the 1st were dispatched to Jeddo, Eckley, Drifton, Hughland and Ebervale. One company of the 9th was deployed to Jeansville, and another, together with the Telford Zouaves, was stationed in Audenried. The remainder of the force stayed in Hazleton as a reserve. This deployment provided security for the most important points in the area.

On Monday, April 12, the M'Clellan Rifles were sent to head off trouble in Pittston, where they remained for eleven days. The 1st Regiment was returned home on April 27. Nine companies of the 9th, plus the Telford Zouaves, remained on duty until May 11, when all but three companies were sent to Wilkes-Barre. Finally on Tuesday, May 18, all Guard troops were relieved and sent home.

The Schuylkill fields remained quiet until June 2, when a number of collieries reopened. Work was disrupted the next day. A crowd exchanged gunfire with a sheriff's posse at a colliery in Mahanoy City, and a coal breaker between Shamokin and Mt. Carmel was burned.

General Sigfried immediately sent Companies F and G of the 7th Regiment to report to Sheriff Werner in Mahanoy City, and Companies B, D, H, I and K of the 7th were sent to the sheriff at Shenandoah. At both places large crowds were in the streets, businesses were closed and the citizenry frightened by the mob. The arrival of the Guard calmed their fears, and, although roving bands of dissidents kept the troops busy, order was restored.

On June 4 the General got word that a party of strikers intended to "visit" Shamokin. Company D, 8th Regiment, from Harrisburg, was sent to Shamokin early next day, causing the mob to disperse without incident. The unit was relieved from duty June 7. The Ashland Dragoons was placed on duty for one day in June because of a threat to a Shamokin colliery.

Troops of the 7th Regiment were relieved on June 12 by request of their regimental commander, Colonel Alexander Caldwell, and were replaced by five companies of the 8th. The Colonel based his request for relief on the fact that his men had responded promptly to the call of duty and had been performing it for ten days at great personal financial loss.

Companies D and I of the 8th went to Mahanoy City, and A, F and G were sent to Shenandoah. The Guard remained on duty until June 23 when, in spite of protests from local civil authorities, the troops were all sent home.

General Osborne remarked in his report: "The sacrifice which men like these make, for the public good, ... by neglecting important private interests, is not ... fully appreciated."

No civil strife marred 1876 in the Commonwealth. The National Guard of Pennsylvania was for the first time assembled at one place and time for the International Exhibition. The entire force, less the First Division, was encamped at Camp Anthony Wayne, Fairmount Park, Philadelphia, for ten days, starting August 3. The First was kept busy extending military honors to visiting soldiery.

General Latta in his report complained that there were still too many Major Generals, and that the sparse population in many parts of the State could not support all the divisions authorized. He recommended a reduction to four divisions, with such brigades as would "best serve convenience and efficiency."

Company F of the 4th Regiment was ordered to provide security around the jail in Mauch Chunk (present-day Jim Thorpe) on June 21, 1877, while four members of the "Molly Maguires" were hanged. The duty was performed without incident.

The depression of 1873 continued into 1877 with unemployment increasing and many men who were employed forced to take large reductions in pay. In his book, *The Slavic Community Strikes*, Victor R. Greene says that mine operators in the anthracite region of Pennsylvania lowered wages by ten to fifteen percent twice in 1876 and again in 1877. On June 1, 1877, according to Wayland Dunaway's *A History of Pennsylvania*, the Pennsylvania Railroad Company reduced the wages of its employees for the second time in four years. The cut led to what Dunaway describes as "the most extensive strike in railroad history... ."

Before order was restored in Pittsburgh, some four thousand Guardsmen and six hundred Regulars were employed. Damage to railroad property in Pittsburgh amounted to almost $3,000,000. Some 1,600 railroad cars and 126 locomotives were burned, and the railroad hotel, its elevator and the Union Station were all destroyed. Colonel Frederick C. Hitchcock, in *History 13th Regiment, National Guard of Pennsylvania*, states that some 25,000 to 40,000 men were idled in the Lackawanna Valley because railroads were not transporting coal. In Áugust a mob of 5,000 men attempted to loot the city of Scranton.

Violence erupted in Pittsburgh on July 19, when a mob halted all freight train traffic out of that city. Major General Alfred L. Pearson, commanding the Sixth Division in the Pittsburgh area, was ordered to duty. He first called on the 19th Regiment, then on July 20 ordered the 14th and 18th Regiments to duty. However, the General was unable to muster enough troops to influence the situation, so Major General Robert M. Brinton's First Division, from Philadelphia, was ordered to Pittsburgh. Enroute, the division was instructed to pick up two Gatling guns and 20,000 rounds of ammunition at the arsenal in Harrisburg.

Some 550 men of the First Division left Philadelphia at 2 A.M. two days later, stopped at Harrisburg, as ordered, and arrived in Altoona at 9:30 in the morning. After eating a hasty meal of coffee, bread and jam, the contingent proceeded on its way, augmented by fifty men from the 2d Regiment and forty-seven from the 6th. The force finally arrived at Union Station, Pittsburgh at 2 o'clock in the afternoon.

While the First Division was enroute to Pittsburgh, the combined strengths of the 14th, 18th and 19th Regiments rose to about 600 men. The 18th was on duty at Torren's Station, while the 14th and 19th, along with Breck's Battery, occupied the Union Depot, clearing the tracks of rioters near that point.

When General Brinton's First Division arrived, each man was issued an additional ten rounds of ammunition, making twenty rounds, in all, per man. The men were also fed some sandwiches and coffee. The division was split, with the Second Brigade being detailed under General Pearson's orders, but against the desires of General Brinton, to guard a train which the Pennsylvania Railroad intended to run out and to protect the railroad in that area. Another regiment was directed to guard a crossing, leaving General Brinton with the 1st Regiment, Weccacoe Legion, Washington Greys and Keystone Battery (pulling the two Gatling guns by hand).

This force moved from the Union Depot to 28th Street, where the 14th and 19th Regiments were supposed to have cleared rioters away from the railroad. Such was not the case. The 19th was out of position, and both the 14th and artillery battery were surrounded by thousands of men, women and children, who milled around the troops. The 14th had become ineffective and the commander of the 19th feared that many of his men sympathized with the rioters. As a result, both regiments were no longer effective.

In the meantime, as the 1st Regiment pushed its way along, assisted by the sheriff, the dense crowd of dissidents became more belligerent. The 1st deployed in a three-sided square in order to press the mob back. As the command came to "charge bayonets," the troops were attacked by a shower of stones. Three soldiers were felled by the flying rocks, pistol shots rang out from the mob, members of the crowd tried to disarm some of the soldiers, and finally some of the troops fired without orders. Before the firing could be stopped a number of rioters had been felled by the gunfire. Fifteen soldiers had also been shot and six others felled by stones. Guard casualties in this skirmish amounted to five percent of the number of men present.

After Generals Pearson and Brinton discussed the matter, they decided to occupy the offices, buildings and roundhouses for the night, in order to prevent the mob from burning them.

The 1st Regiment manned the brick roundhouse, while the Weccacoe Legion and Washington Greys guarded the two main entrances from the railroad side. The 2d and 6th Regiments composing the Second Brigade of the First Division, occupied the offices and upholstery shops near the roundhouse. The Gatling guns and Breck's Battery (from the Sixth Division) covered the 26th Street gates.

The 14th and 19th Regiments were placed in the wooden frame transfer offices and sheds along the tracks between 23d and 24th Streets.

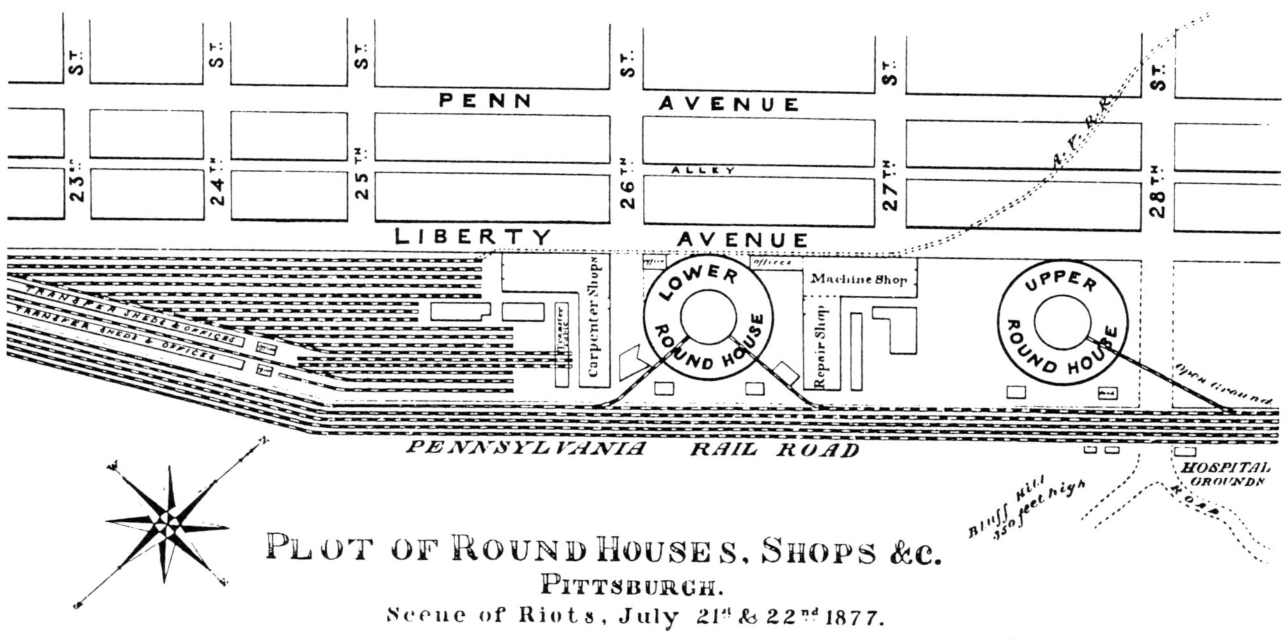

Plot of Roundhouses, Shops &c.

The First Division remained besieged in the roundhouse and outbuildings all night on July 21. No food, water, ammunition, or reinforcement ever reached them. During the siege two men were wounded at point blank range by rioters firing pistols. Other rioters attempted to bring a captured 12-pounder cannon into action against the roundhouse. They refused to obey an order to disperse, and eleven of their number were shot down by a volley of musketry. The mob broke every window in the office building and fired so rapidly into the telegraph room on its first floor that the floor had to be abandoned.

Rioters broke down the iron gates and were halted only by musket fire and the advancement of the Gatling guns. Finally the crowd set fire to the 28th Street roundhouse and adjoining buildings and tried to burn the Guardsmen out of their positions at the 26th Street roundhouse and buildings by running burning railroad cars down the tracks. Some cars were derailed, and the fires caused by those which struck the buildings were brought under control by the defenders.

General Pearson (Sixth Division) went during the night in search of food and ammunition for the First. Meanwhile, the Adjutant General sent a message to General Brinton saying that he was arranging to send ammunition and reinforcements, which should arrive by five or six o'clock the next morning. He also advised that, if escape became necessary, General Brinton's men should withdraw east toward Torrens and Colonel Guthrie's 18th Regiment.

When no relief came the next morning, General Brinton decided to break out of the roundhouse. He formed the exhausted division with the 1st Regiment leading, followed by the remainder of the First Brigade. Next came the Gatling guns. The 6th Regiment, Second Brigade, brought up the rear. Since there were no horses to move the two guns of Breck's Battery, they were spiked, making them useless to the mob.

The column moved out onto 25th Street, a block further west than the mob had expected, then proceeded east on Penn Avenue. The crowd attacked the rear of the column, firing into the marching Guardsmen. The march continued in this way for more than a mile, when the troops halted and fired into their attackers. The Gatling guns also opened fire. This sudden, violent retaliation scattered the crowd, and they never reassembled.

In two days of sometimes savage action four Guardsmen had been killed and another fatally wounded. Many others had been injured. Civilian

Locomotive Bells from the Roundhouse defended by the 1st Regiment.

Destruction wrought by the Rioting in Pittsburgh, 1877. Note the ruins of a roundhouse.

dead were estimated at between sixteen and twenty-five.

The First Division was ordered to Altoona on Monday, July 23, but got only as far as Blairsville, where it met some 300 more men from the Division who had been trying to make their way to Pittsburgh. Lacking transportation for the combined force, the command camped at Blairsville for five days when it returned by train to Pittsburgh, escorting Governor Hartranft and Adjutant General Latta.

On August 1st the division was initially ordered to Philadelphia but was redirected to Scranton. The trip was slowed by rioters, who damaged the track. Skirmishers had to be used. Finally, a car with a Gatling gun mounted on it was placed ahead of the engine. The division was relieved from duty in Scranton by the Fourth Division on August 4. More National Guard troops were sent to Pittsburgh, including the Eighth, Ninth and Tenth Divisions. The last of these men were relieved from duty August 7.

The strike did not confine itself to Pittsburgh railroads but was general in rail yards and railroad stations throughout the Commonwealth. Severe confrontations or actual rioting occurred at Altoona, Huntingdon, Harrisburg, Philadelphia, Wilkes-Barre, and in the Lehigh and Schuylkill valleys, including Allentown and Bethlehem. Exchanges of gunfire in Scranton between a mob and a sheriff's posse resulted in the deaths of three rioters. In Reading a fight with the insurgents resulted in the injury of over 200 Guardsmen out of a force of 253, and eleven dead and fifty to sixty wounded among the mob.

The timely deployment of the Third Division, consisting of Wyoming Artillerists, McClellan Rifles, Telford Zouaves, Scranton City Guards (formed during the strike) and the 9th Regiment, into Wilkes-Barre on July 23 headed off most of the trouble in that city. Although strikers stopped all trains entering Wilkes-Barre on August 1, the city did not experience severe violence.

General sympathy by many people in the Commonwealth with the strikers interfered with the assembly and deployment of many elements of the Guard. Companies, regiments and even parts of entire divisions were delayed or diverted from reaching destinations by widespread civil disorder and wanton seizure and destruction of railroad property. In truth, some Guard members found themselves in a dilemma created by their private views on the strike and their sense of duty as members of the National Guard. With few exceptions they performed their duty most creditably.

The Fourth Division, consisting of the 7th and 8th Regiments, was deployed to Harrisburg on July 23 to guard the State Arsenal. A part of the division had to detour Reading because a mob had already torn up track and burned the bridge over the Schuylkill River. Harrisburg's Company D (City Greys) of the 8th Regiment was the first unit deployed, having been posted at the Arsenal July 22.

The First Troop, Philadelphia City Cavalry, in attempting to join the First Division, got as far as Altoona. They could go no further because of rioters and decided to return to Philadelphia. Along with a detachment of forty-seven men from various Philadelphia units, the thirty-two man troop started back toward Harrisburg. On July 23 the Troop was found at Progress, near Harrisburg, and was attached to the Fourth Division. On July 26 it was ordered west to join the First near Pittsburgh.

Members of a mob in Harrisburg broke into and looted the city's gun stores. The citizens of Harrisburg, in disgust, then organized a posse, under command of the sheriff. Forming regular patrols, they soon restored order, arresting riot leaders.

The 7th and 8th Regiments, stationed at the Arsenal, were never called upon to intervene in the city. On July 25, the 8th was detailed to guard the railroad bridges at Rockville, Dauphin and Marysville. On the 31st, the entire Fourth Division was redeployed to Pittsburgh. After an uneventful tour there, the command was relieved on August 9 and 10.

General James A. Beaver, commanding the Fifth Division, arrived in Altoona on July 22 with about 130 men of his command and 200 more from the First. While the General tried to secure engines with which to send First Division troops on to Pittsburgh, strikers seized the engines from their crews, ran them into the roundhouse, blockaded the gate to the railroad yard and spiked the switch so that it could not be moved. Repeated attempts by the Guard to break through and retake the engines were defeated. Finally General Beaver withdrew his men, placing the Philadelphia contingent in the cars which they had used on their trip to Altoona and deploying his own troops near the railroad shops.

On July 26 and 27 four companies of the 12th Regiment and one company and part of another of the 5th (both regiments part of the Fifth Division) arrived to reinforce the command. These units had either forced themselves onto trains, or had marched overland to reach Altoona.

Governor Hartranft with the Tenth Division passed by rail through Altoona on July 27. Strikers ran a stolen engine up the mountain in advance of his train. It was believed that the dissidents intended to run this engine down the mountain and ram the Governor's engine. However, the strikers' engine ran out of steam and was later found on a siding. The Fifth remained on duty at Altoona until released on August 9.

Meanwhile, the Seventh Division (15th and 17th Regiments), scattered across thirteen counties of northwestern Pennsylvania, was ordered to assemble on July 20. Assembly was severely hampered by lack of rail transport and crowds of belligerent men in various towns. Most of the companies had to march overland much of the distance to their rendezvous. One Guardsman died of overexertion on the march.

Companies B, C and I, 17th Regiment, were left in Erie by request of the mayor to help him maintain order. The remainder of the Division reported to Pittsburgh on July 28, performing security and railroad patrol duty until August 1, when it was sent to Wilkes-Barre, escorting Governor Hartranft. (The First Division was sent from Pittsburgh to Scranton the same day.) The Seventh moved in two train sections. Although the first section was subjected to showers of rocks and coal which were thrown at it, the second section had to halt frequently in order to remove obstacles from the tracks. The division remained in Kingston and Wilkes-Barre August 2-4, securing the railroad locally and protecting employees of the Lehigh Valley Railroad as they worked.

On August 4 the division moved to Scranton, relieving the First. During their stay in that city, the Division, among other tasks, provided protection for members of the posse who had fired into the mob in Scranton, as mentioned earlier.

The assembly of the Ninth Division, consisting solely of the 13th Regiment, was also delayed by lack of transportation and bellicose crowds. Company A had the worst experience. As they marched to the depot in Butler on July 21, a large mob tried to stop the troops by hurling stones at them. Although the men had no issued ammunition, some of them had supplied themselves with a few rounds of "buck and ball," a combination of buckshot and a musket ball. When the crowd began throwing rocks, one of the Guardsmen fired in the direction of the mob. Although no one was hurt, the rock throwing ceased immediately, and the Company boarded the train.

The Company Commander discovered that the water pipe between the water tank and boiler of the engine had been cut. The enterprising commander had it repaired, and the Company proceeded on its way. This unit, along with several others, was halted at Butler Junction, twenty-seven miles north of Pittsburgh, where they remained until the next day, when the Division Commander sent them all back to their unit assembly areas. The Division remained scattered in various company rendezvous until July 26, when it was ordered to Blairsville Junction. On the day following, the Division joined the forces accompanying the Governor to Pittsburgh, arriving there on Saturday morning, July 28. The division remained on duty in and around East Liberty and Torrens Stations until relieved on August 7.

General J. R. Dobson's Tenth Division (Washington Troop, Delaney Guards, Griffin Battery and 11th Regiment), except for the Delaney Guards, which were sent to help the sheriff at West Chester, and Company H, 11th Regiment, which was unarmed, all assembled at Paoli on July 23. On the next day the Division was ordered to Philadelphia by the Governor. The same day General Bolton of the Second Division telegraphed General Dobson to loan him all the troops he could and to send them to Reading with "plenty of ammunition." General Dobson also received an order from the Adjutant General that day directing him to send three companies to Reading. When General Dobson telegraphed the Adjutant General, informing him of the Governor's orders, he was told to comply with the Governor's directive. Meanwhile, General Winfield S. Hancock, in Philadelphia, telegraphed General Dobson that he had heard nothing from the Governor about the Tenth reporting to him (General

Hancock). He then went on, in effect, to refuse the help of the Tenth. General Dobson, no doubt a perplexed man at this point, telegraphed the Adjutant General about General Hancock's message and requested permission to take the Division to Reading. General Latta replied that Reading was quiet and that the Tenth should bivouac for the night. On July 26 the Division joined the Governor's train which was headed for Pittsburgh and arrived the next morning.

The Tenth was placed near 31st Street, a few blocks from the roundhouse which had been defended by the First a few days previously. One company of the Tenth was without uniforms and were serving on duty in civilian clothes. Some of the strikers thought they could take advantage of this. The plan was for some of the strikers to infiltrate, mingle with the civilian-clad Guardsmen and make as many friends as possible. At a prearranged signal these infiltrators were to spike the Guard's cannon and other strikers rush the position. However, one of the infiltrators was apprehended, and the plan was discovered. As a result, when a large body of dissidents gathered, expecting to be able to rush into a camp of unsuspecting Guardsmen, they found to their dismay that everything was in order, the troops on the alert, and that their plan had been thwarted.

At 6 in the morning of July 29, under escort by details of men from Companies E and I of the 11th Regiment, trains were started east, opening the Pennsylvania Railroad for the first time in two weeks.

The Sixth Division vindicated itself when recalled to duty on August 1. Over one thousand men assembled, and the command was sent to "guard the road and property, from below Nanticoke, to Kingston." General Pearson placed his men in a series of pickets and guard posts throughout the area, securing coal breakers and bridges across the Susquehanna. On August 4 the 20th Regiment reinforced the Sixth Division. On August 10 elements of the Sixth Division relieved the Seventh in Scranton, extending their area of operations to do so. The command was finally relieved from duty on September 3.

The entire National Guard of Pennsylvania was activated for varying periods during the strikes of 1877, but a need was perceived for additional troops. Accordingly, on July 25 the Governor accepted the offer of a regiment recruited from the Veteran's Corps, 1st Regiment, National Guard of Pennsylvania. The regiment, designated the 20th, consisting of more than 600 officers and men, was recruited, uniformed, armed and equipped in two days. They were enroute to join the First Division in Pittsburgh by mid-day on the twenty-eighth. The 20th was later moved to Scranton, as noted above, where it was relieved by another emergency reigment, the 1st Regiment Volunteers, on September 20.

The Governor realized he needed a force for long term service in the Wyoming Valley. The 1st Regiment Volunteers was therefore organized of volunteers from National Guard regiments which had already served on strike duty. Authorized by orders of August 17 and 19, the regiment was formed and ready for duty by mid-September. The term of service was set at a maximum of three months.

Another emergency regiment, the 21st, recruited from veterans who were members of the Grand Army of the Republic, had reached the size of a four company battalion. The remaining six companies were ready for muster, when the need for additional troops no longer existed. Those already in service were mustered out. During the emergency three additional companies were recruited for the 4th Regiment and four for the 16th in order to bring these commands up to the authorized ten companies per regiment.

General Latta in his report for 1877 renewed his recommendation to Governor Hartranft that the National Guard of Pennsylvania be reorganized, suggesting, for the first time, a single division and "but one major general, with three or four brigades." The arrangement of troops, he continued, would then conform "to the organization of a division of an army... ."

The experience of the widespread civil unrest of 1877 demonstrated more forcefully than ever how cumbersome and unwieldy the multiple-division structure really was. There were too many generals, staffs and officers for the number of troops in the force. Many "divisions" had only sufficient troops to warrant a colonel in command. The large number of generals, staffs and officers divided responsibility, contended General Latta, and worked to the disadvantage of the Guard in discharging duty. If any "good" came out of the

HEAD-QUARTERS NATIONAL GUARD OF PENNSYLVANIA,
ADJUTANT GENERAL'S OFFICE,
HARRISBURG, *June 15, 1878.*

GENERAL ORDERS,
No. 2.

The following act of the Legislature is published for the information and government of all concerned :

AN ACT

Amending the provisions of the act, entitled "An act for the organization, discipline, and regulation of the militia of the Commonwealth of Pennsylvania," approved May fourth, eighteen hundred and sixty-four, and the supplements thereto.

SECTION 1. *Be it enacted by the Senate and House of Representatives of the Commonwealth of Pennsylvania in General Assembly met, and it is hereby enacted by the authority of the same*, That section one of the act entitled "A further supplement to the act of fourth May, one thousand eight hundred and sixty-four, entitled 'An act for the organization, discipline, and regulation of the militia of the Commonwealth of Pennsylvania,' approved the fifteenth day of April, one thousand eight hundred and seventy-three," which provides as follows, namely : " That in time of peace, the National Guard shall comprise an aggregate not exceeding ten thousand officers, non-commissioned officers, musicians, and privates, and shall consist of two hundred companies, fully armed, uniformed, and equipped, to be distributed among the several military divisions of the State, according to the number of its taxable population ; but the Commander-in-Chief shall have power, in case of war, insurrection, invasion, or imminent danger thereof, to increase the force beyond the said ten thousand, and organize the same, as the exigencies of the service may require," be, and the same is hereby amended, so that the same shall read as follows : " SECTION 1. That in time of peace, the National Guard shall consist of not more than one hundred and fifty companies of infantry, five companies of cavalry, and five batteries of artillery, fully armed, uniformed, and equipped, to be allotted and apportioned in such localities of the State, as the necessity of the service, in the discretion of the Commander-in-Chief may require, and organized into such divisions, brigades, regiments, battalions, and independent companies, with power to make such alterations in the organization and arrangement thereof, from time to time, as he may deem necessary : *Provided*, That the major generals of the line shall never exceed one, nor the brigadier generals five. But the Commander-in-Chief shall have power, in case of war, invasion, insurrection, riot, or imminent danger thereof, to increase the said force and organize the same, as the exigencies of the occasion may require : *Provided*, That whenever an officer shall be re-commissioned, within six months after the expiration of his original commission, in the same grade, or in a lower grade than that in which he has served in the National Guard, his new commission shall bear even date with, and he shall take rank from the date provided for in his former commission."

SECTION 2. That section two of the said act, which provides as follows, namely : " The organization of the National Guard shall conform generally to the provisions of the laws of the United States, and the system of discipline and exercise shall conform, as nearly as may be, to that of the army of the United States, excepting that the minimum standard of a company shall be forty non-commissioned officers and privates," be, and the same is hereby, amended, so that the same shall read as follows : " SECTION 2. The organization of the National Guard shall conform generally to the provisions of the laws of the United States, and the system of discipline and exercise shall conform, as nearly as may be, to those of the army of the United States, and the Commander-in-Chief is hereby authorized to make changes and alterations therein, but such modifications shall conform, as nearly as practicable, to said laws, system, discipline, and exercises, excepting that the numbers of said companies shall consist of not less than fifty, nor more than sixty, non-commissioned officers and privates, and each non-commissioned officer and private shall, at the time of enlisting, sign two enlistment papers, according to a form prescribed by the Adjutant General, one copy of which shall be retained by the company commander, and one forwarded to, and filed in, the office of the Adjutant General ;

The General Order Setting Forth the Act of Legislature creating the Division.

terrible events of 1877, it was the realization that the National Guard of Pennsylvania *had* to be reorganized and made more responsive and efficient.

On June 12, 1878, the Pennsylvania Legislature passed an act completely reorganizing the National Guard of Pennsylvania. Section 1 placed the National Guard, in time of peace, at "not more than one hundred and fifty companies of infantry, five companies of cavalry, and five batteries of artillery ... organized into such divisions, brigades, regiments, battalions, and independent companies" which the Commander-in-Chief (the Governor) desired. He also was given the discretion to alter the organization and arrangements any way he saw fit. Section 1 further provided "That the major generals of the line shall never exceed one, nor the brigadier generals five."

Section 2 required that the organization of the National Guard of Pennsylvania should "conform generally to the provisions of the laws of the United States, and the system of discipline and exercise shall conform, as nearly as may be, to those of the army of the United States... ." It also set the number of enlisted men in each company at not less than fifty, nor more than sixty. Each recruit was required to undergo a physical examination.

General Latta, not Governor Hartranft, conceived the idea of a single division for the National Guard of Pennsylvania. Although the enabling act provided for divisions, it also restricted the number of major generals to one, thereby dictating a single division. The division could not conform to tables of organization of the Regular Army, as some latter-day historians have claimed, because the Regular Army of 1878-79 had no regularly constituted divisions. The largest formation then was the regiment.

Pursuant to the Act, all ten divisions were disbanded on June 12, 1878. The new First Brigade was organized August 28. The Second, Third, Fourth and Fifth Brigades were organized September 23.

Commander, territory and organization of each brigade was as follows:

FIRST BRIGADE
BG George R. Snowden
(Philadelphia)

First Troop Phila City Cavalry
Black Huzzars
Keystone Battery
1st Regt, Infantry (10 companies)

2d Regt, Infantry (6 companies)
State Fencibles (battalion)
Gray Invincibles (company)

(The Artillery Corps Washington Greys and Weccacoe Legion were permitted to remain temporarily unattached, and the commanders were allowed thirty days from the receipt of the reorganization order to recruit their commands each to a battalion of four companies.)

SECOND BRIGADE
BG Frank Reeder

(Counties of Northampton, Lehigh, Berks, Bucks, Montgomery, Delaware, Chester, Lancaster, Lebanon, Dauphin, Perry, Cumberland, Franklin, Adams and York)

Washington Troop
Griffin Battery
4th Regt, Infantry (6 companies)
6th Regt, Infantry (6 companies)
8th Regt, Infantry (6 companies)
11th Regt, Infantry (7 companies)

THIRD BRIGADE
BG Joshua K. Sigfried

(Counties of Susquehanna, Wayne, Bradford, Tioga, Potter, Cameron, Clinton, Lycoming, Sullivan, Wyoming, Lackawanna, Pike, Luzerne, Columbia, Montour, Union, Snyder, Northumberland, Monroe, Carbon, Schuylkill)

Ashland Dragoons
Wyoming Artillerists
7th Regt, Infantry (9 companies)
12th Regt, Infantry (9 companies)
13th Regt, Infantry (9 companies)

FOURTH BRIGADE
BG James A. Beaver

(Counties of Juniata, Mifflin, Centre, Huntingdon, Clearfield, Blair, Fulton, Bedford, Somerset, Cambria, Indiana, Westmoreland, Armstrong, Fayette, Greene, Washington, Allegheny)

Sheridan Troop
Knap's Battery
5th Regt, Infantry (8 companies)
10th Regt, Infantry (7 companies)
14th Regt, Infantry (9 companies)
18th Regt, Infantry (8 companies)

FIFTH BRIGADE
BG H. S. Huidekoper

(Counties of Beaver, Butler, Lawrence, Mercer, Crawford, Erie, Warren, Venango, Forest, Clarion, Jefferson, Elk, McKean)

15th Regt, Infantry (7 companies)
16th Regt, Infantry (8 companies)
17th Regt, Infantry (7 companies)

On March 12, 1879, the new governor, Governor Henry M. Hoyt, himself a former Brigadier General, appointed John F. Hartranft, the former governor, as "Major General, Division Commander National Guard." Thus was born the division which now bears the number twenty-eight. The governor reappointed Brigadier General James W. Latta as the Adjutant General. The keystone was prescribed as the designated badge of the National Guard of Pennsylvania by the Adjutant General on August 27, 1879.

As an experiment, the National Guard of Pennsylvania was assembled in two separate encampments in 1880, during which the commands underwent their annual inspections. Each camp was one week long. Attendance varied, with most men being present on the date set for inspection of his unit. The average attendance finally came to sixty-seven percent of the troops on the rolls. This was satisfactory, taking into account the fact that the men were not paid for service. Only their food and cost of transportation was provided by the Commonwealth. As a result of this experiment, the Adjutant General concluded that there should be an annual encampment of at least one week's duration, with all Guardsmen attending.

The Army Advisor system had its beginning in August 1880, when Lieutenant Colonel G. A. DeRussy, 3d Artillery, was detailed by the Department of the East, United States Army, to inspect the brigades of Pennsylvania's Guard encamped in Fairmount Park, Philadelphia. His report was very detailed and generally very complimentary.

Although the men were not paid for their annual encampments, they did receive pay for the day of

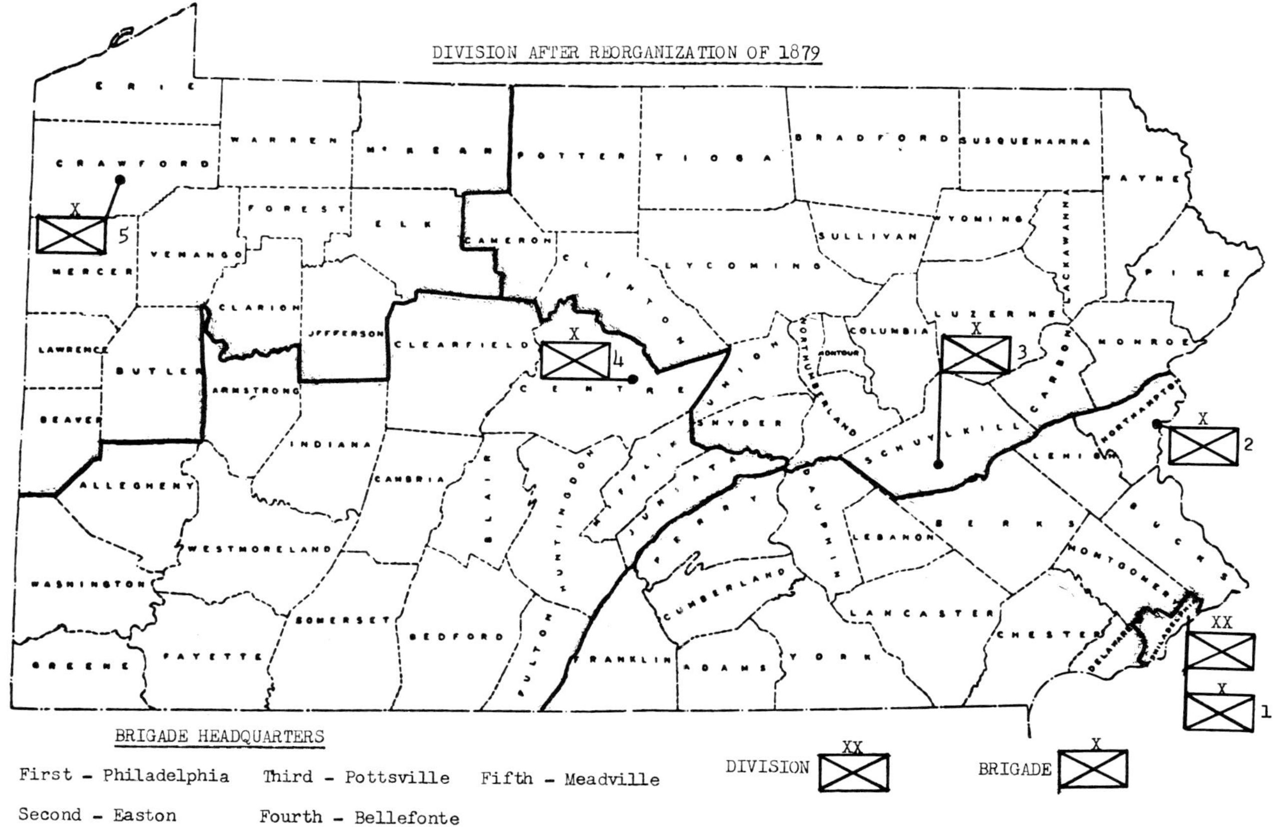

Division After Reorganization of 1879.

annual inspection. The following chart shows some typical authorizations:

Major General	$20.83
Brigadiers	$15.28
Colonels	$ 9.73
Lt Colonels	$ 8.33
Majors	$ 6.95
Capt (Mounted)	$ 5.55
Capt (not Mtd)	$ 5.00
1LT (Mounted)	$ 4.45
1LT (not Mtd) & 2LT (Mounted)	$ 4.17
2LT (not Mtd)	$ 3.89
SGM & 1SG	$ 3.00
Sergeants	$ 2.00
Corporals	$ 1.75
Pvts and Musicians	$ 1.50

Over 7,500 men of the Division took part in the inaugural parade in Washington for President James A. Garfield on March 4, 1881. The troops received many compliments for their excellent discipline and soldierly bearing.

A State law in 1881 required the Guard to hold an annual encampment not to exceed eight days, that the annual inspection take place during this time, and that troops would be provided transportation and subsistence, plus a per diem allowance for up to five days of the encampment.

Section 7 of the act provided for a fine of up to $25 if a soldier failed to report for duty. If the fine was not paid within twenty days, the soldier could be jailed for up to thirty days.

The Division was reduced in 1881 to a three brigade force. The First Brigade, commanded by Brigadier General George R. Snowden, was located in and around Philadelphia. The Second Brigade, under Brigadier General James A. Beaver, covered the remainder of the eastern half of the state. Brigadier General J. K. Sigfried's Third Brigade was stationed in the western half of the Commonwealth. The 7th, 11th and 17th Regiments and Ashland Dragoons, along with six companies belonging to other regiments, were all disbanded.

The Adjutant General's report for 1882 included a comprehensive report by Lieutenant Colonel E. Wallace Matthews, Ordnance Officer and Chief of Artillery for Pennsylvania's division. Among a number of suggestions, the Colonel recommended that all batteries include Gatling guns and but two cannon. He stated that, since the Guard was called most often to riot or other civil disorder duty, a machine gun was much more effective, maneuverable and intimidating than an artillery piece. Further, he explained, the cost of maintaining batteries so armed would be much less than those equipped with heavier cannon requiring horses and associated equipment. He recommended a battery of four Gatling guns (.45 caliber) and two naval 3″ breech-loading rifles (a superior, lightweight field piece, he contended), plus at least sixty rifles or carbines for the men.

A member of the State Fencibles 1870-1880.

[M.]

ORDERS AND CIRCULARS.

HEAD-QUARTERS NATIONAL GUARD OF PENNSYLVANIA,
EXECUTIVE CHAMBER,
HARRISBURG, *March 12, 1879.*

GENERAL ORDERS,
No. 1.

I. The following appointment is hereby announced:
Major General John F. Hartranft, Division Commander National Guard.
He will be obeyed and respected accordingly.
II. The following Staff appointments are hereby announced:
Brigadier General James W. Latta, Adjutant General.
Lieutenant Colonel D. Stanley Hassinger, Assistant Adjutant General.
Colonel Hartley Howard, Inspector General.
Colonel Clarence G. Jackson, Quartermaster General.
Colonel Thomas J. Smith, Commissary General.
Colonel Louis W. Read, Surgeon General.
Colonel A. Wilson Norris, Judge Advocate General.
Colonel John S. Riddle, General Inspector of Rifle Practice.
Lieutenant Colonel William Ross Hartshorne, Aid-de-Camp
Lieutenant Colonel Edward B. Young, Aid-de-Camp.
Lieutenant Colonel Charles M. Conyngham, Aid-de-Camp.
Lieutenant Colonel Nathan A. Pennypacker, Aid-de-Camp.
Lieutenant Colonel Elisha A. Hancock, Aid-de-Camp.
Lieutenant Colonel Galloway C. Morris, Aid-de-Camp.
Lieutenant Colonel David F. Houston, Aid-de-Camp.
Lieutenant Colonel Albert W. Taylor, Aid-de Camp.
Lieutenant Colonel B. Frank Eshelman, Aid-de-Camp.
Lieutenant Colonel John Lowrie, Aid-de-Camp.
Lieutenant Colonel Walter W. Ames, Aid-de-Camp.
Lieutenant Colonel J. Ford Dorrance, Aid-de-camp.
They will be obeyed and respected accordingly.

HENRY M. HOYT,
Governor and Commander-in-Chief.

General Order appointing Major General Hartranft as Division Commander.

A Regular Army inspector, in his report of the Division's annual encampment of 1882, recommended that encampments be extended to fifteen or twenty days. In the shorter training period, he continued, too much time was devoted to relearning for progress to be made. He observed that Guardsmen were deficient in the rudiments of military subjects, a situation which could be remedied by instruction during drills between camps.

By 1886 some of the deficiencies noted by the Regular Army inspector a few years previously had been corrected. Battalion and company drill was now considered to be very satisfactory, but the troops still needed more instruction in sentinel work and skirmish drill.

In 1886 only the First Brigade conducted officer and non-commissioned officer schools and had an examining board which compelled officers to be thoroughly qualified for their assignments. The Adjutant General in his report remarked that the First's schools provided good instruction. He suggested that new enlistees undergo at least two months of preliminary training in the basics and the school of the soldier. Colonel Thomas J. Hud-

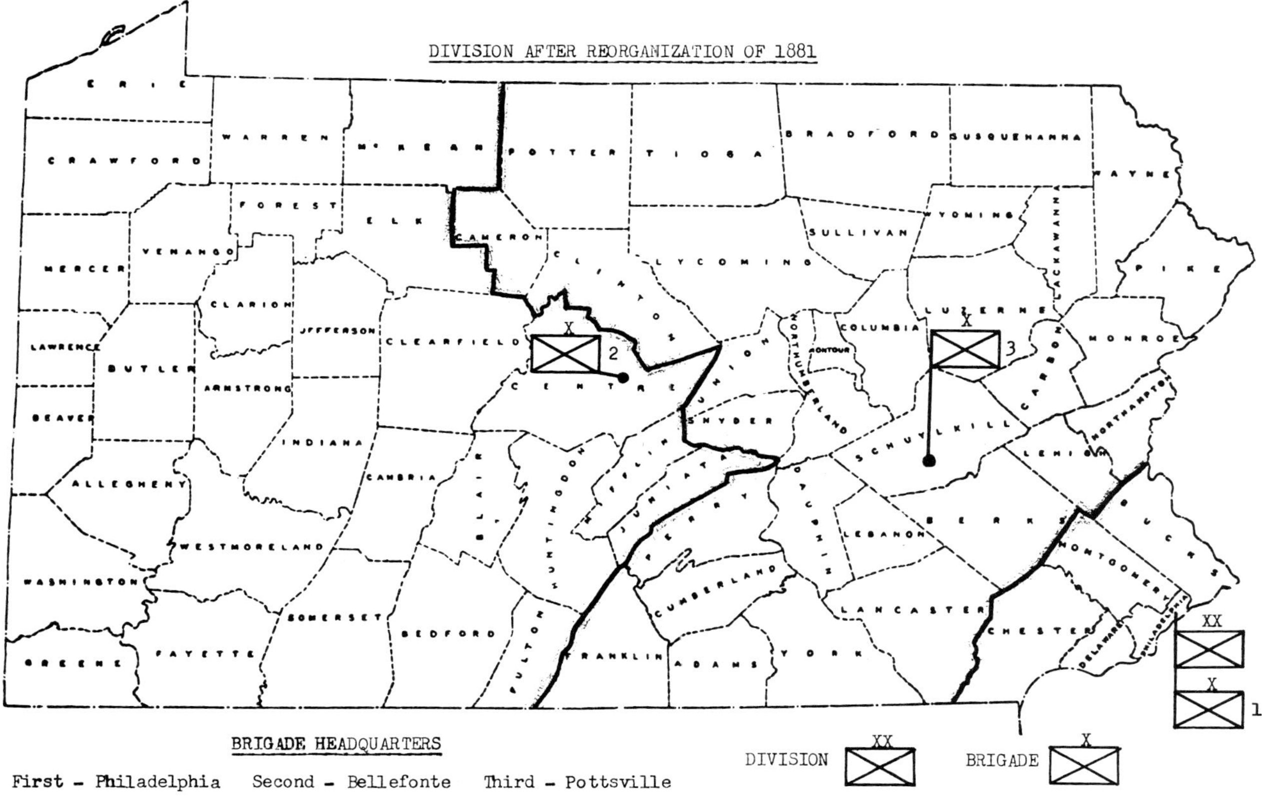

Division after Reorganization of 1881.

son, Pennsylvania's Chief of Artillery, repeated his earlier recommendation that the Parrott and Phoenix cannons of the Pennsylvania artillery be condemned as "absolutely unserviceable and positively dangerous to use."

An act of the State Legislature in 1887 eliminated all similar legislation pertaining to the National Guard which had been enacted in the previous thirty years and, over-all, improved and modernized the Guard. Major provisions were:

a. Increases in the annual appropriation from $200,000 to $300,000;

b. Reduction of the term of enlistment from five to three years and provision for reenlistment bonuses;

c. Conveyance to the State of title to armories in the Commonwealth in those instances where State funds had been applied in purchase;

d. An increase in the period of the annual encampment to fourteen days, at the Governor's discretion.

e. Provision of pay to the men for every day in camp, rather than the maximum of five days pay previously authorized.

f. Provision for a State code of military justice.

The increases in compensation, reduction in the term of service and reenlistment bonus stimulated enlistments and retention of personnel.

In 1886 the owner of the land at Mount Gretna, where the Pennsylvania Guard held annual training, Robert H. Coleman, at his own expense, built a reservoir, cleared portions of the land and extended water pipes throughout the area.

In the 1970's between 1,000 and 2,000 visitors have watched the Division parades during annual training. The figure can be compared with the last day of the encampment of 1887 when General Phillip Sheridan reviewed the Pennsylvania Division. Over 25,000 visitors were present. Attendance at camp that year was ninety-four percent.

The First Brigade Surgeon, in his report for 1888, complained about "the abominable shoes" which the men wore to camp. Shoes were not issued, so the men furnished their own. Judging from this Surgeon's report, the variety and condition of shoes was not suitable for camp duty and marching. He recommended that the Guardsmen be issued the regulation army shoe.

A new cavalry company was formed in Harrisburg in 1888. Known as the "Governor's Troop," it attended its first camp that year at Mount Gretna with the Third Brigade.

During the late 1880's many units of the National Guard either owned or rented their armories. The rental system sometimes led to embarassing situations. As an example, the borough of Phoenixville rented a building to Battery C for use as an armory. One day the borough authorities evicted the battery and "put into the street violently all company and State property," as the report stated it.

On May 30, 1889, Johnstown was inundated by a flood. The 14th Regiment of Pittsburgh, under Colonel Peter D. Perchment, along with Company H of the 5th Regiment (Johnstown), was ordered to camp there in order to assist in preserving order. In addition, the Pennsylvania Guard's Commissary and Quartermaster departments assisted in distributing supplies to flood survivors.

By 1890, qualification boards had been established in the Pennsylvania National Guard, which passed on the fitness of newly elected officers. Many regiments had both officer and non-commissioned officer schools. In some regiments no one was made a non-commissioned officer without first passing an examination.

In 1890 a provisional brigade of Regulars, under Brevet Brigadier General H. Y. Gibson, consisting of two companies of the 11th Infantry, two light batteries from the 2d Artillery and two troops of the 5th Cavalry were encamped with the Pennsylvania Division. These Regulars served as models for the Guardsmen to emulate. The work of Lieutenant William H. Bean, 2d Cavalry, as an instructor of guard duty for the Pennsylvanians was another example of the forerunner of the present-day advisor system.

When the entire Division went to camp at one time, it was customary to group the three Cavalry troops into a provisional battalion of cavalry and the three artillery batteries into a provisional artillery battalion. The Division Commander, Major General George R. Snowden, recommended that an additional cavalry troop and another artillery battery be formed so that a battalion of each could be permanently constituted. He also recommended that the battalion system be established within the regiments of infantry, similar to that being instituted in the Regular Army.

On August 19, 1890, a tornado struck Wilkes-Barre, killing a number of people and causing considerable destruction. Three companies of Colonel Keck's 9th Infantry (Wilkes-Barre) were placed on duty to guard against looting. The civil authorities were gratified by the prompt and efficient assistance of the Guard. The men responded quickly to the call and did excellent work.

One of the Regular Army inspectors with the Division that year's camp, Captain Clinton B. Sears, in his report states: "I know of no other state, unless it be New York, that can put into the field on as short notice so large, so well equipped, and so efficient a body of men," as Pennsylvania.

CHAPTER 6

1892 to Spanish-American War

Uzal W. Ent, Colonel, Hq, 28th Inf Division

IN 1891 THE Regular Army published new drill regulations for a three battalion regiment, requiring the National Guard of Pennsylvania to consider reorganization in order to conform to the new standards (as had already been recommended by the Division Commander in 1890).

For the first time since 1877 elements of the Pennsylvania National Guard were called to duty to suppress disorder and enforce the law, this time in the coke-producing region of Westmoreland County. Both the 10th and 18th Regiments were ordered to Mount Pleasant on April 2 and were on duty within twenty-four hours after the order was given. Most troops were on duty for three weeks, but several companies served fifty-five days. The discreet conduct and disciplined behavior of the troops and their good leadership was considered responsible for the dispersal of large, angry mobs without gunfire.

The Regular Army inspector that year, Captain James Chester, reported that the equipment of the Pennsylvania National Guard's artillery was worthless and that the guns were good only for firing salutes. His report confirmed what the Pennsylvania Guard's Chief of Artillery had already reported several times in previous years.

On July 2, 1892, the authorities at the Homestead plant of the Carnegie Steel Company dismissed the last of its employees, declaring that henceforth the Homestead Steelworks would be operated as a non-union mill.

For the previous six months the Company and the men, represented by the Amalgamated Association of Iron and Steel Workers, had been at odds over a new contract which involved wage scale changes. The differences between Company and Union could not be resolved. Although the Amalgamated counted only a small minority of the workers on the rolls of the union, the other employees at the plant collaborated with it in its efforts to get a contract which would permit small wage increases for at least some of the men. On May 29, Henry Clay Frick, the Carnegie representative at Homestead, issued an ultimatum to Amalgamated: accept the Company wage proposal by June 24 or be dealt with individually. This left the Union with three choices: surrender; disband; or strike.

Shortly after issuing the ultimatum, Frick ordered a twelve foot high wooden fence erected around the plant. The boards were two inches thick, and the fence was topped with three strands of barbed wire. Three-inch holes were bored every twenty-five feet at about shoulder height all along the three miles of fence. Platforms were erected at intervals, upon which searchlights were mounted. Behind the fence and platforms a system of pipes was laid so that high pressure hoses could be connected and used, if necessary. The workers thought that the barbed wire was electrified but generally laughed at the whole idea. They composed a poem about this elaborate structure, called "The Fort That Frick Built":

> "There stands today with great pretense
> Enclosed within a whitewashed fence
> A wondrous change of great import,
> The mills transformed into a fort."

A meeting between Frick and the Union on June 23d came to naught. The twenty-fourth, ultimatum day, passed without incident. On the day following, Frick wrote to Robert Pinkerton for 300 guards, because he intended to start up Homestead on July 6 with non-union employees.

Meanwhile, the mills operated at full-blast until June 28, when lay-offs began. On the night of the thirtieth the workers voted to go out on strike. Although the plant continued production until July 2, on that date the last workers were dismissed. Notice was given that the Homestead plant was now non-union and that no further negotiations would be carried on with Amalgamated.

On July 4, Frick demanded that County Sheriff McCleary place deputies inside the Homestead works to secure it. An effort to accomplish this mission by sending a dozen deputies to the plant was thwarted by a large body of workers who met the deputies at the railroad station.

Sheriff McCleary went to the Homestead plant the next day, where a committee from the Union took him on a tour, showing him that the works were undamaged and under effective guard by workers. The union committee offered him up to 500 of their people as sworn deputies, under bond, to watch the mill and keep out trespassers. McCleary refused the offer. He pointed out that the firm had the right to secure its own property. However, he reached an agreement with the committee that fifty deputies, selected by the sheriff, would take over guard duties at the mill.

The sheriff knew that some 300 Pinkertons in the employ of Carnegie were due to arrive on July 6. With less than a day to gather a posse, he was unsuccessful in securing enough men willing to serve. Perhaps a few more days time would have permitted the sheriff to form a posse and averted the bloody, bitter battle which took place on the sixth.

The Pinkerton force of 316 men was, unfortunately, made up of a mixture of drifters, hoodlums, criminals on the run, and a few college students between semesters, the lot mixed with a few Pinkerton Agency regulars. Upon their arrival in Pittsburgh the men were placed on two barges and propelled along the river to Homestead by tugboat. Alerted to their coming, a crowd estimated at over 10,000 people broke into the Homestead plant and met the barges as they came ashore. Many of the crowd were armed with rifles, pistols and shotguns.

As a group of Pinkertons tried to land, one of the crowd laid himself across the gangplank. Pinkerton Captain Frederick H. Heinde, leading the landing, stooped down as if to move the prone man, when the latter rose up and shot Heinde in the thigh with a pistol. This shot triggered a general gunbattle, in which more than thirty Homestead men and fifteen Pinkerton men were killed or wounded in less than three minutes.

The Pinkerton people withdrew into their barges and were besieged there by angry strikers for the rest of the day. Strikers sent a burning raft down the river, hoping to set the barges on fire; they brought out some cannon but could not depress the muzzles sufficiently to hit the barges, although the cannon fire succeeded in decapitating one of their own number. Approaching the barges in boats, they threw dynamite, blowing a hole in one barge, then fired into the opening with rifles. The rioters even attempted to blow the barges up by releasing gas from a main then firing a rocket into the gas pocket.

With a large number of their party dead and wounded and survivors hiding amid the shambles of their barges and pools of blood, the Pinkerton men finally surrendered. As they came ashore, the Pinkerton force was set upon by the mob. One man had an eye put out by an umbrella; several more were killed, some bludgeoned to death by the crowd. Not a man escaped without some injury.

Amid the chaos and bloodshed the sheriff, as

well as Frick, attempted to get Governor Robert E. Pattison to call in the National Guard. Pattison at first refused to involve the Guard as "private watchmen" for the Homestead works. Instead, he blamed Sheriff McCleary for not having taken stronger measures to gather a posse and head off the affair. However, the sheriff still could not raise a posse. In effect, the Union Committee administered the town of Homestead.

Finally at 10 in the evening of July 10, in response to another plea by the sheriff, Governor Pattison activated the entire Division. The Second and Third Brigades, along with the First Troop, Philadelphia City Cavalry, were dispatched to the Homestead area. The First Brigade was assembled at Mount Gretna by 11 in the evening of July 11. The Second and Third Brigades arrived at Munhall Station in Homestead at one o'clock the next day.

The 4th, 10th and 14th Regiments, along with Battery C, armed with two Gatling guns, formed a provisional brigade under the command of Colonel Hawkins of the 10th Regiment. This brigade peacefully occupied high ground across the Monongahela River from the Homestead works and set up a signal station.

The Second and Third Brigades, each of four regiments, upon arrival at Munhall Station, occupied hills overlooking Homestead from the south. Signal communication was opened with the provisional brigade on the far shore. In all, some 5,378 men of the Division were thus assembled without incident from the civil populace. Another 2,600 men from the First Brigade were in reserve at Mount Gretna.

The strikers feared the coming of the Guard but decided to meet them as friends. They sent a deputation to the Division Commander, George R. Snowden, requesting that the National Guard join them in a parade and tendering the cooperation of the advisory committee in maintaining order. General Snowden rejected both offers, stating that he did not recognize the Committee as a legal governmental body and that the National Guard would do its duty to the Commonwealth and the law. He told the union representatives that everyone could cooperate with the Guard by going peacably about his own business.

A provost guard was formed, under Lieutenant Colonel Charles S. Greene, consisting of three companies from the 18th Regiment and one company of the 15th. This force was shortly expanded into a regiment, and at times, to a regiment plus a battalion. Guard details were posted along the boundary between the town of Homestead and the steel works, as well as at Munhall Station. In addition to signal stations, communications between the elements of the Division on both sides of the river included the steam boat "Little Bill" and a railroad bridge which ran from the north shore into the steel plant grounds. The cavalry, with the addition of the Sheridan Troop and Governor's Troop, was organized into a provisional squadron and daily patrolled the streets of Homestead.

The citizenry still recognized the *de facto* government set up in Homestead by the union Advisory Committee. On several occasions the provost guard was sent to capture local policemen and release prisoners who had been illegally apprehended on orders of the Committee. On the other hand, the Guard protected constables who were serving lawfully drawn warrants.

The relationship between the local civilians and Guardsmen was generally peaceful. Few incidents occurred. At the outset troop morale was good, but as the tour of duty lengthened, the troops became bored. On occasion some men went to town and drank too much, but no serious disciplinary problems arose until the Private William L. Iams incident.

Iams belonged to Company K, 10th Infantry. On July 18, a dissident named Alexander Berkman attempted to assassinate Henry Frick in the latter's office. The attempt was unsuccessful, although Frick was shot twice in the neck and was stabbed with a sharpened file in the hip, right side and near the left knee. Berkman was apprehended and jailed. The remarkable Frick had his wounds treated and remained at work!

Private Iams was near his regimental headquarters when he heard the news of the attack. "Three cheers for the man who shot Frick!" he shouted. His regimental commander, Lieutenant Colonel J. B. Streator, hearing the shout, but not knowing who had been responsible, assembled the regiment. When the Colonel asked the assemblage who had called out, Private Iams stepped forward and admitted he had done so. When asked by the Colonel to apologize, Iams refused,

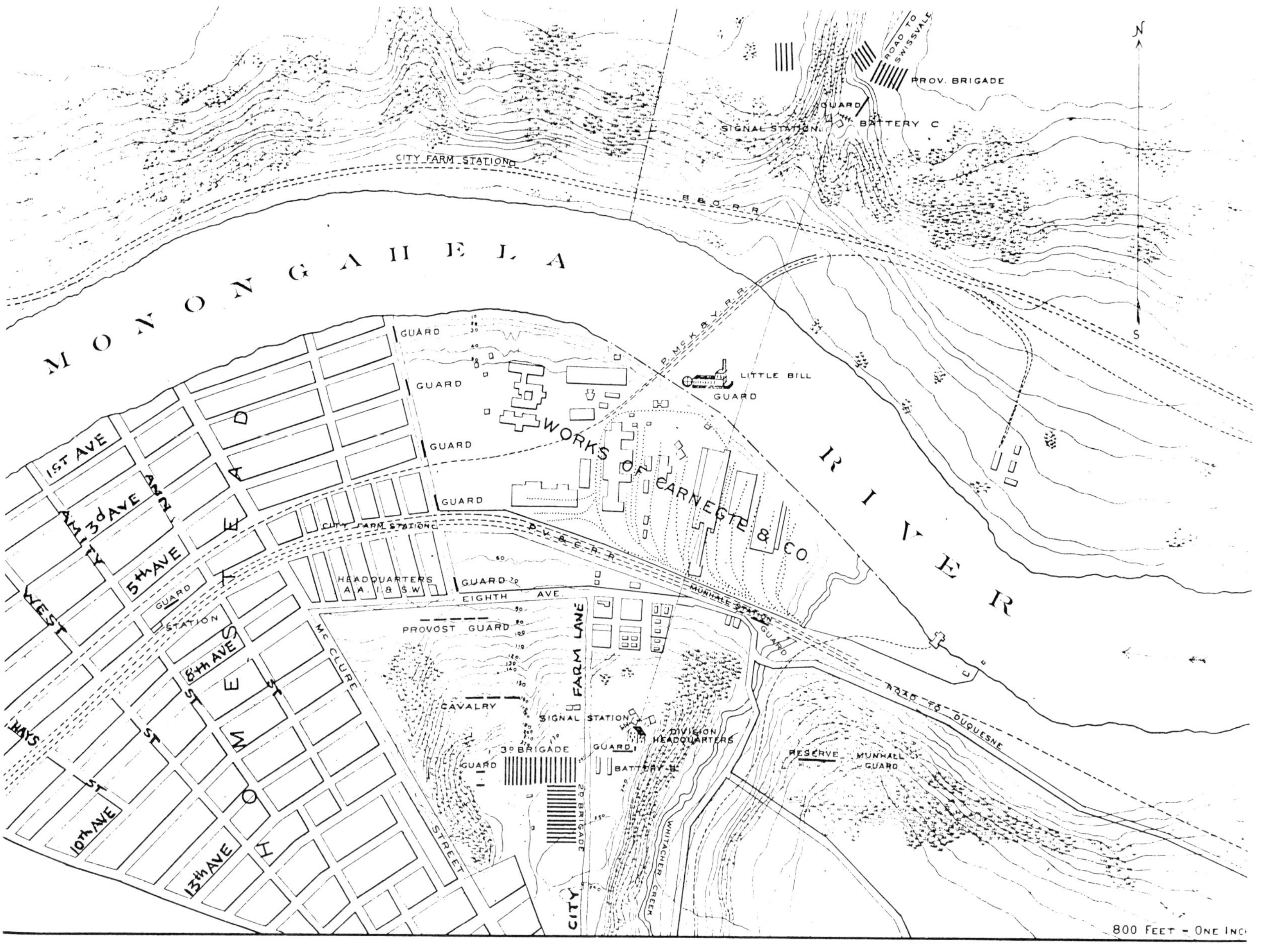

MAP SHOWING HOMESTEAD CARNEGIE MILLS, AND DISPOSITION OF THE NATIONAL GUARD OF PENNSYLVANIA.
TO ACCOMPANY REPORT OF MAYOR A. LAWRENCE WETHERILL, DIV. ADE. CA. NOV. 30TH 1892

Carnegie Mills and the Dispostion of the National Guard of Pennsylvania, November 30, 1892.

exclaiming, "I hate Frick!" He was put in arrest.

The regimental commander and the acting brigade commander, Colonel Hawkins, regarded Iams's call for support for Frick's attacker as an act of mutiny and treason. No doubt influenced by the bloody encounter which had taken place a few days previously between the Pinkerton force and strikers, both officers agreed to string Iams up by the thumbs, under the eyes of a surgeon. After ten minutes the surgeon offered to have him cut down. Iams refused. He was left hanging another eight minutes, when he was ordered released by the surgeon.

Although Private Iams was not injured physically by his punishment, such Draconian measures had fallen out of general use. General Snowden, the Division Commander, also viewed Iams's action as mutinous and traitorous. He also felt that the "usages of war," as he put it, were appropriate and ordered Private Iams discharged in disgrace. The lieutenant who delivered the message from the General to Colonel Hawkins created the impression that Iams should be drummed out of camp before the assembled brigade, and that one half of his head be shaved for the event. Accordingly, Iams, dressed in a suit of overalls with half his head shaved, was drummed from the camp to the tune of *The Rogue's March*. Although Iams filed suit against Streator, Hawkins and the surgeon, he lost the case, and the officers were vindicated. The judge ruled that they took punitive action appropriate in order to maintain discipline in a force which had been called to active duty to quell a riot and promote public order and that the Private's call to cheer Frick's attacker was insubordinate and a detriment to good order.

Most commands on duty at Homestead served until near the end of July, when they were sent home. The First Brigade, at Mount Gretna, was discharged from active duty on July 19. The 5th Regiment was relieved on the Ninth of August. The 15th and 16th were gradually relieved from duty, a few companies at a time. Finally, on October 13, the last of the Division left Homestead.

The hard service at Homestead finally made a large number of tents and other equipment unserviceable. Practically everything which had been issued to the men, except their weapons and basic uniform, was already old—in some instances, over fourteen years old—and was unfit for service. Only the haversack and cartridge boxes were replaced. Some units bought new campaign hats and leggings. The non-standard shoes with which the men supplied themselves also proved a problem because of the hard service. They either did not stand up or were bad for the wearers' feet.

As a result of the duty at Homestead, the Division Commander recommended that the strength of the Division be increased, suggesting twelve regiments of twelve companies each. (This would have permitted each regiment to organize into three battalions, of four companies each, along Regular Army lines.) He also suggested an engineer company and a "signal corps" be formed, as well as another troop of cavalry and one more artillery battery. The cavalry could then be organized into a four-troop squadron and the artillery into a battalion of four batteries. The reorganization proposal, however, was rejected.

A Pennsylvania law of 1889 authorized the formation of four companies of naval militia, to be known as the Naval Battalion of the National Guard of Pennsylvania. At the end of 1892 applications had been submitted by two companies in Philadelphia.

Except for the call up of the 5th and 16th Regiments and the Sheridan Troop for riot duty in the Punxsutawney-Walston mine area in 1894, the period from 1893 was a fairly stable one for the Pennsylvania Guard.

In 1893, two divisions (companies) of naval militia were formed into a battalion in Pittsburgh. These units were attached to the First and Second Brigades, respectively, of the National Guard of Pennsylvania. The Division also paraded in the presidential inaugural parade in Washington, through a driving blizzard of snow and ice.

The artillery received new weapons in 1893, so that each battery consisted of two 3.2-inch breech loading steel rifle guns, two .45 caliber Gatling guns and two 3-inch muzzle loading cannon. The Regular Army inspector during the encampment of 1895 rightfully took the Pennsylvania Guard to task for this mixture of weapons in each battery, pointing out that it took three different manuals (procedures) to serve and fire each set of weapons and three different kinds of ammunition. He recommended a six-gun battery of each type weapon.

In 1894 the Division encamped as a division at

Orderly Room and NCO Barracks, First Regiment Infantry, 1890's.

Gettysburg. A troop of cavalry and a battery of artillery from the Regular Army encamped near the Pennsylvania cavalry and artillery. These Regular troops provided much valuable instruction to the Guardsmen. For the first time a group of men destined to become the first signal company were also in camp with the Division. One officer and nineteen men composed this group. They received excellent training from a Regular Army Signal Corps sergeant.

Men were paid on a scale of from $1.50 to $3.00 per day, depending on rank, for each day of camp, with a .25¢ per day reenlistment bonus for each reenlistment. Drills were held at least once a week and sometimes twice a week in some units. Each company stored a thousand rounds of ammunition in its armory in case of emergency. Many infantry regiments stored additional ammunition at their headquarters. Having a stock of ammunition immediately at hand, plus the fact that camp equipment stored in Harrisburg could be sent out rapidly by train, helped make possible the rapid and effective deployment of the Guard.

The Adjutant General's report for 1894 provides a detailed outline of unit organization of the National Guard of Pennsylvania:

Cavalry Troop

CPT	1	1SG	1	CPL	8	Wagoner	1
2LT	1	QM SGT	1	Trumpeters	2	Privates	28-36*
1LT (Asst Surg)	1	Commissary SGT	1	Blacksmiths	2		
2LT (QM)	1	SGT	5	Saddler	1		

Artillery Battery

CPT	1	1SG	1	Artificers	2
1LT	2	QM SGT	1	Wagoner	1
2LT	1	Commissary SGT	1	Privates	45-56*
1LT (Asst Surg)	1	SGT	4		
2LT (QM)	1	Musicians	2		

Infantry Company

CPT	1	SGT	4
1LT	1	CPL	8
2LT	1	Musicians	2
1SG	1	Privates	35-45*

*Engineer Company***

CPT	1	SGT	4
1LT	1	CPL	8
2LT	1	Musicians	2
1SG	1	Privates	35-45*

*Signal Company***

CPT	1	SGT	3
1LT	1	CPL	4
2LT	1	Musicians	2
1SG	1	Privates	28-32*

* Minimum - maximum numbers authorized.
** These units were authorized, but not formed in 1894.

Note that the artillery battery lists no corporals.

Division Staff
Maj General 1

(LTC) Asst AG	1	(LTC) Surgeon-in-Chief	1
(LTC) Inspector	1	(LTC) Ordnance Officer	1
(LTC) Judge Advocate	1	(LTC) Inspector of Rifle Practice	1
(LTC) QM	1	(MAJ) Aide-de-camp	3
(LTC) Commissary	1	Staff NCOs	6

Brigade Staff
Brig General 1

(MAJ) Asst AG	1	(MAJ) Surgeon	1
(MAJ) Inspector	1	(MAJ) Ordnance Officer	1
(MAJ) Judge Advocate	1	(CPT) Aide-de-camp	2
(MAJ) QM	1	Staff NCOs	6
(MAJ) Commissary	1		

Regimental Staff
Colonel 1

LTC	1	(1LT) Adjutant	1
MAJ	2 or 3	(1LT) QM	1
(MAJ) Surgeon	1	(1LT) Asst Surgeon	2
(CPT) Chaplain	1	(1LT) Inspector of Rifle Practice	1
		Staff NCOs	6

Battalion Staff (As part of Regt)		*State Fencibles*			
MAJ	1	MAJ	1	Inspector of Rifle Practice	1
Adjutant	1	Adjutant	1	Chaplain	1
SGM	1	QM	1	Asst Surgeon	1

Strikes, which had broken out in the coke-producing region of western Pennsylvania during the spring of 1894 spread to the bituminous regions in the Punxsutawney - Walston area by early summer. On June 21 the sheriff of Jefferson County sent an urgent telegram to the Governor asking for the Guard.

The Governor ordered the 5th and 16th Infantry Regiments and the Sheridan Troop, all under command of Brigadier General John A. Wiley, to the scene. The entire force arrived in Punxsutawney at 6:25 in the morning of June 22. Upon arrival the cavalry was placed forward as a screen and the command marched off toward Walston, the scene of most trouble. As the troops approached the area, the cavalry was withdrawn and infantry skirmishers sent forward. As the men advanced overland across steep and hilly ground, through fallen, rotten and burn-blackened timber, they found that the riotous gangs had disappeared, leaving behind a small cannon and other evidence of defensive preparations.

The Guard organized itself in the area. Aside from the need to rescue the sheriff and some of his deputies, who arrived in nearby Adrian by train on June 24, and the dispersal on the 26th of a mob consisting mostly of women, who attacked a trainload of workmen, duty was fairly routine and free of trouble. The one exception, apparently, as reported by the surgeon of the 16th, was that the coffee supplied to the troops was of "an exceedingly cheap grade."

In 1895 the Adjutant General, Brigadier General Thomas J. Stewart, (for whom the Thomas J. Stewart medal is named) called again for the addition of another artillery battery and cavalry troop so that the artillery and cavalry could be organized into battalions which would report directly to Division Headquarters. He also recognized the need for an engineer company and signal unit.

The warehousing of equipment at the Harrisburg Arsenal was improved by storing the equipment by regiment and invoicing a complete list of everything in each regiment's stock. The subsistence department studied the possibility of keeping on hand two or three days' emergency rations in each armory. Stocking equipment by regiment and the positioning of rations at each armory, along with the weapons and ammunition already there, would expedite the assembly and deployment of the Guard in future emergencies.

Captain Alexis R. Paxton, 15th United States Infantry, in his report of his observation of the National Guard of Pennsylvania in 1896, said that "the superior condition of the National Guard of Pennsylvania is primarily due to the people of the State, who take deep interest in its success, and prove that ... by making liberal appropriations for its maintenance and by making wise laws for its discipline... ."

In 1897, on the eve of the Spanish-American

Pennsylvania State Arsenal, Harrisburg, Pennsylvania in the 1890's.

War, a brigade of the Pennsylvania Division was once more summoned by the Governor for onerous strike duty, this time near Hazleton.

Trouble started in mid-August when the laborers of the Lehigh and Wilkes-Barre (L and WB) mining company, below Hazleton, went on strike in protest against the "tyrannical methods" employed by one of the superintendents. Unrest centered among the Slavic and Italian workers. Just as this problem was being resolved, some 500 Slavs struck the Van Wickle Company on August 27th over the issue of pay discrimination. The strike among Slavic and Italian laborers spread to the Anglo-Saxon workers of the L and WB and of the Lehigh Coal and Navigation Company, who were terrorized by the other strikers.

On a number of occasions, to reinforce their demands and show strength in numbers and purpose, large crowds of strikers paraded behind American flags, intimidating everyone they met.

Troopers on Duty in Hazleton, 1897.

A demonstration of this type led to the "Lattimer Massacre."

On September 10, 1897, a mob of persons of Lithuanian, Slovakian and Polish backgrounds marched from Harwood, intending to shut down the Pardee Company's Lattimer Mine. They were met at the mine by Sheriff James Martin and an armed posse. He ordered the marchers to disperse. In the ensuing scuffle the sheriff either fell or was pushed. He ordered his men to fire, whether at the strikers, or over their heads, it was never clear. The posse fired into the mob of unarmed marchers who ran, shouting and screaming, for cover. Eleven people were killed outright. Eventually the toll reached nineteen dead and thirty-nine wounded. In the confusion the sheriff and his deputies escaped.

The sheriff telephoned the Governor, who summoned to duty the Third Brigade, a portion of which was located in the county where the strike was centered, along with the First Troop Philadelphia City Cavalry. Brigadier General Gobin, Brigade Commander, was placed in command of the force. The Brigade consisted of the 4th, 8th, 9th, 12th and 13th Regiments, Battery C and the Governor's Troop.

These units were all deployed by late evening of September 10. The 4th and 8th Regiments, together with one-half of Battery C were posted near Audenreid; the 9th was posted at Hazle Park; the 12th placed east of Hazleton, and the 13th was ordered to Lattimer. The two troops of cavalry and the remaining half of Battery C were encamped in Hazleton, near the Lehigh Valley Railroad yards, two blocks from where Brigade Headquarters had been established. The 4th Regiment was moved on the 13th to the town of Drifton, six miles north of Hazleton. The Brigade was scattered a distance of nine miles, from north to south. Over 2,500 men were thus deployed to cover the twenty-seven square miles of troubled area.

This extended deployment presented the Brigade Commissary, Major Simon B. Cameron, with a serious problem in issuing fresh, untainted food to the various commands. He solved the problem by assembling a special train. Rations were issued from a refrigerator car and a box car to the different regiments as it passed along the track. Only the 4th Regiment had to travel to the railroad for its supplies, a distance of less than a mile. In this manner, rations were issued to the regiments daily between 9 in the morning and Noon.

In addition to normal camp equipage, each company brought 1,000 rounds of ammunition. Some regiments brought additional stocks. The Brigade Ordnance Officer reported that an additional 40,000 rounds of rifle, 2,000 rounds of carbine, and 2,000 rounds of pistol ammunition, plus 200 blanks, and fifteen cases of shrapnel ammunition for the 3.2-inch cannons were all stockpiled in a magazine guarded by the 8th Regiment. Although the troops marched, patrolled, and on occasion, had to disperse large crowds of angry citizens, not a shot was fired by the National Guard, although the men habitually carried ammunition for their weapons.

There were two unusual aspects of this strike, according to contemporary accounts. The first of these was that the "immigrants"—the Italian, Polish, Slovakian and Lithuanian workers—went out on strike, but the Anglo-Saxon elements did not until they were terrorized into doing so by the "foreign" elements. The second unusual aspect was that on several occasions mobs of women attacked laborers who attempted to return to work.

These "Amazons," as several Guard commanders referred to the women in these disturbances, were led by "Big Mary" Septek. Mrs. Septek, known for her forceful character, ran a boarding house near the Lattimer Company Store. When word came that some men had returned to work, "Big Mary" recruited a group of Slavic women to chase these men back home. On one occasion men of Colonel Frank J. Magee's 8th Infantry intercepted a crowd estimated at about two hundred and fifty women on their way to the workings at McAdoo. The troops dispersed the crowd. The following day, September 20, two hundred women, led by "Big Mary," armed with clubs, bars of iron and stones, and encouraged by a crowd of men and boys accompanying them, attempted to drive workers away from the mine works near Lattimer. Captain Eugene D. Fellows, commanding Company F, 13th Infantry, then Officer of the Day, in the words of the regimental commander's report:

> summoned a company and with the relief at the guard tent, proceeded at double time to get into position between the open pit and washery and the crowd of advancing women... . Two more companies were sent immediately to cooperate with him... . The soldiers stopped the women about 500 feet from the washery and edge of pit, and ... drove them back... . The reenforcements ... scatter[ed] the crowds of men and boys and ... [kept] them on the move and [did not] allow them again to gather.

Attempts such as those of the women to keep the various mines shut came to naught, and peace was eventually restored. The artillery battery was relieved from duty on September 24, and all other National Guard troops had returned home five days later.

The weather was cold during the two-week tour, and the constant need for troops to be on duty and patrol literally destroyed the one uniform that each man had. Uniforms were subsequently supplied without cost to the units, because the old ones had been worn out prematurely in service to the Commonwealth.

The United States declared war against Spain on April 21, 1898. Two days later, the President issued a call for 125,000 volunteers to fight the war. Pennsylvania was asked for ten regiments of infantry and four batteries of artillery, totalling 10,800 men. On April 25 the Governor of Pennsylvania ordered the Division to assemble at Mount Gretna "on Thursday, April 28, 1898."

In driving, bone-chilling rains the Division carried out the Governor's order. The Consolidated Morning Report of the Division for April 29th shows that over ninety-seven percent of all assigned personnel were assembled—over 8,900 men. Additionally, although Pennsylvania's assignment did not include cavalry, the Secretary of War and the Army Adjutant General agreed to accept the Commonwealth's three cavalry troops and count their numbers against the quota of 10,800.

In 1898 there was no law which compelled members of the National Guard to go on Federal active duty. Everyone had to be "mustered in" to the Federal service by answering "Yes" to the call of the muster roll. A "No" answer meant discharge and return to home. In his book on the history of the 4th Regiment, Pennsylvania Volunteers, F. H. Reichard described the procedure:

> The regiment was marched to the parade grounds and halted in column of companies; an officer called the roll to see whether all the enlisted men were present; then the Governor and the mustering officer of the Regular Army took their positions in front of the company; the mustering officer then read the following declaration... :
>
> 'The necessity of the situation does not require that any member of the National Guard of Pennsylvania shall consider himself bound by such membership to enlist in the service of the United States, if such enlistment shall impose upon him personal sacrifices not made necessary under the limited call of the President, or hardships upon those who are dependent on him for support. Nor shall any non-enlistment be in anywise considered an avoidance of duty or be to the prejudice of men who, willing to endure anything for their flag and country, are not called upon to render service than can be renered by those upon whom the demands of home and family do not rest so heavily, and who await the opportunity to serve their country.'

"The roll was called again," he continued in the account, "and each man answered 'Yes' or 'No,'... ." Some 1,014 men of this regiment answered "Yes"; twenty-one replied "No." In like fashion most commands registered overwhelming affirmative responses.

Between May 6 and July 22, the entire Division was mustered into Federal service. Some last minute recruiting took place. Records show that although less than 9,000 men reported to Mount Gretna in April, over 12,000 officers and men were mustered into service in a seventy-five day period.

In addition, all regiments sent recruiting parties home in order to fill up the companies to wartime strength of three officers and 106 enlisted men. None of these recruiting parties had any difficulty in filling their quotas. As the 10th Regiment's history records it, there were so many applicants from which to choose, recruiters could afford to be very selective in enlisting the 248 men needed. Although these recruits for the 10th were enlisted on July 25, they were detained in Honolulu until

Bugler, First Pennsylvania Artillery in Puerto Rico.

"The First Shot" The First Pennsylvania Artillery in Puerto Rico.

late November and did not join the regiment until December 2.

Only three regiments, the 4th, 10th and 16th, along with the three artillery batteries and three cavalry troops, ever left the United States. Mustering-in dates, summary of service, and dates mustered out for all elements of the Division are outlined in Appendix A at the rear of the book.

The United States found itself with an army much larger than necessary to defeat Spain. Further, Spanish capitulation on all fronts was more rapid than was expected. The war was short.

The 16th arrived in Puerto Rico July 28, followed on August 2 by the 4th, plus the Philadelphia City Troop and Light Battery B. The Governor's Troop and Sheridan Troop, along with Batteries A and C, joined their comrades six days later. All of the cavalry and Batteries A and C were sent back to the United States on September 3, following the 4th Regiment, which had departed for home on the 1st. Battery B left Puerto Rico on the eighth. The 16th Regiment remained on the island until October 10, when it also left for home.

Although these commands spent little time in Puerto Rico, they were employed, almost from the outset, in advancing on enemy positions and occupying captured territory and towns. Battery B, for example, was in position on the road to El Cayey, with guns loaded to fire, when word came to cease hostilities.

The 16th Regiment, in its advance, seized the towns of El Coto and Juan Diaz without a fight and sent a patrol into San Isabel. Spanish troops withdrew from the latter two towns upon the approach of the Americans.

On August 7 the Regiment exchanged its old rifles for the new Krag-Jorgensen, a weapon well received by the troops. Two days later the 16th distinguished itself in a hard-fought battle with the Spaniards near Coamo. A strong Spanish position blocked the road. On orders from Major General James H. Wilson, commanding the American troops, the 16th moved from a reserve position on the night of August 8 and took a trail north of the blocked road, bypassing the enemy position. The Regiment bivouacked in the hills, continuing the march the next morning, traversing a dozen miles until they discovered a large body of the enemy along the road below and to their right. Attacking the enemy with part of his command, Colonel Willis J. Hulings sent four companies to outflank the Spanish and attack down the road. Before this maneuver was completed, however, the Spanish surrendered. Nine of the enemy lay dead, sixty-

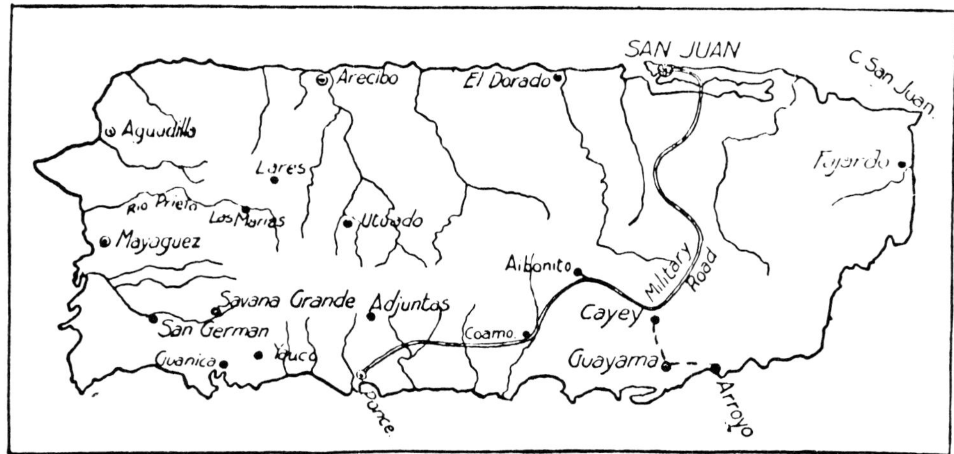

Sketch Map of Puerto Rico.

nine more were wounded and 211 became prisoners. The 16th lost one man fatally wounded, and five others received lesser injuries.

According to the regimental history there is evidence that Colonel Alexander L. Hawkins, or some politician "pulled strings" to have the 10th Pennsylvania ordered to the Philippines. At any rate, on May 17 the Regiment was diverted from going to Chickamauga Park, Georgia, and ordered to the Philippines as part of General Wesley Merritt's command. The Regiment arrived by train in San Francisco on the 25th, where it trained and fitted.

Embarking on the steamship *Zealandia* on June 14, the men sailed for Hawaii in a convoy carrying the expedition. They arrived in Honolulu on the evening of the 24th. The next day the entire expedition, as guests of the people of Hawaii, dined in the Royal Gardens while the Royal Hawaiian Band serenaded them. (The 10th Pennsylvania also destroyed the 1st Nebraska in a baseball game, 16-0.)

Continuing on their journey, the force finally entered Manila Bay on July 17, and the troops went ashore four days later, encamping six miles south of Manila.

A line of trenches was established by the Americans about 1,000 yards from the Spanish, running east from Manila Bay to a macadam road which ran between Cavite and Manila. American regiments took turns manning these trenches on twenty-four hour tours. On July 31 it was the 10th Pennsylvania's turn. Major H. C. Cuthbertson, in command of the fortifications that day, placed Companies A, C, H and I in the trenches, along with a battery of artillery from Utah. Companies D and E were put in reserve 200 yards in rear of the line. Company K was located about the same distance in rear of the right of the trench line, but east of the road. Company B was guarding a road two miles further to the rear.

The day was quiet, but that night a fierce typhoon sprang up and the troops were deluged by torrents of wind-driven rain. About 11 in the evening the Spanish commenced firing artillery from

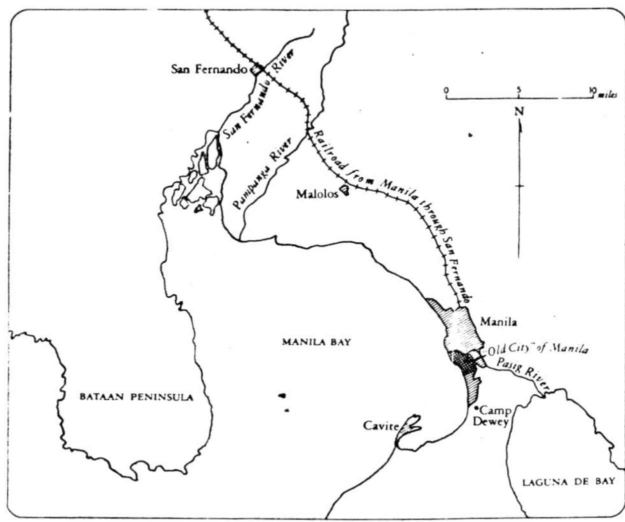

The Philippines: Central Luzon and the Manila area.

a fort in their lines. After a half hour's harmless bombardment, according to regimental reports, Spanish infantry advanced toward a point east of the American trenches and started to fire by volley into the Pennsylvania troops. Major Cuthbertson rapidly brought Company K into a line east of the road and extended his defensive position by moving Companies D and E onto line to the right of Company K. These three companies numbered about 200 men. A hot rifle fire was exchanged with the Spanish until about 1:30 A.M., when Battery K, 3d U.S. Artillery, arrived and was posted near the center of the 10th Regiment's line. A battalion of the 1st California Infantry arrived about the same time and relieved Companies D, E and K. The Spanish ceased their attack on the American right. About 2 in the morning the enemy renewed its attack on the trench line but broke off the engagement about an hour later. Nine men from the 10th were killed and some forty wounded in the battle. Although some historians claim that the Spanish did not really attack in strength and that the 10th was just suffering the usual jitters of troops new to battle, they did stand their ground and fight, believing they were engaged in a desperate battle against great odds. Believing makes it so. From that day on it was the "Fighting Tenth."

On August 13 the Spanish surrendered, and the 10th entered Manila along with other American troops. The insurgents of Emilio Aguinaldo, who had helped besiege the city, were refused entrance into Manila by the department commander. Aguinaldo, angered by this treatment, gradually occupied trenches and blockhouses around Manila with his troops until the American army in the city was virtually besieged by its late friends and allies. On December 2 Companies A and B, under Major Cuthbertson, were detached for duty as a guard at the hospital which had been established on Corregidor Island.

The situation between Aguinaldo's insurgents and the Americans occupying Manila gradually deteriorated until the night of February 4, 1899, when the Filipino army attacked American positions all along the eighteen mile perimeter. United States troops were on the alert, however, and the attack was repulsed. The sixty-man outpost of the 10th was subjected to heavy rifle fire from near a place called the Chinese Hospital but

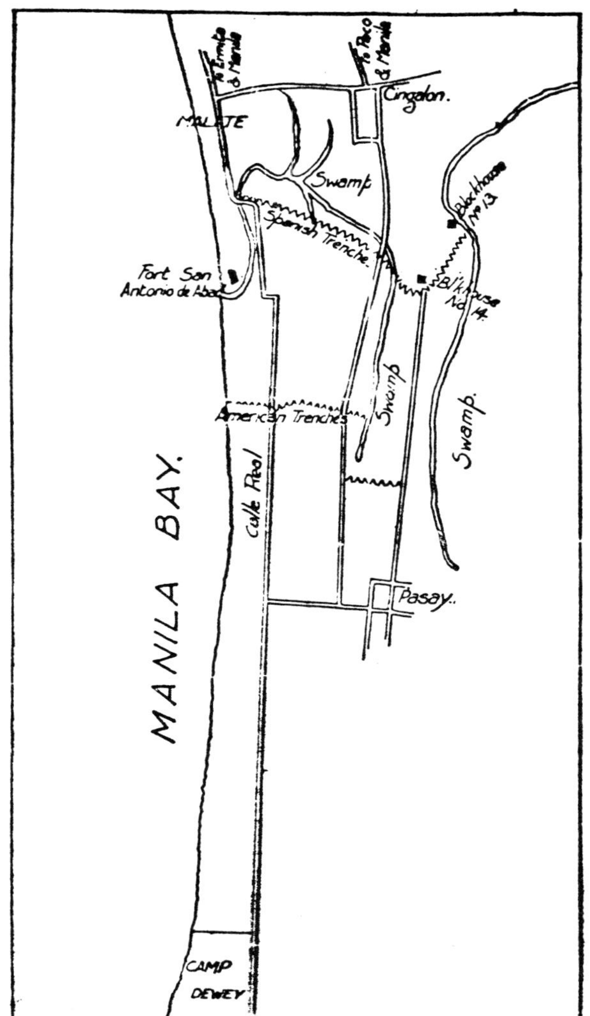

Sketch Map of Area Defended by the 10th Pennsylvania.

was reinforced by the remainder of the regiment. The 10th suffered no casualties. The next morning the Utah Battery shelled the enemy positions and the 10th, "cheering and yelling," charged. As the Regiment approached the Filipino lines, the enemy withdrew.

After securing the Chinese Hospital, the 10th continued its advance up a long hill, on the crest of which was located a Spanish fort around De la Loma Church. Filipine troops in the fort poured a rapid fire against the advancing Americans, wounding Major Everhart Bierer. As the Regiment neared De la Loma, the insurgents retreated toward a stone blockhouse north of the church. The blockhouse, too, was taken by storm. On orders, the Regiment halted and occupied the

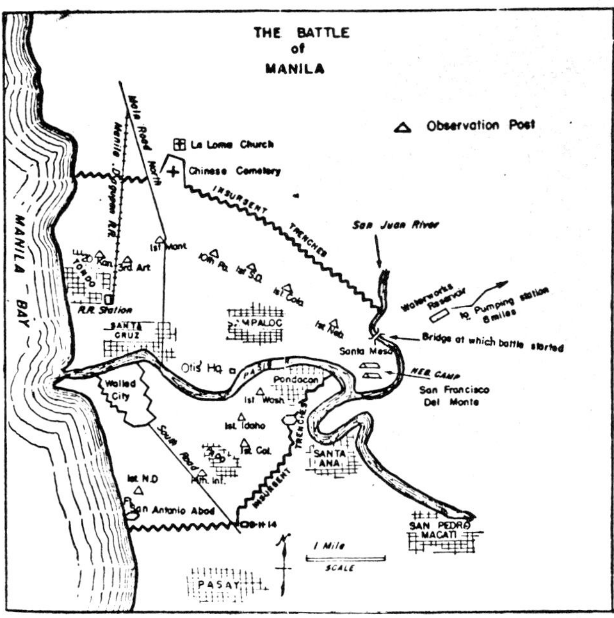

The Battle of Manila.

ground it had taken. In addition to the wounded major, one man was killed and four more wounded this day. The 10th occupied these lines until March 25. On the seventeenth a lieutenant and three enlisted men were wounded in a skirmish with a group if insurgents who attacked the position.

On March 25 the Regiment, deployed in a skirmish line, took part in the American army's general advance against Malolos, capital of Aguinaldo's republic. Fighting throughout the hot day, the Regiment steadily advanced against the defending Filipinos, fording rivers, through dense jungles, over rocky hills and valleys.

The attack was continued the next day, an equally hot and dusty one. Many men were prostrated by the heat. The enemy at one point formed along a wood line behind entrenchments and opened a heavy fire. Artillery was brought to bear, and the infantry moved forward, driving the enemy from his trenches and pursuing him through two towns, capturing several flags in the process. Another man was killed that day, and six more were wounded.

On March 27 the 10th was in reserve, but Company I was sent to help out a detachment of the 3d United States Artillery, which was fighting a large body of Filipinos. In carrying out this mission, a large number of insurgents were captured, and twenty-five were killed. Two men of Company I were wounded.

The advance was continued on the twenty-ninth. The Filipinos continued their "Fight, retreat, fight, retreat" tactic until the Regiment arrived on the banks of an unfordable river near Guiguinto, some five miles south of Malolos. Finding an iron railroad bridge, the 10th crossed and was being followed by the 1st Nebraska Regiment, when the insurgents opened fire from the front and both flanks. The 10th had been ambushed! Colonel Hawkins, the regimental commander, quickly formed the Regiment and fought back, supported by artillery, which was brought into action. In a spirited fire-fight, the ambush party was driven off. The Regiment lost three killed and sixteen wounded, including Hawkins, who was not seriously hurt.

A blinding rain storm and continued hit and run Filipino resistance marked the advance March 30, during which the 10th lost another man killed and two wounded. At noon on the last day of the month the Pennsylvanians entered Malolos. The Ameri-

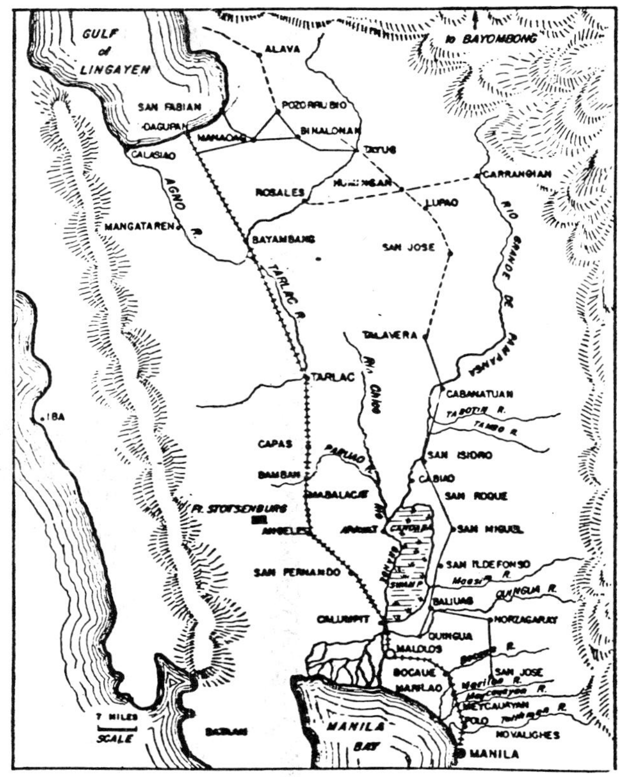

Area North of Manila.

The Well-Groomed First Squad of Company I, 10th Pennsylvania Volunteer Infantry.

Some More "Well-Groomed" Veterans on their Way Home.

Colors of the 10th Pennsylvania Volunteer Infantry after the Spanish American War. This flag was carried into all the battles of the Regiment in the Philippines.

can force halted, and the 10th was posted near the town, where it remained until April 14, when it was relieved and ordered to Cavite.

Colonel Hawkins was appointed Commander of the Military District of Cavite with troops consisting of the 10th Pennsylvania, two batteries of the 1st California Heavy Artillery, a battery of Wyoming Light Artillery and a troop of Nevada Cavalry. The district included Cavite City of about 5,000 people and was later extended to include Corregidor. On May 14 Companies A and B were relieved on that island by Companies E and H.

The regiment continued to garrison Cavite until June 30, 1899, when it was ordered to return to the United States. The Regiment sailed on the transport *Senator* for San Francisco by way of Japan's Inland Sea, with stops at Nagasaki and Yokohama.

On the night of July 18, two days after leaving Yokohama, Colonel Hawkins, who had been in ill health all during his tour of active duty, died on the transport. His remains were finally buried at Washington, Pennsylvania.

The Regiment was mustered out of Federal service August 22, 1899. In its eleven month tour in the Philippines, the 10th lost fifteen men killed or dead of wounds, six dead of disease, seventy wounded and one man missing. The Regiment had borne its assignment well; the men had acquitted themselves with honor and gallantry. Its veterans could point with justifiable pride to their battlefield accomplishments, and Pennsylvania soldiers everywhere could take pride in the reflected glory of their comrades in the "Fighting Tenth."

Hundreds of Pennsylvania soldiers were among the thousands who contracted typhoid fever while in the several large camps established to accommodate the huge volunteer army. Camp Thomas, on the Chickamauga, Georgia, battlefield was one of these. Two others were located in Florida, one at Port Tampa City and the other near Fernandia. All were ill-sited.

Latrines, as specified by the Regular Army, which had laid out Camp Thomas, were too close to mess areas. The water table at Tampa City was so close to the surface that latrine pits could be dug down only two feet. A number of days' rain flooded the drill field and caused the latrines to overflow. The camp at Fernandia was located on high ground overlooking a swamp three quarters of a mile wide. Pennsylvania commanders protested the assignment of these camps, but were overruled until it was too late. Only after typhoid fever had assumed epidemic proportions was anything done.

1899 was a year of reorganization and reinstatement of units into the National Guard of Pennsylvania after the war. It also became necessary to re-equip and re-uniform the entire force. By the end of 1900, however, the Division was reconstituted. The entire force, less Battery A, which was without artillery pieces, went into the annual camp that year "in a thoroughly efficient condition," as the Adjutant General's report put it.

The Division, less men awaiting discharge, numbered 724 officers and 9,387 enlisted men, as

Corporal of the 1st Pennsylvania Volunteer Infantry in the field at Camp Thomas, Chickamauga, Georgia, 1898.

PVT G. Lewis, a cook with Company G, First Pennsylvania Volunteer Infantry.

Masters of the culinary art - The Mess Section, Company G, First Pennsylvania Volunteer Infantry, Camp Thomas, Chickamauga, Georgia, 1898. Note the undershirt with the Keystone on the front and the pet pig held by one of the men.

of September 30, 1899. On March 31, 1901, the division contained 714 officers and 8,952 enlisted men. It consisted of three brigades, with a total of fourteen regiments, four troops of cavalry and three batteries of artillery.

The Adjutant General noted in his report, however, that fully sixty-five percent of the men in the first post-war encampment of 1900 had never been in military service before but that they were high caliber men. Lieutenant General Nelson A. Miles, commanding the United States Army, visited that camp, together with some 25,000 people, who came to see the Division review. In a letter afterwards to the Adjutant General, he noted the "interest, efficiency and soldierly bearing" of the command. He also observed that the Pennsylvania National Guard's practice of conducting annual encampments by brigade and division was responsible for the fact that "Pennsylvania troops have always rendered such excellent service and obtained for them their fine reputation." The record of the Pennsylvania National Guard in the Spanish-American War demonstrates General Miles's remarks to be well justified.

CHAPTER 7

1900 to Mexican Border

Uzal W. Ent, Colonel, Hq, 28th Inf Division

FOR A NUMBER of years, the National Guard had used a rifle range at Mount Gretna without rental. During that time several buildings and other improvements were added to the range. However, by 1901 ground around the range was being sold off, and many private persons were attempting to acquire the land, too. At the suggestion of the Military Board somewhat over 241 acres were finally purchased by the State at $20 per acre to preserve the range for the Pennsylvania National Guard.

In July 1902, the Division encamped on the battlefield at Gettysburg. Secretary of War Elihu Root, among the distinguished visitors that year, was impressed by Pennsylvania's Guard. In his remarks at the conclusion of his visit, he said: "... the National Guard in all States should be brought to the same high standard as that in Pennsylvania... ."

In early 1902 a wage dispute commenced between the anthracite coal operators and their employees, resulting in a general strike. The generally quiet atmosphere which first prevailed broke down in early July. By the end of the month rioting, and on occasion, gun-battles, erupted in many parts of Carbon and Schuylkill Counties. The Governor resisted involving the Guard, hoping that the sheriffs and local citizenry would be able to cope with the situation.

Finally, on July 30, confronted with a riotous situation beyond the capability of local authorities to handle, the Governor ordered to duty the Division Commander, Major General Charles Miller, Brigadier General J. P. S. Gobin, of the Third Brigade, as well as the 8th and 12th Regiments, Companies F and G of the 4th Regiment, and the Governor's Troop. The entire force was assembled at Shenandoah, Schuylkill County, on the evening of the last day of July and remained there preserving order for some weeks. On August 18, the Division Commander had to furnish troops to support the sheriff of Carbon County. Men from the camp at Shenandoah were sent.

In spite of the commitment of National Guard troops to some of the trouble spots, the general situation in the Commonwealth continued to deteriorate. The following chronology shows the spreading disorder:

August 27 - Second Troop of the Philadelphia City Cavalry sent to reinforce General Gobin.

September 23 - 13th Regiment sent to reinforce General Gobin, who was ordered to extend forces in order to support the sheriffs of Lackawanna and Lebanon counties.

September 24 - 9th Regiment sent to General Gobin, who was then ordered to support the sheriff of Luzerne County.

September 28 - Sheridan Troop sent to General Gobin.

September 29 - 4th Regiment ordered to duty with Gobin's force, who then was ordered to extend support also to Northumberland County.

At last, on October 6, the Governor activated the entire Division. Portions of Luzerne, Schuylkill, Carbon, Lackawanna, Susquehanna, Columbia and Northumberland counties were being terrorized by riotous mobs. People were being intimidated and beaten, railroad tracks were being torn up, and lawlessness had become so largescale and covered so wide an area that neither the civil authorities nor the elements of the Guard then on duty could rectify the situation.

The Division Commander was charged with protecting from harm all men desiring to return to work and with guarding "trains and other property from unlawful interference." He was empowered to arrest anyone engaged in "acts of violence and intimidation" and to hold them prisoner until they could no longer endanger public peace.

The turbulent area was divided into three brigade districts. The First Brigade, under Brigadier General John W. Schall, was assigned Carbon County, the southern half of Luzerne County and a small section of Schuylkill County with headquarters at Tamaqua. The First Troop was placed at Manila Park, Panther Creek Valley, along with elements of the 6th Regiment. The remainder of the 6th was placed in and around Lansford and Summit Hill. The 1st Regiment was disposed in and around Hazleton; the 2d in New Philadelphia and St. Clair, and the 3d was located at Minersville. A section of guns from Battery A was assigned to each regiment.

Colonel Willis J. Hulings, the acting commander of the Second Brigade (until the return of Brigadier General Wiley from Fort Riley, Kansas) established headquarters at Ashland, along with Battery B and the 5th Regiment. The Sheridan Troop and 18th Regiment garrisoned the Shenandoah region. The 10th went to Shamokin, the 14th to Mahanoy City and the 16th to Mount Carmel. This disposition covered the troubled areas in Northumberland and Carbon counties and a portion of Luzerne.

General Gobin established the headquarters of his Third Brigade at Wilkes-Barre. He also placed the Second Troop, Philadelphia City Cavalry, Battery C, and the 9th Regiment in that city. The Governor's Troop was deployed to Olyphant, along with the 13th Regiment. The 4th Regiment was split between Nanticoke and Plymouth, the 8th was sent to Duryea and the 12th was deployed to Providence.

On October 10 Division Headquarters was moved from Harrisburg to Pottsville. By October 21, the strike was over, but it appeared prudent to keep most of the troops on duty. A few units were relieved on October 24, but commencing November 5, the National Guard was gradually withdrawn from the strike areas and sent home. The last units were relieved on November 12.

One civilian was shot by a guard when he refused to halt at the guard's command. One Guardsman was shot in a saloon fight. Colonel Theodore F. Hoffman, commander of the 8th Regiment, died of exposure. It cost the Commonwealth almost $1,000,000 to employ the Division for the strike. This tour of strike duty was marked by alarms and occasional confrontations with crowds and mobs, but, in spite of the deaths noted above, was generally free of the violence characterized by the strike of 1877. Like all other duty, however, it was marked by several memorable events.

One day Colonel Wendell P. Bowman, commander of the 1st Regiment, was riding on a horseback tour of his outlying outposts. Seeing a mob on the verge of violence, he drew his pistol from its holster and boldly charged into the crowd, ordering the strikers to disperse. Surprised by the Colonel's bold action, the mob slowly melted away.

One of the companies of the 9th was on its way back to its quarters early one October morning, after an all-night patrol. Suddenly they heard shrill screams and loud cries from a nearby colliery. The noise could mean only one thing to the commander—another mob was in action. He ordered his men to double-time, hoping to arrive at the scene before too many people were hurt. The double-time became a race, for the noise increased as the unit drew closer to the scene. As the company made a "column left" around a large building, they beheld the source of the "unmistakable" noise. Before them was a yard full of mine mules, braying at the morning sun. Officers and men halted and commenced laughing so loudly that the mules stopped braying. The company was treated to a breakfast of ham and eggs, generously provided by the colliery cook. But the company officers were later roundly "chewed out" by the regimental commander "for accepting the hospitality of one of the parties to" the strike.

A number of significant events marked 1903. The General Assembly appropriated $37,000 to procure a permanent campground for the Division. New rifles were issued to the troops, and Congress passed the now-famous "Dick Act."

The principal provisions of this act were to declare the National Guard to be the "Organized Militia"; that all other able-bodied men of the nation would constitute the "Reserve Militia"; within five years after the law went into effect "the organization, armament and discipline of the Organized Militia" would have to conform to Regular Army standards; the United States would arm and equip the Guard without charge; units of the Guard would be required to have twenty-four drills (or target practices) per year; instructors from the Regular Army would be detailed to the Guard; inspections by both the Regular Army and Guard officials was mandated; Guardsmen on Federal active duty would be subject to the same regulations as the Regular Army and receive active duty pay; and National Guard commands could be called to Federal active duty for up to nine months. If the unit, at the expiration of the nine months, elected to volunteer for service in the Volunteer Army (for the war or other national emergency for which such an army would be needed) as a unit, it had to be accepted "as is." It could not be broken up and its men used as a pool of individuals for the volunteer army. The Act also abolished brigade bands, established regimental bands, and provided for a hospital corps as part of the Medical Department.

An additional troop of cavalry was formed in 1903, but the Adjutant General called for one more to be organized in the western part of the State. Although four Engineer companies, a Signal company and a Hospital Corps of five companies were all authorized, none of these type units had been formed by the end of 1903.

The Regular Army observer during the encampment of the Division at Gettysburg in July 1904, Colonel James Regan (Commander of the 9th United States Infantry) had this to say, in summing up his report:

> The National Guard of Pennsylvania is a fine, earnest body of citizen soldiers and one of which not only the State, but the Nation should be proud. The Major General commanding, the Governor [and] the Adjutant General ... evince the greatest interest in everything pertaining to the Guard. If the Nation could call out this Division just as it is today, it would give an excellent account of itself... . Every time I saw the Division ... it was a marvel to me how so many citizens could be brought together from multitudinous callings and from every town and city in the State and maneuvered with such precision and order. Every credit is due the citizens of the Guard for the sacrifice they make and the good work they have done and are doing.

Another troop of cavalry was formed on May 11, 1904, at New Castle. Designated Troop F, it brought to six the total of cavalry troops in the Division.

The Adjutant General in his report for 1905 noted that a number of commands were authorized but either were not in existence or only partially manned. For example, by 1905 the Division was authorized four engineer companies, twelve troops of cavalry, five artillery batteries of light artillery, a "signal corps" of one company and a "hospital corps," of five companies. However, only six cavalry troops and two light batteries were actually in existence. (It was not until 1908 that any change in this state of affairs came about. Two engineer companies were organized that year, along with a signal corps company in Pittsburgh on September 12.)

The General Assembly in 1905 enacted a bill authorizing payment of an annual uniform and equipment allowance of forty dollars for commis-

Staff of the First Regiment Infantry at the Annual Encampment, Gettysburg, PA July 23-30, 1904.

sioned officers who were "required to be mounted" and thirty dollars for all other commissioned officers. However, the officer had to spend at least that sum in order to qualify and was required to supply proper vouchers showing the payment. If the officer spent more than the allowance, the overpayment could be included in the voucher for the next year.

The Commander of the 13th Regiment, Colonel Frederick W. Stillwell, in his report for 1905, lamented the stress placed by inspectors on "guard duty, discipline, condition of clothing, rifles, books and papers," and that they based the efficiency of the unit on a total of the ratings on these matters. He declared that marksmanship should have an important part in determining the merits of the various commands. "Certainly in time of war a soldier's ability to kill his man is paramount in determining his efficiency. The present system of rating has a tendency to make us try to find out what the inspectors want, then give it to them ... ," he continued.

A provisional brigade of the National Guard of Pennsylvania participated in the Inaugural Parade for President Theodore Roosevelt on March 4, 1905. The 1st Regiment, plus four companies each from the 4th, 8th, 10th, 12th, 14th and 16th Regiments, the bands of the 10th and 12th Regiments and the Hospital Corps, 12th Regiment formed this command. General Gobin commanded the Special Parade Unit.

An act of Legislature of May 11, 1905, brought a State Armory Board into existence. Consisting of the Governor, Adjutant General and five persons appointed by the Governor (three of whom had to be National Guard officers), the Board was charged with providing, managing and caring for armories of the National Guard of Pennsylvania. The term for each appointed Board member was five years. Section 8 provided that no more than $20,000 could be expended for an infantry armory nor more than $30,000 for an artillery battery or cavalry troop armory. The sum of $250,000 was appropriated for the Board's first year of operation.

When the new State Capitol Building in Harrisburg was dedicated on October 4, 1906, a provisional brigade of the Guard, consisting of a regiment from each of the brigades, under the command of Brigadier General John A. Wiley, took part in the ceremony. In all, three companies each came from the 1st, 2d, 3d, 4th, 6th, 9th, 16th and 18th Regiments; two companies each from the 5th, 8th, 10th, 12th, 13th and 14th Regiments. The bands of the 1st, 16th and 12th Regiments, along with the Hospital Corps of those regiments also participated.

Gobin, by 1906 a Major General, in his report complained that the Regular Army observers that year in camp showed by their criticisms "an absolute want of knowledge of the conditions and workings of the National Guard... ." He went on to observe:

> The capabilities of officers are judged ... from the standpoint of the hours experience, forgetting or ignoring the fact, that these officers have fully as much duty to perform in organizing their commands at their various homes, keeping the ranks filled, and their men efficient and enthusiastic, as is their work for the brief period, when they appear at the head of the command. This is widely different from commanding a body of men supplied by the War Department... .

His observation later in the report that the newly-formed State Constabulary (State Police) would "doubtless relieve the Guard in many instances from riot duty," was most prophetic, for that has subsequently been the case many times.

The Division encamped at Gettysburg in July 1908. The Division Commander in his report praised the conduct of the men of the command but condemned the conduct and appearance of Regular Army soldiers in the town of Gettysburg. (Some Regular Army troops were also in camp, to assist in instruction for some of the Guard commands.) The Regular Army troops did not meet the standards and orders of the National Guard. The Regulars scoffed at the idea that the National Guard had any authority in the matter. When appeals to the Regular Army Officers present did not alter the situation, the Guard Provost Marshal's men resorted to force. One Army man received a bayonet wound, and several others were hit on the head. This seemed to settle the question of jurisdiction but "made it very unpleasant" for the Division Commander.

In 1908 the War Department arranged for joint maneuvers of the Regular Army and the Organized Militia. Camps for maneuvers were established in different parts of the nation. The one for the northeastern United States was at Pine Plains, New York (site of present-day Fort Drum). The Governor designated Pennsylvania's 3d Regiment Infantry to represent the Organized Militia of the Commonwealth. The regiment left Philadelphia on July 5, arriving in Pine Camp the next day.

Four officers from the Regular Army met the regiment and worked with the command during the entire encampment, assisting, advising and umpiring the regiment during maneuvers. For the maneuver the regiment became part of the First Infantry Brigade, under Colonel William Paulding, 24th United States Infantry, along with the 2d Connecticut, 4th Maryland and First Corps of Cadets of Massachusetts.

Drills and lectures from 7:00 in the morning to 8:00 at night opened the first day of camp. On Wednesday the troops were turned out on a few hours' notice, in heavy marching order toward the "enemy." After encamping for the night, the brigade attacked the enemy. In a similar fashion, the regiment took part daily in attacks, defenses, marches under combat conditions, outposts, rear guard actions, and the like.

Although the Regimental Commander, Colonel William G. Price, reported that ninety-five percent of the enlisted men he had personally canvassed stated that it was the most instructive camp they had ever known, he rightly remarked that the troops should have received some instruction before going on the maneuvers so that they would have been better able to understand and perform the duties expected of them. Fifty percent of the enlisted men had never been to camp before.

By 1909 the Division had 147 companies of infantry (fourteen regiments), six troops of cavalry, two batteries of field artillery, two engineer companies and a signal company. The Pennsylvania Guard was authorized 968 commissioned officers and 13,886 enlisted men. Because of strength shortages and the fact that many of the authorized separate units had not yet been organized, the actual strength stood at 722 officers and 9,558 enlisted men. At year's end the position of Division Commander was vacant when Major General John A. Wiley, the previous commander, retired on August 28, 1909.

A reorganization was mandated by Congress in 1908, under which "the organization, armament, and discipline of the organized militia [was required to] be the same as that which is or may be prescribed for the Regular and Volunteer Armies of the United States."

Accordingly, the signal, engineer, artillery and cavalry units were detached from the brigades and attached to the Division Headquarters. The 5th Regiment was disbanded and its companies split among the 8th, 10th and 12th Regiments. The Adjutant General was made Chief of Staff and the Department of Adjutant General, Inspector General, Judge Advocate General, Quartermaster, Subsistence, Pay and Ordnance were all provided for in the reorganized Division Headquarters. The reorganization took effect January 1, 1910. Prior to that, however, a provisional brigade of Pennsylvania Guardsmen participated in an inaugural parade, this time for William Howard Taft on March 4, 1909.

On January 1, 1910, in addition to the changes previously noted, a separate brigade (a fourth brigade) was formed of the 4th, 6th and 8th Regiments. This brigade, plus the 14th Regiment, which was left as a separate regiment, reported directly to the General Headquarters of the National Guard of Pennsylvania. For the first time since 1879, then, the Pennsylvania National Guard was no longer composed of a single division. Brigadier General C. B. Dougherty was appointed Division Commander in early 1910.

"Automobile trucks" were secured by the State Adjutant General, Brigadier General Thomas J. Stewart, as transportation for all camp equipment for the annual encampment of 1910. They proved to be more economical than wagons and were generally well received by commanders and members of the Quartermaster Department.

On May 15, 1910, the six troops of cavalry were reorganized into two squadrons—"A" and "B." Squadron "A" consisted of the First and Second Troops, Philadelphia City Cavalry, and Troop A. Troop G was authorized, but not yet in existence. Squadron "B" consisted of the Governor's Troop, Sheridan Troop, and Troop F. One additional troop, yet to be organized, was also authorized for this squadron. Each squadron was to be commanded by a major. He was given an adjutant, an officer to act as combination quartermaster and commissary, and a sergeant major. (Troop G was added to the rolls on June 16 and the fourth troop of Squadron "B," known as Troop H, was formed on November 2, 1911.)

A camp of instruction was conducted at Mount Gretna May 19 to May 23 for company officers and battalion officers of infantry organizations. (Field and regimental officers could attend if they temporarily waived all questions of rank and would "participate on an equal footing with all in squad and company drill.") The course was designed to prepare the officers for the annual encampment and included drill, care of the rifle, first aid, handling the Government ration, company paper work, map reading, extended order, patrols, advance and rear guard, outposts, attacks and defense, and preparation of field orders. Some 403 officers attended the course.

The camp of instruction was again held in 1911, from May 14 to May 19 with over 390 officers attending. The course, on a more sophisticated basis, included military topography and a whole series of classes specifically for "field and staff officers." (The course was considered so valuable, that it was repeated in 1912.)

In 1912, the men of the Pennsylvania Guard were issued new cotton olive drab uniforms conforming to the Regular Army summer uniform. Continuing a practice of some years, under War Department directive, the Pennsylvania National Guard sent a portion of its troops "to participate in a Joint Encampment of the United States Army forces and the Organized Militia at Mount Gretna" July 6-13, inclusive. The Second Brigade and Second Squadron of Cavalry, plus Battery C, were detailed to this camp.

For the first time a fully equipped field hospital was established at the Division's 1912 encampment at Gettysburg. The hospital was used to impart valuable training to medical personnel of the Division.

Major General C. B. Dougherty, Division Commander, commented in his 1912 report that:

> ... under the provisions of ... Federal Law ... combining the Regular Army and the Organized Militia in the first line of the Nation's defense it becomes absolutely necessary to not only take a larger and more comprehensive view of the service which the citizen soldiery may be called upon to perform, but it becomes absolutely essential that efforts be made to improve the forces comprising the Organized Militia for the demands which may sometime be

> made upon it... . the Guard is progressing; ... a greater interest is being inculcated in the minds of the officers and enlisted men, and ... all the preparations for instruction will bear good fruit and be of inestimable benefit to the Guard as a whole.

The National Guard may no longer be called the Organized Militia; otherwise, General Dougherty's comments still apply.

The commander of the Signal Company, Captain Fred G. Miller, reported that his company ran schools three nights a week so that men of his company could receive instruction without losing time from other civilian night schools which many of them attended. The company commander had two former non-commissioned officers of the Signal Corps present at the armory in the evenings in order to render assistance to the men of the unit. His training program also required his officers and non-commissioned officers to write papers on various subjects. The unit also conducted regular drills. In all, the Company had an ambitious training program.

Brigadier General C. M. Clement, Third Brigade, reflected the opinion of many other leaders when he remarked in his report that "the beneficial effects of the Camp of Instruction were plainly manifest in the improvement shown both in drill and during maneuver; ... ."

General Orders Number 1, of the State Adjutant General, dated January 8, 1912, established a war recruiting system conforming to directives from the War Department. The system called for a recruiting officer from each local organization who would begin to recruit his command to war strength if his unit were called into Federal Service; but who would remain behind with a "suitable detail of enlisted men" and continue to recruit, joining the unit only just before it left the Mobilization Camp. Each regiment and separate battalion was required to establish a party to recruit for the command during the remainder of the war, and each State was ordered to maintain a State mobilization camp, through which all recruits would have to pass for their initial physical examination, equipment, and whatever initial training time would permit.

The Annual Clothing Allowance for 1912 included $150 for each regimental band, $325 for each signal, engineer and infantry company, and cavalry troop; $500 for each battery of artillery, and between $35 and $50 for the staff non-commissioned officers at various levels. It was made clear that the money was for the purchase of clothing items for enlisted men.

The Division encamped by brigades in 1913. The Third and Fourth Brigades camped near Selinsgrove, on high ground overlooking the Susquehanna River, the First Brigade at Mount Gretna and the Second at Indiana.

Two singular features marked the camps of the Third and Fourth Brigades. Each during its encampment had a pontoon bridge constructed across the Susquehanna to be used in training, and both brigades planned to use Guardsmen who were licensed aviators to fly airplanes to take photographs from the air, to carry messages, and to reconnoiter. Both plans demonstrated the imagination and advance thinking of the leadership of these two commands.

Major Charles S. Farnsworth, who was detailed by the War Department in 1913 as an Inspector Instructor of the Pennsylvania National Guard, read a paper before the National Guard Association of Pennsylvania in January 1913 on "Recruiting and Drill Attendance in the Organized Militia." His comments are worthy of summation because they bear a remarkable resemblance to programs currently in vogue. He addressed the matter of drill attendance first, noting that attendance in 1912 averaged thirty-two percent. His plan of action was designed to attack this problem and, at the same time, increase unit strength. The six facets of his program which, he admitted, had come from an article by Lieutenant D. M. Greene, Indiana National Guard, were:

One - Inform the general public concerning the country's probable need for trained soldiers in large numbers and the objects and aims of the Organized Militia. At the same time get prominent local men personally interested in the local organization.

Two - Inform firms, corporations, individual employers and parents of all desirable men concerning the interest of the organized militia, of the units themselves, and that the State desired their cooperation.

Three - Provide similar information to labor unions, social organizations and "socialistic societies."

Four - Maintain a card file in each unit, giving the names, addresses, "occupation, place of work, age, relatives and associates, character of amusements, habits, personality" and other useful information about likely candidates for enlistment.

Five - Have a follow-up system to secure individuals listed on the cards, using hand bills posted in public places, letters sent to individuals, inviting them to the armory, special announcements and invitations to company meetings or entertainments, or personal solicitation by men of the company directly with the prospective candidate.

Six - Get boys of 15 years of age interested in the Guard by inviting them to observe drills, target practices, or to go along with small parties of the Guard unit on patrols, or foster their participation in games (sports). Utilize Guard officers and noncommissioned officers to address groups such as the local YMCA, Boy Scout troop(s), and high school boys.

Major Farnsworth did not agree with paying men to recruit, but he felt that anyone engaged in recruiting should be paid for any expenses incurred.

The use of sub-caliber devices in artillery and other large caliber weapons is not new. The United States Ordnance Department in 1913 offered a whole series of targets, representing enemy artillery pieces, caissons, machine guns, and men, standing, kneeling or prone, along with wooden sticks to support the figures, nails, and sleds to carry the targets representing guns and caissons, plus 600 feet of rope or sash cord for hauling the miniature targets. Some thirty targets of guns and caissons, twelve of machine guns and fifty each of the men, along with sticks, nails, sleds and rope were available "at a cost of about $11.00."

On June 15, 1914, the Civil War and Spanish American War battle flags of Pennsylvania regiments were transferred to the rotunda of the State Capitol building in an impressive parade and ceremony. A veteran of each regiment carried the flag of his command in the parade to the rotunda.

The cavalry of the Pennsylvania Guard was reorganized into a regiment in May 1914, designated "First Regiment Cavalry." Troops I (Sunbury), K (Lock Haven), L (Bellefonte) and M (Lewisburg) were all added to the rolls by transferring four infantry companies from the 12th Regiment. The regiment was divided into three squadrons, each with four troops.

General Orders 51, War Department, dated July 2, 1914, required an increase of seven men in each company of infantry and engineers, and troop of cavalry of the Organized Militia, in order to meet minimum strength requirements. Conversely, signal company authorizations were reduced. Type A units lost twenty-one men; Types B and D two men and Type C units one man. (The record does not indicate which type unit Pennsylvania's signal company was.)

Battery D was formed in Williamsport that July, from Company I, 12th Regiment Infantry.

Captain Elmer K. Rupp, Company D, 18th Infantry, began presenting a dollar to each unit member who brought in a recruit. The plan stimulated recruiting in his company and was approved by the Regimental Commander. State funds were sanctioned for payment of the honorarium. (In the late 1950's, $5, paid from other than State funds, so stimulated recruiting in the 28th Division Headquarters Company that the fund allocated for the purpose was soon exhausted.)

War Department plans in 1914 called for four Regular divisions and twelve National Guard divisions to be the first line in case of war. According to Major General John F. O'Ryan, commander of the New York National Guard, as of November 1914, the Regulars could field only two divisions and the Guard but three more. New York's 6th Division and Pennsylvania's 7th were two of the three. (Divisions were at that time authorized 22,000 men.)

The *State Army and Navy Journal* noted in its November 1914 edition that the cost of feeding Pennsylvania Guardsmen at camp that year had exceeded the twenty-eight cents per man per day allowed by the United States Army. The bill for one brigade was enumerated thus: Groceries, $1,197.24; meat, $2,799.26; bread, $556.14; coffee, $225.72; onions $50.16; potatoes and beets, $310.72.

Ironically, as the cavalry arm of the Pennsylvania National Guard was increasing in size, the number of young men with previous experience with horses and even suitable horses, themselves, were decreasing in numbers. So Captain Charles

C. McGovern, Troop H, reported at the annual convention of the National Guard Association in 1915. (Matters didn't get any better. A veteran of the 103d Cavalry told this author that in the late 1920's and early 1930's the Guard received condemned United States Army horses for their use.)

In early 1915 a major reorganization of the Pennsylvania Guard was in the offing, and speculation was rampant. A new military code was enacted by the Pennsylvania Legislature which made the grade of staff officers conform with grades of comparable staff officers in the Regular Army; abolished the election of officers above the grade of Second Lieutenant; provided for twelve troops of cavalry, enough for a regiment; a brigade of field artillery (two regiments of six batteries each); a battalion of three companies of engineers; a two-company signal battalion; four field hospitals, and enough infantry companies for twelve and one-half infantry regiments. Officers were to be appointed in the future. The immediate result of this abolition of the election system was a marked increase in the number of officers attending the officers' instruction camps at Mount Gretna.

By early 1915 Battery A, located in Bethlehem, was added to the artillery units of the Commonwealth. Its addition brought the number of batteries to four. The four batteries were brought together at Tobyhanna that year for an artillery camp of instruction.

A plan, previously considered, to convert some infantry into artillery batteries, so as to form an artillery regiment surfaced again in October 1915. By December two more batteries had been added to the four already in existence, and the First Regiment Field Artillery, headquartered in Harrisburg, was organized. The new batteries were E and F, both of Pittsburgh.

The roster of the National Guard of Pennsylvania for February 1916 included also a field battalion signal corps, composed of a headquarters, a wire company and a radio company. The entire battalion was located in Pittsburgh, reorganized as such effective January 27.

Civil disorders in Allegheny County in the spring caused the Governor to call out the 18th Infantry and a provisional squadron of cavalry. The appearance of the troops in the affected areas served to cool the situation and avert further bloodshed.

Pay scales had improved over the years for Guardsmen. First sergeants, in 1916, were paid $11.25 per day; battalion sergeants major $10.00; sergeants and cooks $7.50; corporals and artificers $5.25; privates and musicians $3.75.

Mexico was in the throes of a revolution and General Francisco (Pancho) Villa, arch-revolutionary, was on the loose. In March 1916, he and a column of his men raided Columbus, New Mexico. Eight members of the 13th United States Cavalry and nine American civilians were killed. A "punitive" expedition into Mexico, led by General John J. Pershing, was opposed by both sides in the Mexican revolution. While Pershing's column was south of the border, Mexican forces moved close to the Texas boundary. On May 5, both Glen Springs and Boquillas, Texas, were raided by the Mexicans. The Texas, New Mexico and Arizona National Guard forces were mo-

Company M, First Regiment - Ready for Mexican Border Service.

bilized and were moved into positions along the border by the 11th.

The 5,260 Guardsmen of these states, although welcome additions to American forces along the border, were not considered sufficient for the task. Since the situation between the United States and Mexico was so poor and in danger of further deterioration, the President on June 18, decided on a general mobilization of the National Guard of the United States.

A new National Defense Act, approved by the President on June 3, 1916, embodied many provisions important to the mobilization and utilization of the National Guard. Key points were: 1. It increased the Regular Army to 175,000 men for a period of five years and also authorized the formation of an Army Reserve. 2. It declared that the Militia would consist of "all able-bodied citizens of the United States between the ages of 18 and 45. 3. It divided the Militia into three classes: the National Guard, the Naval Militia and the Unorganized Militia.

The National Guard, however, was mobilized under the terms of the Act of 1903, leading to considerable confusion—both for the Guard and for the Regular Army. This confusion lead to misunderstandings, charges of mismanagement, waste and inefficiency. Aside from all this, the Guard was mobilized, and Pennsylvania Guardsmen were among the troops deployed to Texas.

The Division staff assembled at Mount Gretna on June 22, the site selected for assembly of the command. The field hospital units, engineers and ambulance units arrived at Gretna June 23, the infantry the next day, followed by the cavalry and artillery on June 25. Everyone had to undergo a physical examination. Many were rejected for active service, but the remaining men were all mustered into Federal Service by July 4. The Division Commander and his staff were mustered in the following day.

The Division, designated the 7th, in accordance with a War Department plan numbering the four Regular Divisions and the twelve National Guard Divisions, was led by Major General Charles M. Clement. His Chief of Staff was Colonel George Van Horn Mosely, formerly an Army General Staff Captain. Brigadier General William G. Price, Jr. commanded the First Brigade, consisting of the 1st, 2d and 3d Infantry Regiments. Brigadier General Albert J. Logan led the Second Brigade, com-

The Pennsylvania Railroad takes our boys to Texas.

Pennsylvania soldiers on duty in Texas.

posed of the 10th, 16th and 18th Regiments. Brigadier General Christopher T. O'Neil's Third Brigade contained the 4th, 6th and 8th Regiments. Colonel John P. Wood commanded the 1st Cavalry; Colonel William S. McKee, the 1st Artillery; Captain Ray W. Fuller led companies A and B, Signal troops; Major James F. Edwards commanded Ambulance Companies 1 and 2, as well as Field Hospitals 1 and 2.

By July 25, the Division was assembled in a camp outside of El Paso which was subsequently named Camp Stewart, in honor of the long-time Adjutant General of Pennsylvania, Brigadier General Thomas J. Stewart. The main street of the camp was called Pennsylvania Avenue and was close to four miles long.

On July 12 a battalion of the 2d Infantry and one from the 10th were ordered to Marfa Station in the Big Bend district. This outpost was over one hundred miles from the railroad, and roads were almost impassable. On July 16, Company B of the Engineers was sent to Nogales, Arizona.

On July 27 Major General Frederick Funston, commanding the Southern Department, ordered the Division reorganized to comply with the National Defense Act of 1916. Company F, 4th Infantry became Company C, Engineer Battalion.

Service on the Mexican border was arduous but proved not too dangerous for the Pennsylvanians. Tough training, combined with reorganizations, prepared the men, unknown to them, for combat in the Great War. Marches of up to one hundred miles, artillery practice with live ammunition, rifle practice, and finally, full scale maneuvers, involving brigades of infantry and regiments of cavalry, supported by artillery, marked service "on the Border."

On September 19 orders were received to return the men of one brigade to their homes. A North Carolina brigade replaced it.

In the meantime, back in Pennsylvania the Fourth Brigade, consisting of the 9th and 13th Regiments and a separate battalion of four infantry companies, fretted because they had not been called to Federal Service. The brigade went to their annual camp at Mount Gretna, where the idea was conceived to transfer both regiments to the artillery. Permission was granted, and the men went home from camp, and recruiting for the "new" units began. In the midst of the recruiting campaign, however, orders came cancelling the 13th Regiment's conversion to artillery. Instead,

In the field on the Mexican Border.

that regiment was ordered to the border, replacing the 2d Regiment, which had been designated to convert to artillery.

Thus mobilized, the 13th returned to Mount Gretna on August 15, remaining there until October 5. During this period over four hundred men failed the physical examination for entry into Federal service. Men from two companies of the separate battalion (a remnant of the old 12th Regiment) were added to the ranks of the 13th, replacing some of the loss. The Regiment finally joined the Division at El Paso in late September.

The saga of the 9th Regiment, sister of the 13th in the Fourth Brigade, was different. The 9th, under Colonel Asher Miner, was successfully converted to the "Third Pennsylvania Artillery." As such, it too joined the Division at El Paso on October 5.

On September 21, with the departure of one brigade, the remaining troops at El Paso were formed into a war strength division of 24,000 men. (It should be noted that this was not a "Pennsylvania" division, but one made up of all troops present.) The newly-formed First and Third Artillery Regiments were formed into a brigade, under command of Brigadier General William G. Price, Jr.

On October 3 the 1st Regiment started for home in a severe snow storm. Later that month Lewis machine guns were issued to various units, creating machine gun companies in each regiment. Other changes in regimental organization, dating from late 1915, include the addition of a regimental headquarters company, a machine gun company, a supply company and a sanitary detachment.

In November exercises included a convoy problem for the entire Division and several maneuvers, pitting brigades, supported by cavalry and artillery, against other brigades similarly supported. November 14 the First Artillery left for Pennsylvania, followed on December 18 by the 18th Regiment. Between January 2 and January 19, the remainder of the Division, except for the 8th and 13th Regiments, the 3d Artillery and Company C, Engineers, all returned home. The remaining commands were all released from active duty by March 1917, but as will be seen in the next chapter, the men of the Pennsylvania Division would have little time at home before they would be called upon to serve their country once more—this time in the "War to End All Wars."

CHAPTER 8

World War I

Truman Eyler, Gettysburg, Penna.

THE YEAR 1917 was a momentous one for the Pennsylvania National Guard. Many units returned from service along the Mexican border and demobilized in the early weeks of the new year, conscious of a sense of well-being derived from months of training and conditioning. However, they were not to remain inactive for long.

The story of the entry of the United States into the war on the side of the Allied powers on April 6 is a complicated one. The main concern of the National Guard was that this was the sort of national emergency which they were intended to meet.

Various regiments of the National Guard were called back into service as early as March 25 in anticipation of the declaration of war. Indeed, the 13th Pennsylvania Infantry had only entrained for its return to Pennsylvania from the border on March 21, and the regiment was informed en route that its mustering-out orders had been rescinded in view of the likelihood of war.

The declaration of war on April 6 had several immediate effects on the Guard. The first of these was that many of the units were called up and thus continued their training and their acclimatization to full-time service. The second effect was that many of the units were scattered about the Commonwealth in an effort to guard against an anticipated wave of sabotage by agents of the Central Powers and pro-German sympathizers. The numbers of the Guard afforded only weak protection for the myriad of strategic sites and installations; however, no wave of sabotage occurred, and the few incidents that did develop were dealt with effectively.

More regiments were called up for service on July 15. These were swiftly recruited up to full war strength by voluntary enlistment. The Pennsylvania National Guard Division as a whole was drafted into federal service on August 5, and further training soon ensued. The first detail of officers was sent to Fort Sill, Oklahoma, on August 12 for training in the use of machine guns, automatic rifles, musketry, hand grenades, and the bayonet. These officers would later return to the Division and act as instructors for their units.

During this period the War Department in Washington took a hand in the fate of the regiments of the Pennsylvania National Guard. The old division that had served on the Mexican border had been composed of three infantry brigades (each of three regiments), an artillery brigade (also of three regiments), a regiment of cavalry, and auxiliary signal, engineer, and hospital troops. In May of 1917, this was altered somewhat by the addition of a military police battalion, two more field hospitals, two more ambulance companies, augmentation of the engineers to regimental strength, and the organization of divisional headquarters troops. Yet another set of tables of organization was set forth on August 8, and this had far-reaching effects on the Pennsylvania regiments. The number of infantry brigades was reduced from three to two, and the number of regiments in each brigade was also reduced from three to two. A machine gun battalion was added to each infantry brigade, and a third was assigned as a divisional battalion. A number of other formations were also added to the divisional structure, including a trench mortar battery, a train headquarters, an ammunition train, a supply train, a sanitary train, an engineer train, a mobile ordnance repair shop, and a medical department. (It should be noted in passing that most of these support units were equipped with wagons and horses or mules. Relatively few trucks were supplied for the division's use.)

The August tables also required certain changes in the strength of units in order to compensate for the reduction from nine regiments to four. The basic strength of an infantry regiment was set at 3,750 officers and men, who were divided into three battalions. Each rifle company, composed of four platoons and a headquarters section, was increased to a strength of 250 men. The machine gun and supply companies were also somewhat enlarged, while the headquarters became the largest company in the regiment, with the administrative section, the 37 mm. guns, a pioneer platoon, a signal section, and the band.

Before all of these changes were effected, though, the Pennsylvania National Guard moved south. In late August and early September the troops journeyed to Camp Hancock outside Augusta, Georgia. This camp became home to the 28th Division until April of 1918, but the troops viewed it with misgivings upon arrival. Routine back-breaking construction work was the first order of business. The camp was barely half-finished, with incomplete buildings, dismal excuses for roads, and poor or even non-existent drill grounds, obstacle courses, and rifle ranges.

New reorganization plans arrived in the midst of all this work in the form of General Order No. 14 (September 22, 1917) and General Order No. 22 (October 11, 1917). General Order No. 22 overrode certain aspects of General Order No. 14, and attention will be concentrated upon General Order No. 22.

This order changed the Division's designation from 7th Infantry Division to 28th Infantry Division, in accordance with instructions from the Secretary of War and the Adjutant General of the Army. This numerical designation, like all the new lower level designations, was set in accordance with the re-numbering of the formations of the Regular Army, the National Guard, and the National Army (those units formed through the operations of the selective service system). Under this system, division numbers 1 to 25 and regimental numbers 1 to 100 would be reserved for the Regular Army. Divisional numbers 26 to 75 would be reserved for the National Guard divisions (along with the appropriate regimental numbers), while higher numerical designations would be reserved for divisions formed of draftees. Auxiliary formations also received numbers to conform to the system; for example, the supply train of the 28th Division was designated 103d as were most other divisional units. This system was aimed at providing one large, orderly, and anonymous army devoid of regional or unit rivalries and histories. It was a sound plan on paper, but it failed to account for human nature. The 28th Division adopted the nickname "Keystone" and developed a red keystone-shaped shoulder patch, officially adopted October 27, 1918.

The 55th Infantry Brigade was composed of the 109th and 110th Infantry Regiments and the 108th Machine Gun Battalion. The 109th Regiment was based on the 1st Pennsylvania Infantry with additions from the 13th Pennsylvania Infantry. The 110th Infantry was based on the 10th Pennsylvania Infantry and additional troops from the 3d Pennsylvania Infantry. The 108th Machine Gun Battalion was drawn from the 4th Pennsylvania Infantry and the Machine Gun Troop of the 1st Pennsylvania Cavalry.

The 111th and 112th Infantry Regiments and the 109th Machine Gun Battalion constituted the 56th Infantry Brigade. The 18th Pennsylvania Infantry was the base of the 111th Infantry, sup-

France at Last! The 28th Division arriving.

ported by elements of the 6th Pennsylvania Infantry. The 112th Infantry was formed from the 16th Pennsylvania Infantry and elements of the 8th Pennsylvania Infantry. The 109th Machine Gun Battalion was drawn from the 4th Pennsylvania Infantry.

The 53d Field Artillery Brigade was basically formed from the 1st Field Artillery Brigade of the Pennsylvania National Guard. The 1st, 2d, and 3d Regiments became the 107th, 108th, and 109th Field Artillery Regiments. The 103d Trench Mortar Battery was also created, largely from the First City Troop of Philadelphia.

The 1st Engineers of the National Guard became the 103d Engineers, and the 103d Field Signal Battalion was derived from the Field Signal Battalion of the National Guard. The field hospitals and ambulance companies, previously numbered 1 to 3, were re-numbered 109th through 112th and collectively designated the 103d Sanitary Train. Other organizations such as the engineer train, supply train, ammunition train, and military police were also re-numbered 103d and formed from detachments or pre-existing nuclei of units.

Major General Charles M. Clement commanded the 28th Division. His original brigade commanders were Brigadier General Frederick Stillwell (59th), Brigadier General Albert J. Logan (56th), and Brigadier General William G. Price, Jr. (53d). Only Price commanded his brigade throughout the war and demobilization. The other posts were occupied by a succession of officers, and all of these cannot be noted in a brief history such as this.

The breaking-up of the old regiments caused a considerable amount of feeling among the men of the 28th. Hostility to the loss of the old regiments with their traditions and histories was intense, particularly among personnel of the disbanded units. This was held down by keeping the bulk of the regiments together in most cases and submerging them in a new identity. No overt hostility resulted, but at least one writer charged that the reorganization led to uncertainty and confusion. The same writer also used the incident to argue in favor of future plans for a swifter and more carefully-planned mobilization for war.

The general situation of the 28th Division at Camp Hancock was not one to be envied. It had

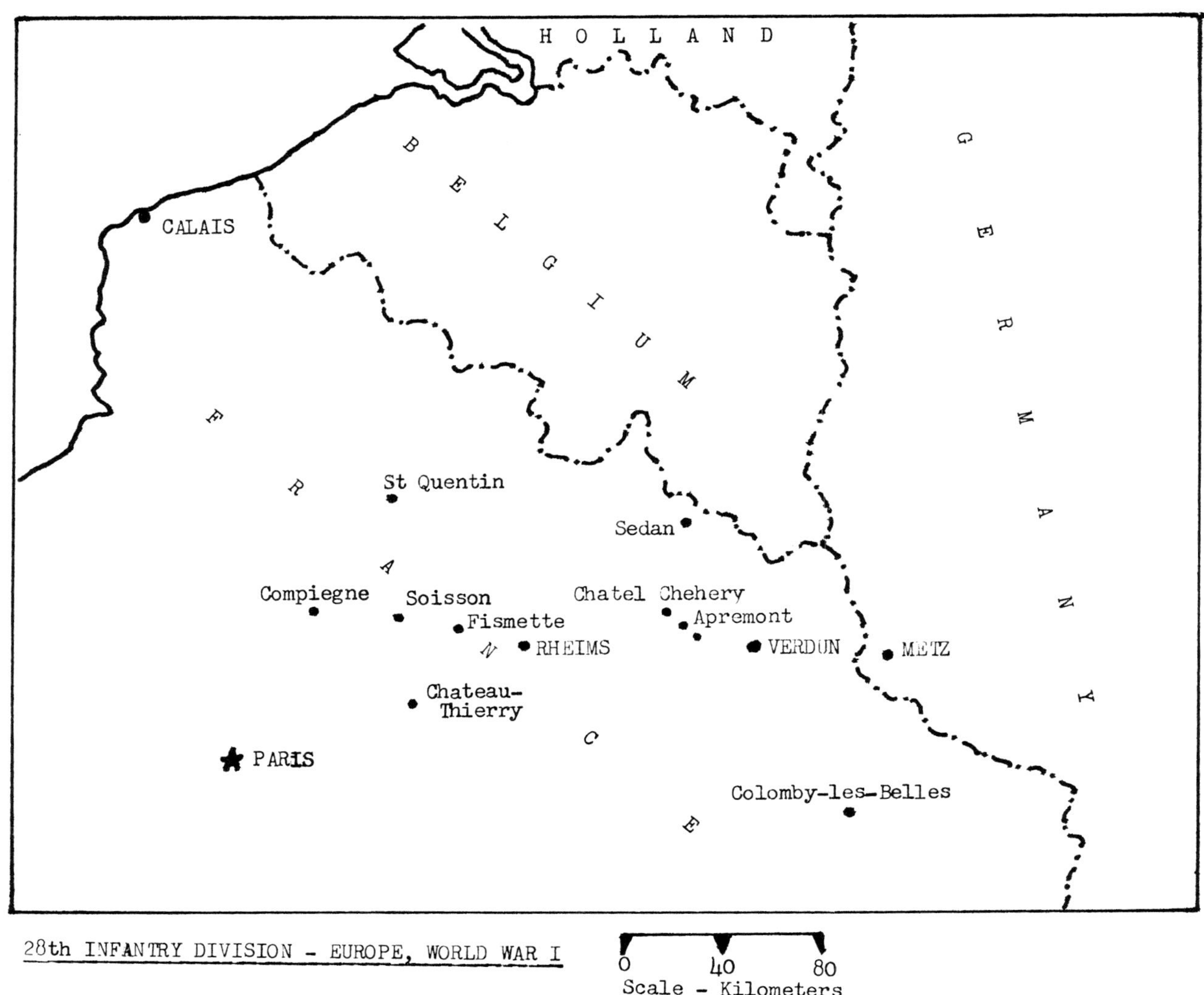

28th Infantry Division - Europe, World War I.

been forced to carve a home out of the wilds and had been shaken by a radical reorganization. On top of these troubles came difficulties with equipment and training. Most of the troops had arrived at camp in summer uniforms, and enough winter uniforms for all did not arrive until winter was in full force. Sufficient blankets for the command did not arrive until January. The troops of the division lacked full equipment until they reached France. At Camp Hancock they had only one bayonet for every three men. The machine gun battalions used wooden guns for training purposes, and there was only one 37 mm. gun in the entire camp. Thus the National Guard bore the brunt of the unreadiness for war.

Despite these shortages and shortcomings, the officers and men of the 28th Division set to work with a will. The first sixteen weeks of training were completed in January, and a program of eighteen weeks was begun. Training included entrenching, target practice, close order drill, hikes and marches, bomb throwing, parades, and reviews. Daily periods of drill ran from 7:30 to 11:30 in the morning and 1:00 to 4:00 in the afternoon. Wednesday afternoons were set aside for cleaning, and regular inspections were held on Saturdays. The evenings consisted of classes and lectures in which patriotism and civics vied with technical subjects for the attention of the troops.

This routine continued into April of 1918, by

which time many soldiers were anxious for overseas service. The careful training paid off in men accustomed to military life and a division from which unfit elements had been culled. The 28th Division was prepared to operate as a unit, and officers and men had come to know and trust each other, an invaluable asset. Changes in the higher command had occurred, however; all of the general officers of the Pennsylvania National Guard, save Brigadier General Price, had either resigned or been disqualified for reasons of health. Major General Charles H. Muir replaced General Clement on December 15, 1917, and remained in command of the Division until October 22, 1918, and later from April 19, 1919 to its mustering-out.

The fears of being left behind disappeared in April when orders arrived for elements of the Division to entrain for camps in New York. The infantry brigades received first priority and swiftly arrived at the assigned camps. Overseas equipment was issued to the troops, and they underwent final inspections and examinations. These tasks took only a few days, and then the troops journeyed to New York harbor where they boarded a wide variety of vessels ranging from military transports to H.M.S. *Olympic*, sister ship of the H.M.S. *Titanic*.

Occasional rough weather constituted the only hazard faced directly by most of the troops on the voyage across the Atlantic. However, the 111th Regiment shared the *Olympic* with two battalions of Regular infantry; one writer noted that boxing was a popular diversion, both with and without gloves. Entertainment of a far different sort arrived early on the morning of May 12 when a German submarine surfaced next to the *Olympic* in search of a target. The alert captain threw the helm over and ran the boat down, awakening the passengers with the bumping and scraping of the hull. When the sub bobbed to the surface in the ship's wake, the deck gunners fired several rounds into it. Escorting destroyers picked up thirty-one dazed survivors from the sub, U-103.

The trip overseas lasted from ten to fourteen days, and most of the troops landed first in England, where the populace welcomed them enthusiastically. The journey across England and the Channel took a few more days and also culminated in another joyful reception. The Pennsylvanians noted the absence of men of military age from the crowds and thus inferred the desperate positions of their allies.

The strategic situation of the Allied powers was critical as the Division disembarked. General Henri Petain had quelled the mutinies in the French armies following the disastrous offensives of April, 1917, but French morale remained fragile. The Italian armies had been engulfed in the catastrophe of Caporetto in October of 1917, drawing off substantial French and British forces. Russia and Romania had both been swept out of the war by the Central Powers and internal instability. The strain of three and a half years of war also wore heavily on Great Britain, still shaken by

In the British training area.

Men of the Iron Division take a break on the way to the front.

the enormous casualty lists of the Somme and Passchendaele. For all of these reasons the Allies determined to stand upon the strategic defensive until the United States could throw her fresh armies into the balance.

The situation of the Central Powers was also desperate. Unrestricted submarine warfare had failed either to starve England or interfere with the transport of American troops to Europe. The Allied blockade, on the other hand, was causing unrest and starvation in the camp of the Central Powers.

However, the collapse of Russia released numerous German formations for action on the western front. These troops were veterans of years of combat, but they were also fresh, having seen few hostilities on their front for many months. Even the addition of these fresh troops gave the Germans a superiority in the West of only about ten percent, scarcely a decisive advantage. General Erick Ludendorff, *de facto* supreme war leader in Germany, realized this and also the necessity for making the first blow decisive. He determined to capitalize on the superiority of the new German offensive tactics. These new infiltration tactics depended upon two factors, a swift, heavy, and paralyzing artillery bombardment and the infiltration of the enemy's front by small bands of highly-trained infantry groups. The combination could reduce an army to chaotic fragments.

Ludendorff decided upon a series of options and launched his first attack on March 21, 1918, in the area of the old Somme battlefields. The objective was to separate the British and French armies, and the initial attack was a stunning success. However, it failed in the face of tenacious resistance. A second German drive on April 9 launched in the area of Ypres and the Lys River met with only limited success. Losses in both offensives were about equal on both sides, a serious situation for the Germans since American troops continued to arrive in ever-increasing numbers.

Ludendorff still planned to smash the British, but he planned to draw reserves away from their front. Therefore a major German assault was launched on May 27 against the Chemin des Dames sector along the Aisne River. It was an apparent success and halted only at Chateau-Thierry, thirty-seven miles from Paris along the Marne River; however, it was a huge salient and dangerously vulnerable to counterattack. Another German drive in June farther to the west failed miserably.

Two final attacks were planned by Ludendorff, who had originally felt that he could afford only one offensive. The first of these was the Champagne-Marne offensive of July 15. It was planned only to draw off Allied reserves, but the German troops hailed it as the *Friedensturm* ("peace offensive") and expected decisive results. This was the first major action of the 28th Division. The second projected attack, the crushing of

28TH DIVISION (55TH BRIGADE)
CHÂTEAU-THIERRY SECTOR, JUNE 28–JULY 14, 1918
CHAMPAGNE-MARNE DEFENSIVE, JULY 15-18, 1918
AISNE-MARNE OFFENSIVE, JULY 18–27, 1918
MAP NO. 1

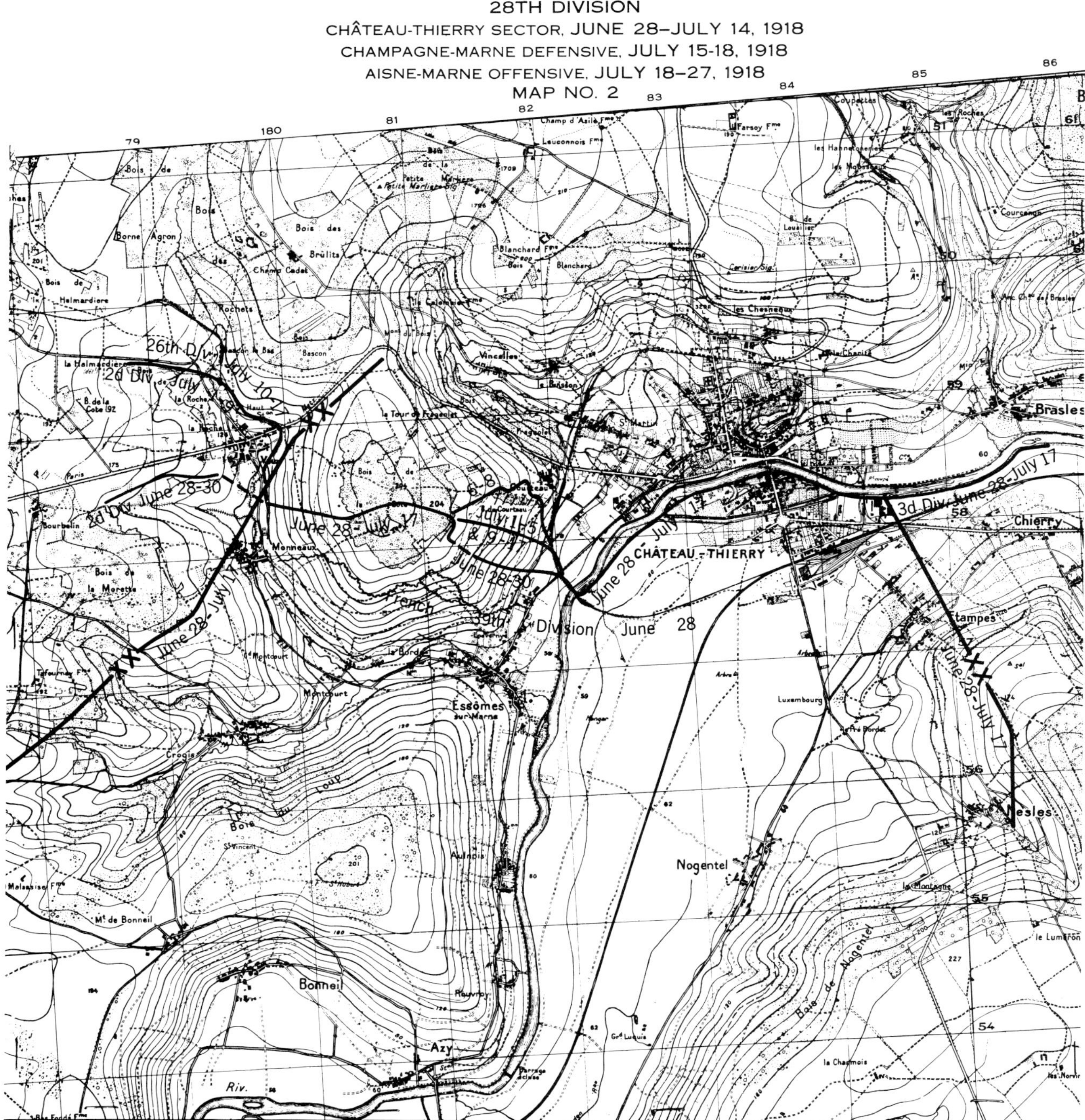

the British armies in Flanders, was never launched because the Allies seized the initiative in the West after the German failure along the Marne.

Upon its arrival in France, the bulk of the 28th Division joined the British for training. The original planning assigned the 28th as one of the American formations to operate in cooperation with the British in Flanders, particularly in light of the desperate situation of the British armies in March and April. Therefore late May found the 28th training with the British 24th Division. The troops had exchanged their Springfield rifles for British Enfield rifles and had received Lewis machine guns as well. Relations between the troops were good, although the Pennsylvanians found the British rations, notably jam and tea, insufficient. The troops of the 28th also stored most of their extra equipment and personal property at this point.

Training in the British sector lasted only a few weeks, and the first indication of a change of assignment was the order to exchange rifles once again. The Division was ordered in June to operate in support of French forces expecting a German attack, and to this end the troops moved by foot, truck, and railroad to the south, renewing their acquaintance with the infamous French *hommes 40, chevaux 8* railroad cars. Most of the Division arrived in the region immediately south of Chateau-Thierry by late June. The troops had been equipped with French automatic rifles and machine guns to make up for the lack of equivalent American weapons.

During the latter days of June and the early days of July some troops of the 28th Division first saw action. Several platoons from the 111th Infantry aided a French counterattack on June 29, and on July 1 seven engineers from the 103d were killed in a gas attack during bridging operations. The troops anticipated the prospect of action eagerly and received an alarm on July 4 with enthusiasm. The Division moved into reserve positions behind French troops, but the alert was called off, and the 28th Division returned to rear areas on July 5, where it remained until July 14.

The German offensive of July 15 was predicted by Allied intelligence operations. The 28th Division moved out of reserve, and most of its units took up positions in the second line of defense south of the Marne and east of Chateau-Thierry on July 14. The 56th Brigade held the left of the position, while the 55th Brigade held the right.

However, not all of the 28th held positions in the second line of defense. Companies B and C of the 110th Infantry and companies L and M of the

An impromptu concert by members of the Pennsylvania National Guard, amongst the ruins of Chateau-Thierry, France, July 24, 1918.

28TH DIVISION
AISNE-MARNE OFFENSIVE
JULY 28–31, 1918

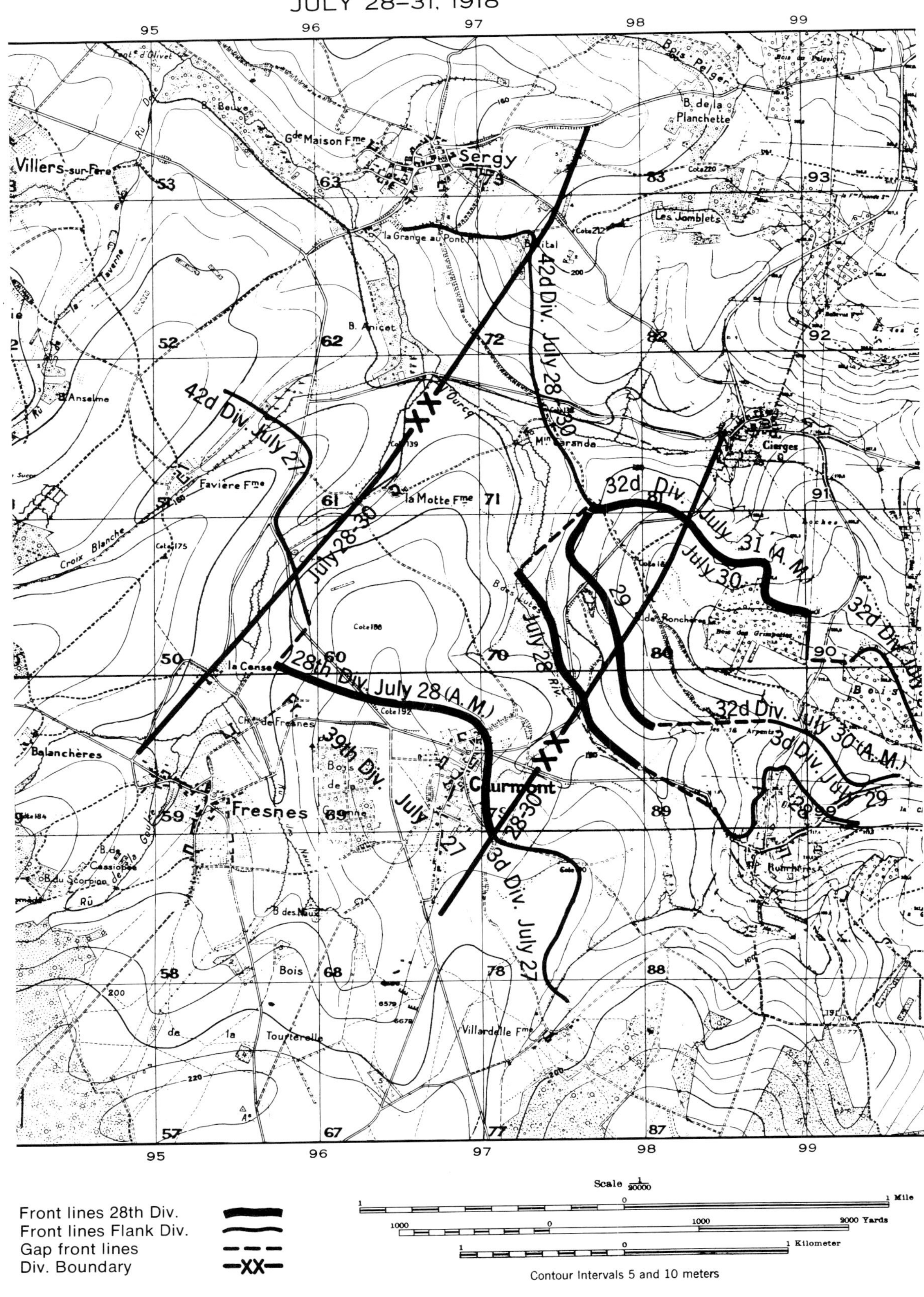

Graves of soldiers of 110th Infantry, 28th Division, buried at Fresnes, France, July 17, 1918. Killed in action during the Chateau-Thierry battle. Lt. Col. W. W. Fetzer; Wagoner Ralph Clark; Sgt. Jos. Malek; Pvt. 1.Cl. Walter H. Distler; Privates James L. Farrell and Warren Striver.

German Aeroplane, brought down by a barrage from the field guns of the 28th Division, near Jaulgonne, France, July 29, 1918.

A Truck Train of the 28th Division passing through the town of Dravegny, France, August 5, 1918.

Captured German machine gun nest in the top of a house, now occupied by the Yanks, near St. Gilles, France, Aug. 8, 1918.

109th took up positions in the front line on instructions from the French command. They were intended as a stiffening force for the French defense and were sandwiched in between the French units. The Pennsylvanians knew no French, and no liaison was provided, a tragic oversight.

The *Friedensturm* was heralded by a fierce German bombardment late on July 14. A fierce infantry assault followed closely upon the heels of the barrage, and the Germans began to infiltrate the French lines. The French abandoned the first line after a brief defense and followed their standard practice of permitting the enemy to expend his strength against a series of positions. However, the Pennsylvanians received no orders and held until they were surrounded. Only at that point did the troops attempt to retire, and all four companies virtually ceased to exist in the ensuing confusion. L and M companies of the 109th contained only 150 out of their original complement of 500 men.

28TH DIVISION
AISNE-MARNE OFFENSIVE, AUGUST 1–6, 1918
FISMES SECTOR, AUGUST 7–17, 1918
OISE-AISNE OFFENSIVE, AUGUST 18–SEPTEMBER 7, 1918

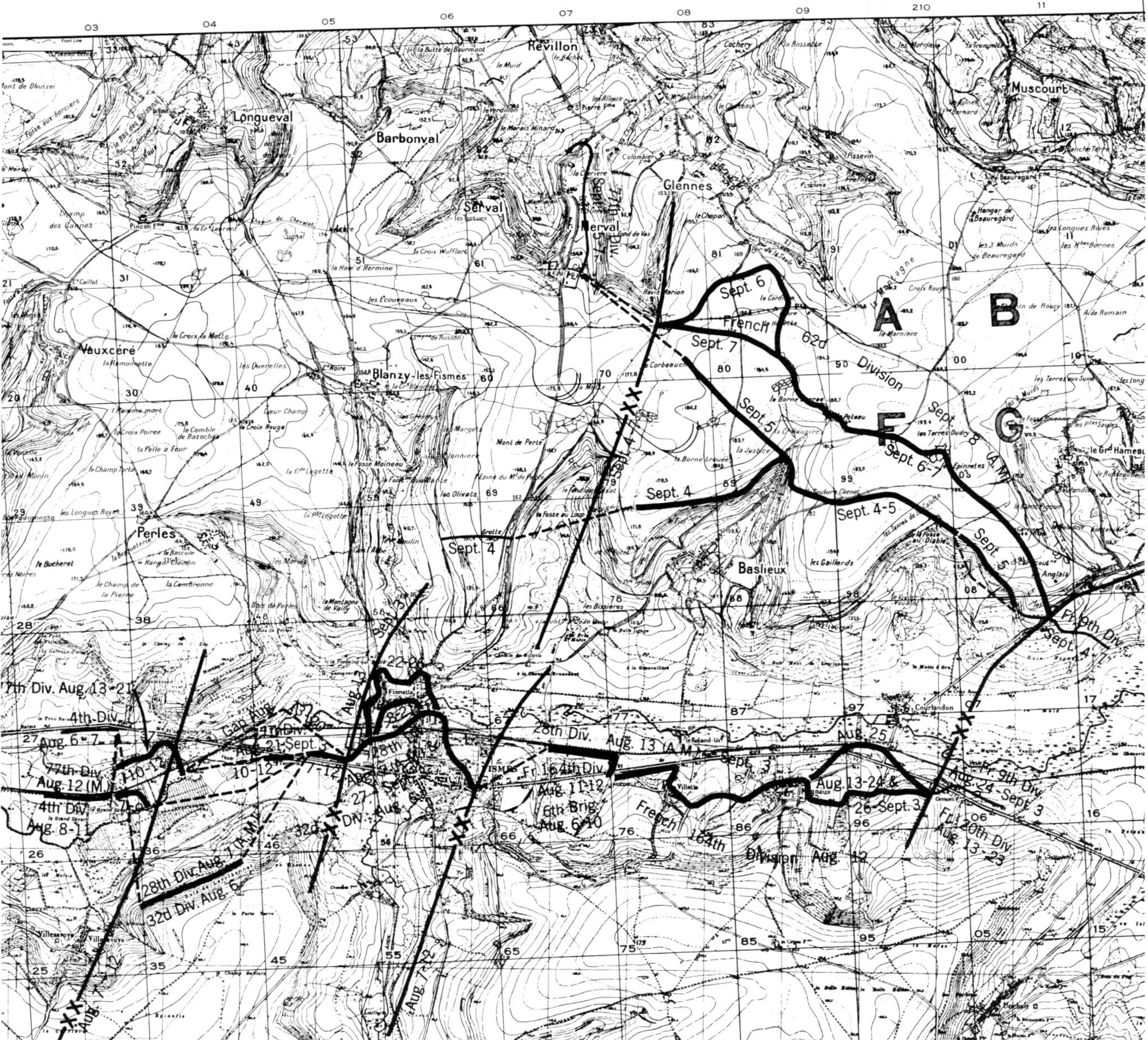

The German assault thus rolled on to the second line of defense and collided with the main force of the 28th Division, where it suffered a repulse with prohibitive casualties. Allied intelligence which predicted the attack and dwindling German resources contributed to the defeat, but the 28th Division, along with other troops of the American Expeditionary Force and the French Army, decisively defeated the final German offensive.

The 28th Division received its nickname of the "Iron Division" in connection with this staunch defense south of the Marne. General Pershing, commander of the American Expeditionary Force, had visited Division Headquarters, and, upon being informed of the stand of the 109th Infantry, had declared, "Why, they are Iron Men." The name was subsequently applied to the entire division.

At this point the Allied command determined to wrest the strategic initiative from the exhausted

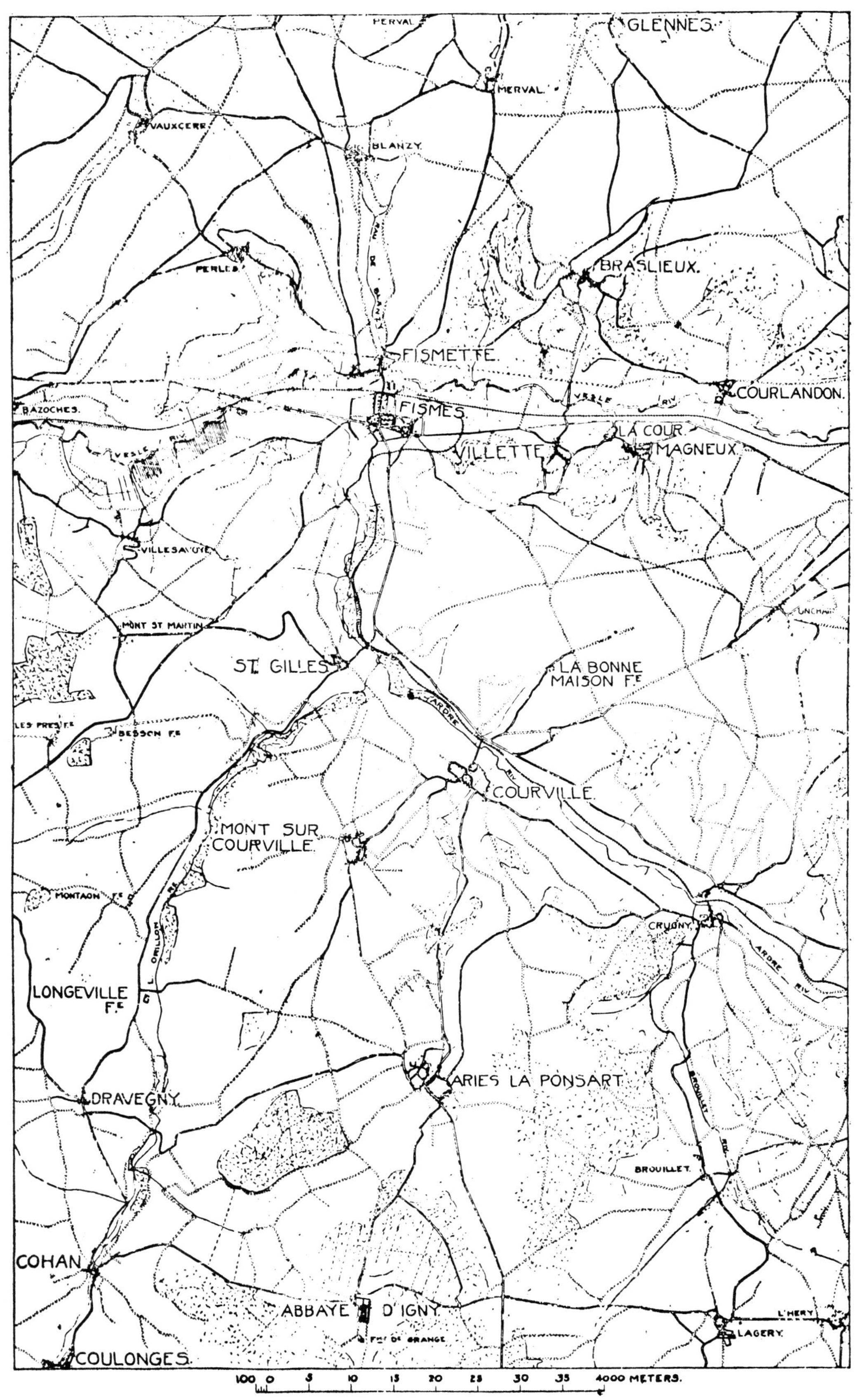

Fismes Sector. The division entered the sector by the Coulonges, Cohan, Dravegny road. August 12, 1918, the sector was extended to the east and the Abbaye D'Igny, Arcis le Ponsart, Courville road was then used to supply the east sector.

The wrecked City Hall, Fismes, France. September 5, 1918. This city was captured by the 28th Division during the drive to the Vesle River, in the early part of August 1918.

and dispirited Germans, and a counter-offensive was launched almost immediately along the Marne sector. The 28th found its true role as a shock assault division in this action. Counterattacks on July 16 cleared the enemy from the south bank of the Marne and gave the division better positions from which to resist any renewal of the German attacks. Further attacks the next day resulted in heavy casualties and little progress, but a German withdrawal was detected on the eighteenth.

The 28th Division joined other American and French divisions in a pursuit of the retreating enemy in the latter days of July 1918. The troops spent a short time refitting and burying their dead and then crossed to the north bank of the Marne. Heavy rains and congested traffic turned the roads into quagmires and impeded the progress of the division.

The Germans conducted a skillful rearguard action on the north bank of the Ourcq River, and the 28th Division launched a series of attacks in the last days of July in an attempt to dislodge the defenders and drive them northward. The 55th Brigade in particular was the assault formation, and the Germans contested the position fiercely as they attempted to safeguard the withdrawal of their stores and artillery north of the Vesle River. French promises of artillery support were honored only weakly at best, and the first attacks collapsed in heavy fire from the unshaken defenders. Further assaults were organized despite serious losses and the failure of units on either flank of the 28th to advance as far as the division, and the German positions fell by July 30 at a cost of over 1,100 American casualties.

The Pennsylvanians moved to the rear areas that night as other troops came to the front lines. The 109th Infantry received replacements of thirteen officers and 945 men to make up for their losses. The 110th received a heavy bombing attack on the night of August 1-2; these passed for

Members of the 28th Division preparing to rid themselves of that elusive little animal called the "Cootie," at the baths erected by the 103rd Engineers of the 28th Div., on the site of the famous Abbey D'Igny, France, August 29, 1918.

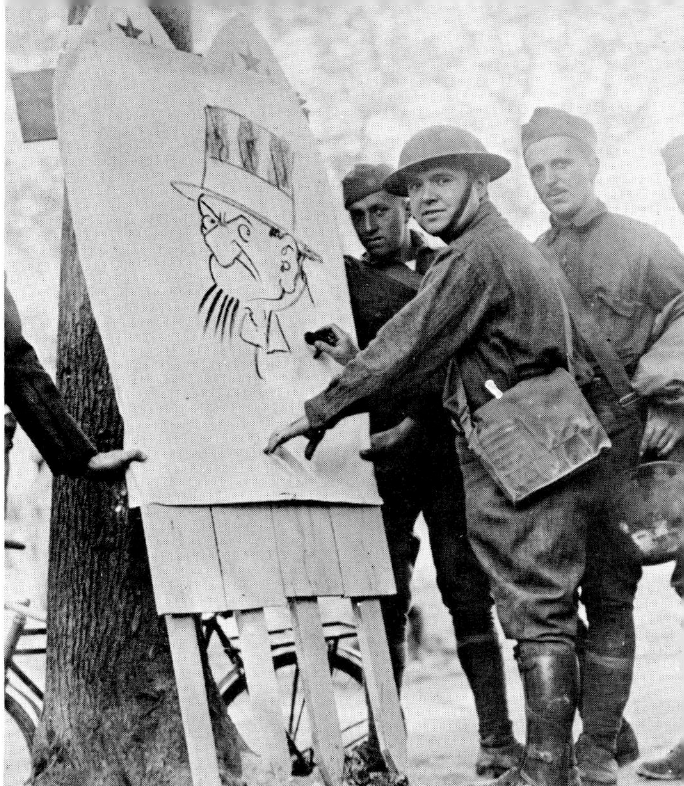

Claude Shafer, of the Cincinnati Post, cartoonist, entertaining the members of the 110th Infantry and the 103rd Engineers at the Abbaye D'Igny, France, while they were awaiting their turn at the baths. Mr. Shafer is the originator of Old Man Grump, featured in the Cincinnati Post. August 29, 1918.

Stone Dam which crossed the Vesle River at Fismes, France, and converted into a bridge by the 103rd Engineers to enable the Infantry of the 28th Division to advance and capture the town of Fismette on the opposite side of the river. This was the scene of the bloodiest battle on the Vesle River, and the 28th Division lost over 7,000 men killed, wounded and missing. Photo taken September 5, 1918 at 1:15 P.M.

The seven colored pictures following this page represent members of the Pennsylvania Militia and National Guard throughout history. Unfortunately, not every command or branch of service could be depicted. Note, however, that each soldier represents a specific command in history, as well as the present-day counterpart element of the Pennsylvania Army National Guard.

First Troop Philadelphia City Cavalry
1774-1781
Troop A, 1st Squadron, 104th Cavalry
28th Infantry Division

First Regiment,
Pennsylvania Volunteer Infantry
1814-1815

111th Infantry

48th Infantry Regiment,
Pennsylvania Volunteers
1861 - 1865

1068th Military Police Company,
213th Support Group

10th Pennsylvania Volunteer Infantry
1898 - 1899

110th Infantry

109th Infantry Regiment, A.E.F.
1917 - 1918

103d Engineer Battalion and
109th Infantry

112th Regimental Combat Team
1941 - 1945

112th Infantry
229th Field Artillery
Co. C, 103d Engineer Bn
Co. C, 630 Tank Destroyer Bn
Co. C, 447th Anti Aircraft Bn

The 28th Infantry Division
Soldier of 1979

rest days for the Division. Early in August the troops marched northward again to the line of the Vesle River. The long hikes in darkness along muddy and congested roads certainly constituted no novel experience for the division.

In their retreat to the north and east the Germans abandoned much of the territory that they had gained from the Allies in the offensive of May and June. They established another defensive line north of the Vesle River, and their artillery fire forced the columns of the 28th to leave the roads as they approached the lines. The Division relieved the 32nd United States Division from positions along the south bank of the Vesle, and the troops soon named the sector "Death Valley" because of the phenomenally deadly German bombardments of high explosive and poison gas. Heavy fighting centered for days around two towns in the sector, Fismes on the south bank and Fismettes on the north bank.

The 53d Field Artillery Brigade finally managed to rejoin the division along the Vesle. The component units had not shipped out from the United States until a considerable time after the infantry had left. The French could supply artillery support, but the German threat required the transportation of fresh American infantry in large numbers. Thus it was not until May that the brigade was scheduled for departure from the United States, and it did not arrive in France until the first week in June.

The brigade had trained long and hard in its service on the border and during its time at Camp Hancock. However, arrival in France meant that the brigade, instead of seeing action at once, merely began another period of training, this time with French equipment and under French supervision. The 107th and 109th Regiments were issued the famous French 75 mm. field guns, while the 108th, designated the heavy regiment of the brigade, received French 155 mm. field guns. All regiments received horse-drawn transport and equipment as well.

"Death Valley" became a familiar area to the gunners as they moved into position in support of the infantry and machinegunners over several days in the second week of August. Despite careful posting of the guns and construction of entrenchments, the artillerymen still suffered heavy casualties because of a combination of skillful

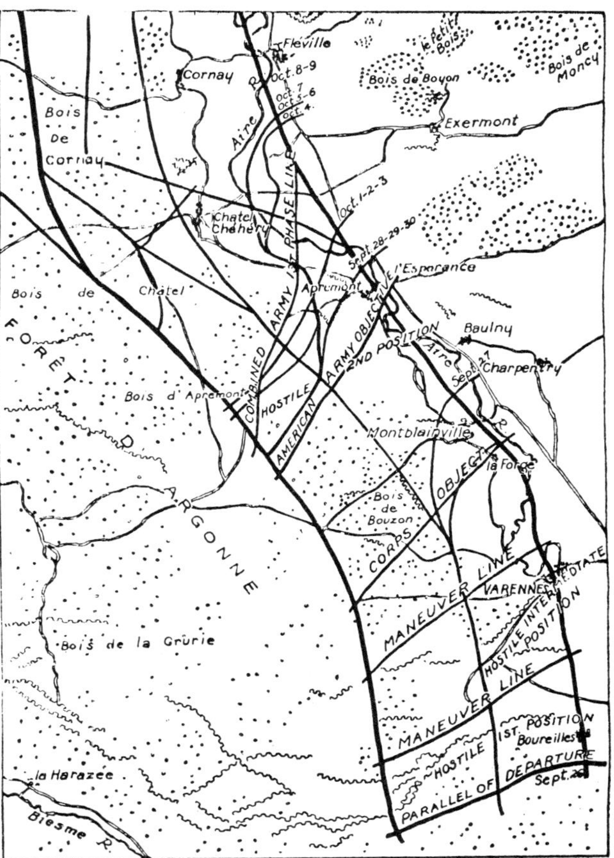

The Plan of Attack for the Twenty-Eighth Division in the Argonne.

Edward D. Green of Co. "A" 110th Infantry, 28th Division, wounded, and being assisted to a dressing station by a captured German prisoner and one of his comrades. Boureilles, France, September 26, 1916.

Meuse-Argonne Offensive

107th Field Artillery, 28th Division crossing bridge just repaired by Co. "A" 103rd Engineers at Boureilles, Aire River, France. September 28, 1918.

Members of Companies "A" & "E" 103rd Engineers repairing a dynamited bridge over the Aire River near Boureilles, France, Spetember 27, 1918, during the Argonne Drive.

Fighting under a gas attack. (Note masks on the men)

German gunners, the use of poison gas, and deadly air raids.

The concentration of the major units of the division probably helped to keep morale high in the heavy fighting. During the last weeks of August and the first week of September the regiments alternated their periods of service in the front lines, occasionally moving to the rear for periods of refitting and receiving replacements.

One particularly sanguinary combat occurred on August 27. The 28th had managed to seize Fismettes on the north bank, and on the day in question it was held by only companies G and H of the 112th Infantry. This weak force tempted a German counterattack under cover of a heavy fog and smoke shells. They swiftly broke into the village, and fierce fighting with grenades ensued. Both companies were virtually destroyed by the lack of support and the heavy attack.

Winning revenge for this tragedy became a task for the other units. The 28th Division was gradually withdrawn from the sector during the first week of September and sent to the rear for a few days of rest. The troops still remained subject to German air attacks. Casualties in the fighting along the Vesle had been very heavy. The 111th had been reduced to half its former strength, and many of the replacements had been in the army for only a few weeks, willing but untrained.

The 28th had a larger task than the pursuit of the retreating German armies beyond the Vesle. In mid-September, having been brought closer to established strength, the troops of the Division began a series of marches and truck rides which carried them to the rear of the Allied lines south of the Argonne Forest. Most of these trips were made in the dark in an effort to preserve the secrecy of the gathering of American troops in that sector. French troops continued to occupy most of the front lines in the sector until the night of Sep-

28TH DIVISION

CLERMONT SECTOR, SEPTEMBER 19–25, 1918

MEUSE-ARGONNE OFFENSIVE, SEPTEMBER 26–OCTOBER 10, 1918

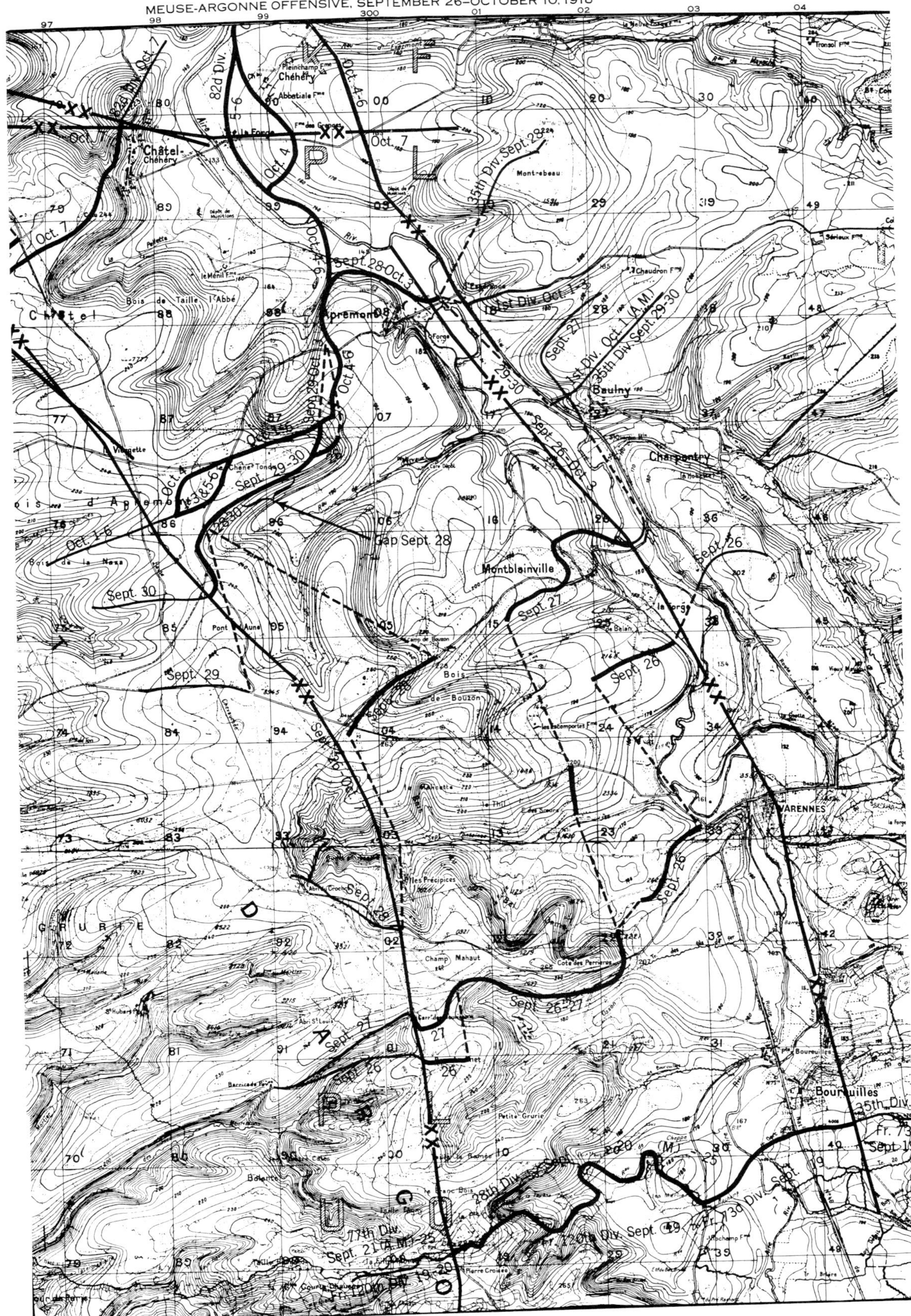

Many men fell wounded.

tember 25-26, and all American officers who entered the front lines for reconnaissance wore French uniforms at the time.

The Meuse-Argonne offensive was intended by the Allied high command as the southern pincer of the great counterattack of the autumn of 1918. German forces were to be pushed north of the Sedan-Meziéres railway, thus causing grave logistical problems. The 28th Division became part of the American First Army, and it was also named as one of the initial assault formations. Its status as a veteran unit and its possession of its own artillery doubtless contributed to this decision.

The preparatory bombardment began early on the morning of September 26, concentrating first upon the enemy wire entanglements and obstacles and then upon the German front line. Then it was transformed into a rolling barrage at 5:30 in the morning.

The infantry went "over the top" at 5:30 on the first day and made excellent progress. Even elements of the artillery moved forward before noon. Formations of tanks assisted in the assault, and the important town of Varennes, a storehouse and transportation center, was seized early in the attack. The advance of the 28th was facilitated by several factors: the veteran state of the Division, the surprise of the attack, and the open terrain in most of the divisional sector. The inexperienced 77th Division on the left of the 28th was considerably hampered by the rough terrain of the forest itself.

Progress continued in the last days of the month with the capture of the Montblainville by dark on September 27 and occupation of Apremont the next day. However, a customary pattern began to reappear soon, and the battle entered a second phase. German reserves entered the battle, and a slugging match soon developed between the tir-

Line of German prisoners captured by the 28th Division in the Argonne Forest, Oct. 4, 1918.

View of the main street of Buxierres, France, October 20, 1918. This town was used by the 109th Infantry, 28th Div. as a rest camp. The road was repaired by the 103rd Engineers.

An old French road house used as Headquarters of the 28th Division, near Euvezin, Nuerthe-et-Moselle, France, Oct. 21, 1918. Note the captured German narrow gauge railroad in the foreground.

ing American forces, disorganized by their advance, and the tenacious defenders. Most of the fighting in October soon degenerated into this slow, exhausting combat.

The famous incident of the so-called "Lost Battalion" in this campaign is worthy of note. This unit had advanced beyond its fellows of the 77th Division and had been surrounded by the alert foe. In order to relieve the pressure on this battalion and the front of the 77th Division, elements of the 28th faced westward and delivered an attack into the forest. The progress of this move soon eased the pressure as the Germans evacuated some of their forward posts to avoid being cut off in turn; thus it can be said that the 28th Division made a major contribution to the rescue of the "Lost Battalion."

Grim and bloody fighting for the next German defensive position distinguished the second phase of the battle. Châtel Chehéry and the ridge of Le Chêne Tondue figured prominently in the actions of the Division. Space does not permit an account of those complicated operations, except to note that the troops of the 28th finally carried the positions at a terrific cost. By dusk on October 6 none of the battalions of the 109th Infantry had more than 300 effectives.

Most of the 28th Division was ordered from the front during the night of October 8-9. It had been reduced to a shadow of its former strength; the 110th Infantry took over 1,000 casualties in the battle, while the entire 109th Infantry had been reduced to less than 400 effectives. Efficiency was further impaired by the incorporation of large numbers of selectees into the division. However, their morale was good, and they gained from association with veterans.

The 28th spent most of mid-October in transit once again. The troops entered the Thiaucourt sector south of the city of Metz as their next assignment. The area had formerly been known as a quiet sector, but the Americans soon changed this with their enthusiasm for night raids and patrols. Nonetheless, it was relatively quiet. The 110th Infantry took only about a hundred casualties during its time in the sector.

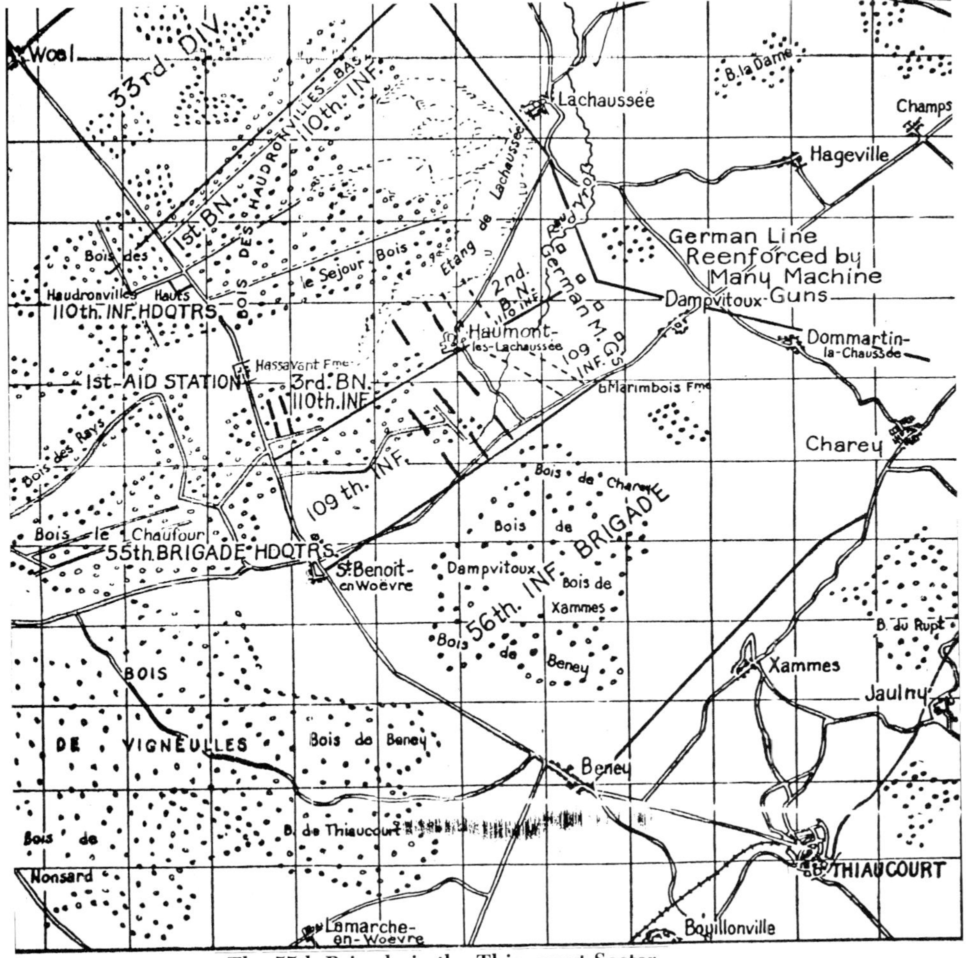

The 55th Brigade in the Thiaucourt Sector.

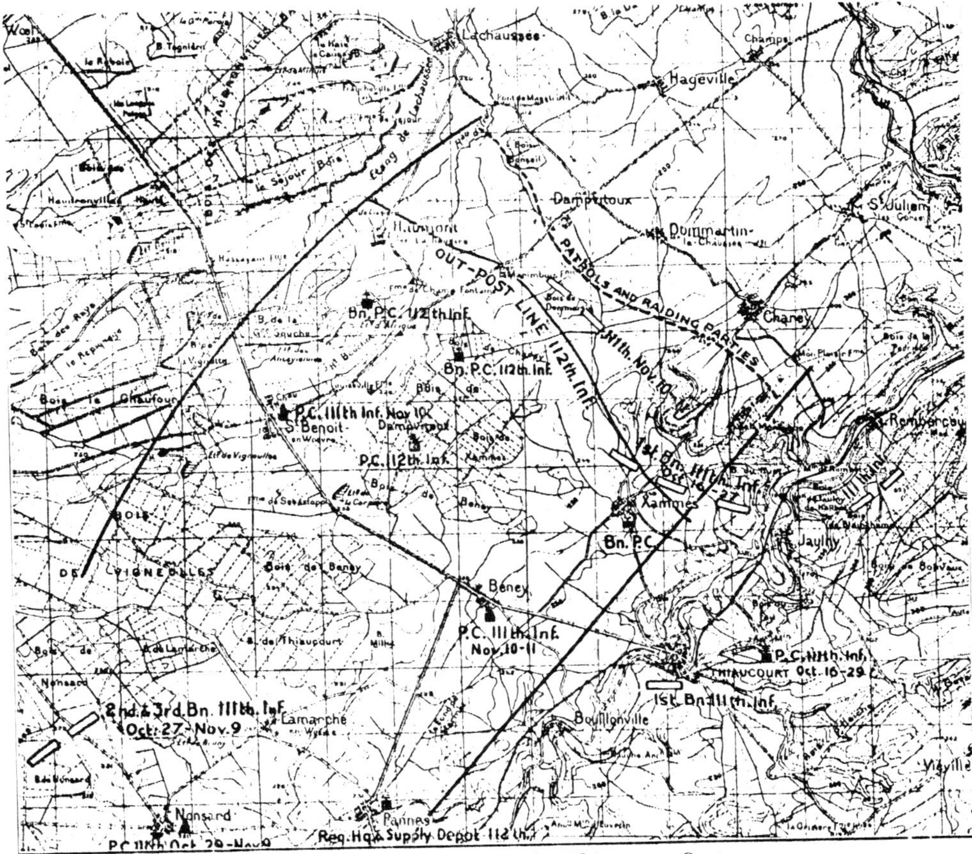

The Fifty-Sixth Brigade in the Thiaucourt Sector

28TH DIVISION
THIAUCOURT SECTOR
AND
WOËVRE PLAIN OPERATION
OCTOBER 16–NOVEMBER 11, 1918

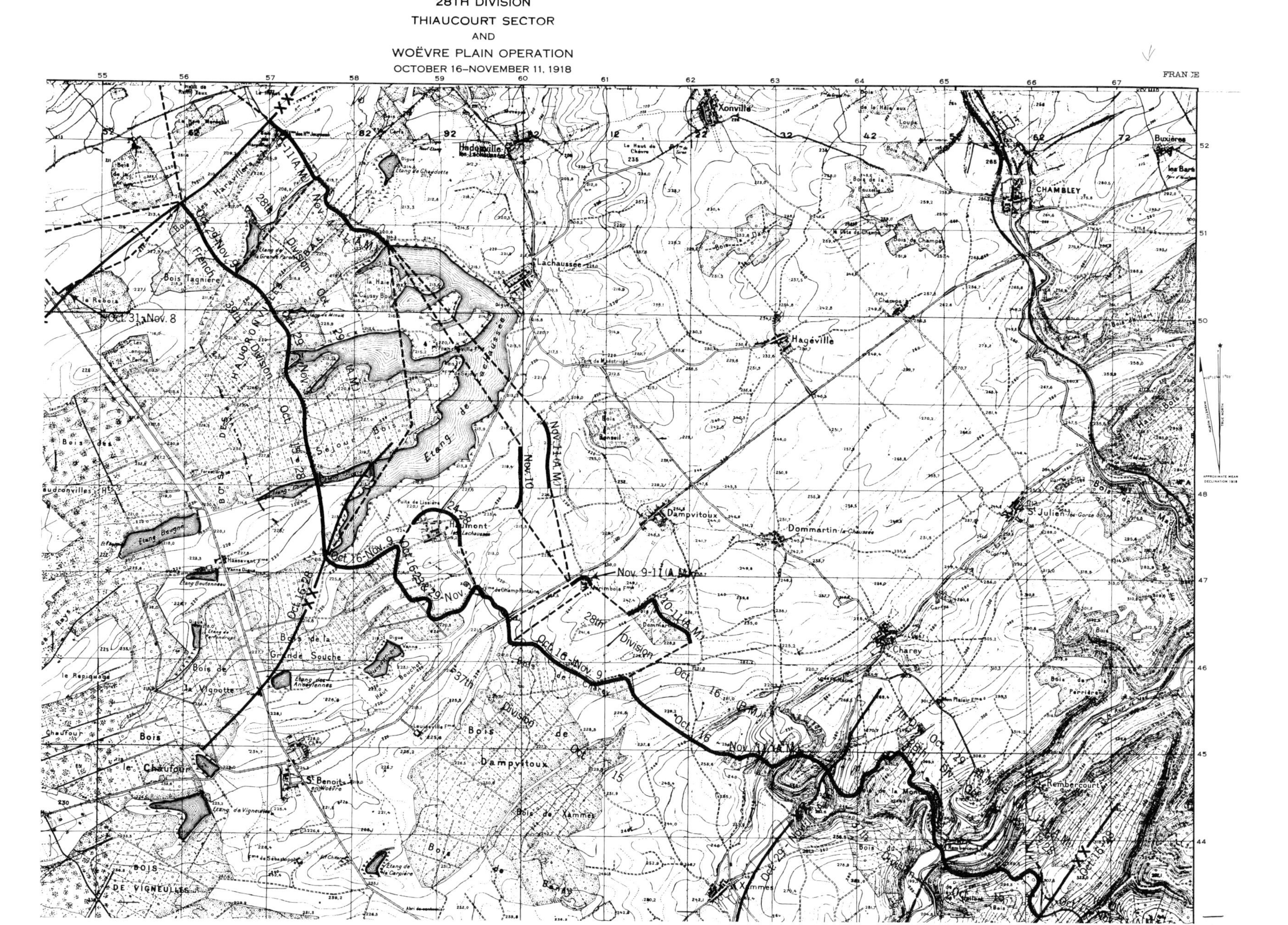

Metz was the target of a projected Allied offensive in the sector, and the 28th was guaranteed a key role in the project. However, the coming of the Armistice on November 11 forced the abandonment of these plans. The 53d Field Artillery Brigade had already been separated from the division and sent to Belgium in conjunction with the 91st American Division. The brigade was doubly distinguished for its record and for being the only American artillery to serve in Belgium.

The troops of the Keystone Division realistically welcomed the Armistice; there were few regrets that the war had ended. Rumors floated about that the Division would join the Army of Occupation, but nothing came of them, and it soon moved into the rear areas. The troops spent much time salvaging the battlefields, building roads, drilling, and holding athletic contests. Boredom replaced the Germans as the great enemy in the winter of 1918-1919.

The 28th changed quarters several times during the winter and early spring. Most of the component units sailed from France in late April, landing in the United States during May. The process of mustering-out varied from one unit to another and will not be detailed here. It suffices to state that most mustering-out was performed at Camp Dix, New Jersey, and that it was largely completed by May 23. Philadelphia had had a great Welcome Home Parade for the Pennsylvania soldiers on May 15. Thus by May 30, 1919, the career of the 28th Division, the "Iron Division," in the Great War was history.

The "Keystone" and the "Y" were there. Men of the Iron Division get coffee from YMCA Doughnut Girls.

The Red Star Liner "*Finland*", used as a transport during the World War, and which brought the 103rd Engineers and 103rd Field Signal Battalion of the 28th Division from France to America. Sailed from St. Nazaire, France on April 20, 1919, arrived in New York on April 30, 1919.

CHAPTER 9

Between the Wars

Allan G. Crist, Colonel, ARNG Retired

EVEN AS THE men of the "Bloody Bucket" were battling the Hun, those in authority in Harrisburg and elsewhere were thinking about what should be done "When Johnny Comes Marching Home Again".

Of course there would be a National Guard again! Not only did the law prescribe one, but for thousands, being a Guardsman was more-or-less taken for granted; it was not only a tradition, but a way of life. The home-town Guard unit (some still called it "the milishy") provided not only in most instances the only permanent, visible presence of "the Military," but a source of civic pride and prestige. So, hardly had the Armistice been signed and the last shot fired in anger when the planning started.

There were more questions about the who, what, where, when and how of the matter than anyone could answer immediately. One of them revolved about how the wartime substitute Pennsylvania Reserve Militia (the too-old, the too-young, and not-quite physically capable) might fit into the picture. As *Our State Army & Navy Journal*, the Guard's quasi-official monthly, expressed itself editorially in its December 1918 issue: "It is hoped that the next Legislature will give deep consideration" to the issue of how veterans could re-enter State service, and, "it is to be hoped that the State may find some means by which both the former Guard and the present Reserve Militia may be amalgamated."

Certainly, some fighting soldiers had seen all they wanted of the uniform, but others were equally eager to get back into the Guard. It was in that same issue that Sergeant Major Frank M. Funk of the 55th Infantry Brigade Headquarters wrote from France: "Will there be room for us? Have provisions been made to receive us again as members of the Pennsylvania Reserve Militia? Will we have an opportunity to enter the lists for promotion and advancement and have the same chance as those who already are members?" Others, too, reflected his concern, and their sincerity was reflected in the numbers who did come back into the Guard when reorganization got under way.

That happy circumstance tended to nullify the effects of an action that soured relations between the Federal Government and Guardsmen nationwide. The latter had assumed that, under the provisions of the National Defense Act which gave Guardsmen a dual State and Federal status, all returning Guard officers, and all enlisted Guardsmen whose State enlistments had not expired, could be held to their State service obligation. However, there was a hitch: the Federal Government had drafted, not just called or ordered, the Guard into service. Therefore, every veteran Guardsman was "home free."

From within the Guard's own ranks came a suggestion that did not set well with the majority. Colonel Franklin Blackstone expressed his feeling that the Guard should have no larger units than battalions; if and when those were combined into provisional regiments, the latter should be commanded by Regulars. The highest State rank a Guardsman could attain under his plan would be that of Lieutenant Colonel. Following his return, and muster-out of United States service as a Major, Blackstone was given command of the re-created 3d Infantry at Pittsburgh in the reorganization. In the years since, his proposal has surfaced periodically, both as to the Guard and other Reserve Forces, but promptly sank.

Two of the unresolved problems initially (and this was nationwide) were just how much a National Guard was needed by the States for their own internal peacetime purposes, and how much by the Federal government for war or national emergency. Closely related was the matter of how the Guard was to be organized and equipped. The pre-war composition was relatively simple because the focus was primarily on the States' needs which were mostly for infantry but in some cases, a certain amount of cavalry and field artillery. For modern warfare, however, more sophisticated types of specialists were an absolute necessity, and the "old" units had to be shaken-up, restructured, re-equipped and retrained. Pennsylvania authorities could not wait until these matters were hashed-over and decided by Congress and the War Department.

The Commonwealth would have a division of some kind—that much was certain—so time was not allowed to stand still. Reappointed, the wartime Adjutant General Frank D. Beary, early in 1919 saw the existing Pennsylvania Reserve Militia as "a splendid nucleus for the expansion and proposed division organization" but acknowledged "a feeling of unrest and uncertainty" among its members "and a doubt as to their future status". However, he saw returning veterans as "ready to take up a portion of the burden" and both elements, along with former draftees, evolving into "a National Guard second to none in the country".

Brigadier General William G. Price, Jr., commander of the wartime 53d Field Artillery Brigade ("undoubtedly the best known figure in line" for the post, said the *Journal*), was pegged to command, effective July 1, on a fulltime basis. Pennsylvania was one of the few States to maintain a modest Division headquarters and staff (one Lieutenant Colonel, two Majors, one Captain, one First Lieutenant and two stenographers) at its own expense.

But first, "the boys" had to come home. Unlike the chaotic post-V-J-Day, 1945, rush to disintegrate the Army, organizations were returned from "the Great War" substantially intact, and a joyous citizenry was eager to give a flag-waving, rip-roaring welcome.

There was enough of the 28th to march an eight-and-one-half-mile route through Philadelphia on May 15, 1919, in what the *Journal* called "the greatest demonstration ever known in America". The publication expansively reported that more than two million had participated in the welcome and that whenever the line halted "the boys were showered with food and flowers". Leading the march was Major General Charles J. Muir, the wartime commander; at his left, Colonel David J. Davis, claimed to be the only National Guardsman to have remained as a Division Chief of Staff in Federal service. There were ovations for the wounded, riding in 114 automobiles. At the end of the column came a caisson bearing a symbolic coffin, upon which was a great floral wreath and a banner bearing the numerals "4,105" representing the men of the 28th who had been killed in action and were dead from other causes. On other occasions in other cities and towns home town folks comparably honored their own returned units.

State authorities had decided for themselves that again Pennsylvania would have a division. Initially it would comprise 7,600 men, growing to between 14,000 and 16,000 by 1921. Soon thereafter the War Department confirmed allotment of a division—but it would keep the 28th designation and the red keystone insignia for the Regular Army. Little did it reckon with the proud veterans and the Pennsylvania citizenry; both reminders of its wartime service were perpetuated in the new organization.

Reorganization began promptly, working from the top down with appointment of three Brigadiers: Richard Coulter, Edward C. Shannon and George C. Rickards.

Washington still had not made up its mind about a postwar divisional structure so, without waiting, the State authorities built upon the prewar regimental designations: 1st, 2nd, 3d, etc., despite the fact that not only the Regular Army but the Guard in other States each had its own 1st, 2nd 3d, etc., Infantry, Cavalry, Field Artillery, or whatever.

There was no dearth of applicants for commissions. By mid-summer, General Price had heard from between 500 and 600 aspirants who had had overseas service. Many, of course, already were officers, but there is room for belief that many were ex-doughboys who coveted the Sam Browne belt, shoulder bars, regalia and perquisites. Officers, moreover, did not have to pull KP or dig latrines and garbage pits! What they found out, however, was that planning, copious paperwork, meetings and other responsibilities cut heavily into leisure and family time. Defections began and by 1921 General Price would be reporting something like 700 resignations. The *Journal* took up the editorial cudgel in 1926 with a plaint that recurs even today: "Guard officers cannot be overburdened with directions from Washington, if the quota of experienced commissioned personnel for the State service is to be maintained."

Washington's postwar defense planners had been optimistic, targeting a nationwide Guard strength of 428,400 by 1924; but less than a year after the Armistice its reports were gloomy: out of 126,000 strength allotted currently to all of the States, only 34,300 had been enrolled. The Militia Bureau in the War Department blamed "widespread indifference on the part of eligible men toward military service; a dislike for strike duty; and hostility of labor unions toward the National Guard." A *Journal* survey among some other States found that in many instances men declined to re-join because officers had been appointed "who did not make their armory life very agreeable by being arbitrary in their methods;" some were unwilling to serve under National Army officers (those identified scathingly at the time as "90-day Wonders.")

Nevertheless, recruiting did make strides; in certain cases substantially all of the personnel of whole platoons of the Pennsylvania Reserve Militia went over to the "new" Guard; about 1,000 of the 7,000 men "on board" by early 1920 had come from that source, which was in the process of being phased-out. Lack of a firm Table of Organization & Equipment notwithstanding, equipment was coming in, too. It was considered newsworthy when a convoy of truckloads of gear passed through Philadelphia en route to Harrisburg. By mid-summer, the Pennsylvania National Guard had about $10,000,000 worth of Federally-owned ordnance, motor vehicles and other materiel.

The new 28th's first two-weeks period of field training was scheduled at the old Mount Gretna camp site for the first two weeks of August, 1920. It promised to be, said the *Journal*, "the greatest in the history of the Pennsylvania National Guard. The equipment will be of the finest and latest regulation and particular attention will be given to the comfort of the men. The rations and mess arrangements will receive great attention by the superior officers." Three Field Artillery Battalions probably would attend "the camp of instruction" at Camp Bragg, North Carolina.

The Division's major components at the outset were the 1st, 3d and 6th Infantry, all at Philadelphia under Colonels Millard D. Brown (soon succeeded by Lieutenant Colonel Jackson W. Study). George E. Kemp and Robert M. Brookfield, respectively; 8th, Harrisburg, Colonel Edward J. Stackpole, Jr.; 10th, Waynesburg, Colonel Edward Martin; 13th, Scranton, Colonel Robert M. Vail; 16th, Grove City, Colonel C. Blaine Smathers, and 18th, Pittsburgh, Colonel William R. Dunlap; the 2nd and 3d Separate Infantry Battalions at Lancaster and Allentown, Majors William C. Rehm and Orlando C. Miller; 1st, 2nd and 3d Field Artillery, Pittsburgh, Philadelphia and Wilkes-Barre, Colonel Churchhill B. Mehard, Howard S. Williams and William S. McLean, Jr.; 1st Engineers, Philadelphia, Major Harry Anderson; Signal Corps, Pittsburgh, Major Thomas P. Rose; Division Trains, Harrisburg, Colonel Maurice E. Finney; 1st Cavalry, Philadelphia, Colonel John P. Wood (soon succeeded by Colonel George C. Thayer); Ammunition Train, Allentown, Lieutenant Colonel Clarence J. Smith, and an innovation for the Guard, a cavalry-type mounted, armed and equipped military police unit at Pittsburgh, under the command of Major Charles C. McGovern.

Despite the *Journal's* somewhat equivocal characterization of the first postwar field training period as having been "as a general whole ... in every way the most successful ever conducted by the National Guard," with half of the actual quota required to constitute the Division present, it reported an "absolute slump" in manning several major units. The 1st Infantry had only two companies organized; the 3d but one minimum-strength battalion with two field officers, and the 2nd Field Artillery "unable to effect any organization." By contrast, the 10th Infantry shone not only at maximum authorized strength, but was one of the first Guard regiments in the nation to receive Federal recognition.

After one of the many wrangles with the War Department over the years, the Guard nationally had achieved a major Congressional victory in a National Defense Act revision which, among many things, provided for a Guardsman rather than a Regular Army officer to head the Militia Bureau. To the 28th fell the honor of supplying the first: General Rickards.

Ready or not, as early as February 1921 the new Guard was subjected to the first in the annual series of Federal inspections, just as the initial organizational structure was being shuffled. The 1st Aircraft Company, offshoot of the old 4th Infantry at Reading, was organized, initially under Captain Joseph Eisenbrown. The 8th Infantry found itself transformed into a second Cavalry regiment as part of a newly-created Brigade. Company locations and regimental assignments were shuffled. By the time the dust settled, a new, nationwide standard Table of Equipment had been adopted, and the wartime model designations had superseded the old—sentiment being respected through authorization to carry the old regimental numbers parenthetically after the new; e.g., 112th (16th Pennsylvania) Infantry.

The new structure introduced such newcomers as the Division Special Troops; Division Train; 28th Observation Squadron; 103d Medical Regiment, and 53d Field Artillery Brigade. At the same time it peeled-off some organizations to fulfill the Army's new pattern of Corps and Army units. Before year's end the new divisional organization had been completed, and the 28th had become the first fully-formed Guard division in the new "Army of the United States"—Regular, Guard and Reserve. By-and-large, that basic structure remained fixed clear through to the succeeding peacetime years.

Today one might wonder how so much had been accomplished in so short a time following a horrible "World's War," with its enormous casualties, military and civilian. The world was sick of slaughter; this had been "the war to end all wars." There had been a great wave of revulsion, a rise of pacifism and anti-militarism, and diatribes against "the merchants of death" whose strongboxes allegedly filled as national treasuries emptied. The international villains had been crushed to earth; the virtuous were triumphant; the question was who now threatened the United States in its splendid isolation. There would be no war; hence, why squander resources on armies and navies?

Such sentiments did have some effect. The Army and the Navy were cut back to a fraction of their war-time size; the original super-ambitious projected strength of the Guard was shrunken; only shortly before Pearl Harbor was it permitted to go as high as 242,000. That meant "line" companies and troops were limited to just three officers and about sixty-five enlisted men. The number was approximately the maximum that many of the smaller towns realistically could recruit, anyway.

True, now and then parading Guardsmen might hear a few voices taunting: "Look at the Boy Scouts" or "Here come the tin soldiers!" Some persons barely concealed their bemusement, trying to figure why grown men would spend so much time "playing soldier."

Yet, there was a widespread, underlying, vague sense that just maybe the United States as the emerging leader of the new world might be caught up some day in events not of its own making; that it might be well to follow Oliver Cromwell's famed dictum: "Praise the Lord, and keep your powder dry." It was a time, too, when Memorial Day, Flag Day, the "Glorious Fourth" and the new Armistice Day were marked by appropriate ceremony. Patriotism was alive and well.

For Pennsylvanians as a whole the 28th "Iron

Group of Troopers, First Troop, Philadelphia City Cavalry in Dress Uniform circa 1920-1930.

Division" was a source of pride and of prestige. Combat-seasoned veterans of the recent war were salted throughout the new divisional ranks from top to bottom; their presence certified that this was no untested "milishy," and younger men coming into military age had no hesitancy in subjecting themselves to such leadership and comradeship.

Why did anyone want to serve? No one was forced to join, and the reasons why so many did—and still do—are many and complex, and they vary from individual to individual. There was basically an ingrained patriotism, although none bandied the word about. In mankind there was, and is, an innate desire to serve, beyond one's own selfish interests—as exemplified by volunteer fire companies, rescue squads, and many other unselfish groups. There was, and is, the gregarious instinct; the "buddy principle" (one Guardsman persuading a friend to sign-up with him) applied.

A Guard unit had some of the characteristics of a civilian club—with military attributes. The wearing of a uniform had, and has, its appeal to many. The inexplicable but almost universal thrill of marching in a disciplined body to the blare of bands behind waving flags played its part. The opportunity to prove oneself capable of "roughing it" to some degree and even exposing oneself to an element of danger incident to field training and maneuvers seemed to have an appeal for some. Guard service, even "drill night," afforded a valid opportunity to do something different; to get away from the day-in, day-out routines of civilian job or profession.

So, through "the Roaring '20's" and the depression-ridden '30's between the two World Wars, the Guard thrived. Untold thousands passed through the ranks of the 28th. A great many tried one "hitch" and then faded back into civilian life, but a substantial number, officers and men, liked the Guard well enough to maintain solid continuity with ten, twenty and more years of service—substantial numbers through both World Wars and even into the Korean War. Guard service sufficiently whetted the interest of others to get appointments to West Point or to enlist in the Regular Army.

Introduction of warplanes and tanks had made their impression on military thinking, but the doctrine of conventional warfare was little changed, and Americans had no reason to muse about a "fifteen-minute war," as later was the case. Accordingly, the hoary policy prevailed of depending upon a peacetime Regular Army, backed-up by a modest-sized, reasonably well-equipped and trained Guard and cadre-type Reserve capable of receiving and training hordes of civilian draftees over a period of months. There was no sense of great urgency or pressure to demand too much training time of part-time soldiers who, it had to be remembered, did not have to join the Guard in the first place.

Weapons, equipment and tactics had changed but little; funds for the conduct of large-scale maneuvers and the transportation of masses of troops to the few available sites were non-existent. Knowledgeable Guardsmen accepted the necessity of tightening-up the training standards from those of an earlier day. Increasing numbers of officers and men took advantage of the greater opportunities to attend various Army Service schools and to spend evening hours on the study of Army correspondence courses and week-night classes at the Armory. General Price's report of January 31, 1924, remarked that the previously-mentioned annual inspections were "a severe strain to officers and men" who were called upon to demonstrate greater knowledge than before.

Still, there could be a "fun" aspect of Armory

drill and field training—time for after-hours conviviality even in the Spartan environment of a pyramidal tent, or an evening's excursion to Lebanon for the few who had "wheels." Torchlight band concerts attracted many. The mid-camp Saturday was "Field Day," pitting unit representatives against each other in a melange of athletic and military-oriented competitions.

There were other events to lend some sparkle to Guard service. One was the mid-May 1922 celebrations of the fourth anniversary of the "Iron Division's" embarkation for Europe. Secretary of War John W. Weeks and the famed General John J. Pershing accepted invitations to attend the Philadelphia events, which included, of course, a parade, a "sham battle" patterned after that of Belleau Wood, and a banquet. Elsewhere there were banquets, reviews in local armories, demonstrations of drilling, machine gunnery, and cavalry tactics, and, of course, band concerts.

Every four years, many Guardsmen could look forward to a trip to the Capitol to march in an inaugural parade, but that for Governor Pinchot's first term, in January 1923, was a disappointment to some who liked long lines of infantry. This time sparse transportation and subsistence funds held Guard participation to close-in units; tractor-drawn 155mm artillery and a handful of light tanks rumbled by; although some cavalry units participated also, their flashing sabers and clattering hooves livening the grim procession.

Men in mounted units had the advantage of being able to do cost-free recreational riding on the basis that the "hay-burners" needed exercise, too. Many units held horse shows, boxing matches, track and field meets. Baseball, basketball and football teams evolved into leagues, with the Guard's Military Athletic League's first indoor track "carnival" in the 108th Field Artillery's Philadelphia Armory on March 15, 1924, considered (in the words of the *Journal*) "a huge success," with 412 entries. Some of the long-established and more prosperous units, inhabiting the larger armories, had their own club rooms, card rooms, billiard rooms, gyms and other well-patronized recreational facilities.

The 28th Tank Company at Norristown, exhilirated by the success of its musical comedy production, publicized an offer to stage the two and one-half-hour show for any Guard unit at the latter's own place for a share of the "gate." There is no record of any takers.

A major event of the 'tween-wars years was a memorable "pilgrimage" to France in May, 1928 for the dedication of Pennsylvania's memorial to the dead of the Great War. A composite battalion comprising a man from every unit constituted a Memorial Escort at the dedication of a bridge at Fismes honoring the 28th, a colonnade at Varennes, and a fountain at Nantillois memorializing the 28th, along with others.

Three years later the Commonwealth sent the 55th Infantry Brigade Headquarters Company, a provisional company from the 112th Infantry and the Governor's Troop to take part in the 150th anniversary commemoration of Cornwallis' surrender to American and French forces at Yorktown.

In 1925, when Philadelphia's Mayor Kendrick got a 120-acre flying field near Hog Island in the Delaware River for use by the 103rd Observation Squadron, the *Journal* commented: "It is to be expected that in the future the air forces will comprise one of the largest branches of the National Guard." The problem was that the unit had to make do in an old police station for its paperwork and technical training and go to the 103d Engineers Armory for drill. It could not get any planes until it got a hangar. Three years later the Squadron was ejected from the police station without knowing where it was going next. On the bright side, it was getting three new aircraft.

It had had its first casualties in 1926 with the deaths of Captain John Batty and Master Sergeant James W. Cheeseman in a nose dive from 150 feet altitude shortly after takeoff at Langley Field, Virginia. The Squadron's next multiple fatality claimed Lieutenant Franklin A. Johnson and Sergeant Paul W. Dernoeden, Jr., in 1931. Returning from maneuvers over Mount Gretna and circling low for a landing of Middletown Air Depot, the craft apparently lost flying speed and crashed.

It was an event when in 1928 a "Falcon" 0-11 was ferried-in from Buffalo in 150 minutes. "All the boys are very enthusiastic over the new ships and the way they perform," reported the *Journal*. The spring sun was drying-out the field and they were getting in more air time. Two officers in a round trip to Middletown tested a new interphone set between pilot and observer that Lieutenant Frank A. Johnson had been experimenting with—its feature being that the fliers could converse without throttling-back the engine.

Many ground-bound units were "disadvantaged" in terms of what euphemistically could be called "armories." For a quarter-century they had made do with such expedients as post office basements, upper floors of public halls, and even dance halls and police stations. A "horrible example" was Latrobe's Company M, 110th Infantry; banned from the floor of a second-story hall for drill because the cadenced feet loosened the ceiling plaster below, it had only a 10-by-20-foot stage available.

The 1927 General Assembly took a nibble at relief measures with a million-dollar appropriation. Contracts were let for new armories at Carbondale, Latrobe and Lock Haven, and plans were advanced for Lancaster, Norristown, Williamsport, Scottdale, East Stroudsburg and Wilkes-Barre.

The 1920's often were referred to in later years as "boom" years, and prospects looked so bright that the military authorities proposed to get on the ballot for the 1928 elections a constitutional amendment that would permit a $5,000,000 armory construction bond issue. However, so many other "big spending" proposals were involved that all sunk under the threat of a too-heavy financial burden.

Those were the days, too, of tent camps, such as currently exists at Fort A. P. Hill, Virginia. Sudden storms on occasion toppled sodden canvas, poles and ropes onto eight-man occupants. In 1926 the first fatal electrical storm ever known at Mount Gretna's encampments killed Private Leonard A. Kerr of the newly organized Headquarters Company, 103d Medical Regiment. Private Kerr's arm was touching the center pole when a bolt hit it; four comrades received shocks.

They were the days when the horse-drawn artillery trained in alternate years at Tobyhanna, the only State locale where cannon could be fired "live," and at Colebrook, near Mount Gretna. No mounted unit had enough horses for everyone; by doing field training at different dates, the cavalry and field artillery could make use of each other's mounts. The 108th Field Artillery, while at Tobyhanna in 1926, had among its official visitors Major General Douglas MacArthur, then commanding the Third Corps Area. Another event was a fly-over the camp by three Martin bombers, three observation and three pursuit planes incident to the nation's sesquicentennial exhibition in Philadelphia.

After years of the relatively dull "O.D." uniform, a movement spread gradually to the return of the distinctive uniforms that many units had sported in by-gone days. Erie's Battalion of the 112th Infantry and Lock Haven's Troop F of the 104th Cavalry were among the first. The Regular Army was dropping the "stand-up" blouse collar in favor of the "roll" type; Guardsmen were authorized to make the switch at their own expense, but cautioned against doing it "on the cheap" by trying to convert existing blouses. The conversion required one-third of a yard of cloth and an unsuccessful attempt to match the wartime issue's three different types and some fifty-seven shades could result in "coats of many colors." All the Guardsmen had to do was wait about three years when the Army started issuing new-style Melton uniforms and the "Pershing" visored cap.

By the "new" 28th's tenth anniversary, things were looking up. General Shannon stated that "the Guard has reached a state of efficiency not before enjoyed by National Guard troops." However, he saw the need of further improvement; the Division could not attain maximum efficiency in peacetime training which failed to contemplate its training as a division—not smaller bits and pieces. Mount Gretna was too small for that, but he looked forward to seeing the whole Division plus the 52nd Cavalry Brigade there for at least one week in 1929, and "undertake movements and maneuvers of an entirely different type from anything that has been undertaken in the National Guard since the War." It was worked out for the various Brigades' dates to overlap one week, and to conduct a three day-two night march in two columns in both forward and retrograde movement in the vicinity of what was to become the Indiantown Gap Military Reservation.

Pennsylvanians had served under 100 organizational flags during the War; and Pennsylvania Guardsmen carried them in the 1929 inaugural parade of President Herbert Hoover. The downpour that soaked the Guardsmen and other paraders may, in retrospect, have seemed omi-

Officers of the 28th Division Train, Mt. Gretna, 1929.

nous of the Great Depression that was triggered by the stock market crash of the following October.

It was at a time, too, when the long stagnation in development of military weapons was ending. It had been announced that the famed Springfield rifle would give way to a .276-caliber semi-automatic capable of firing 200 to 300 rounds per minute; the ratio of three service- and five training-type aircraft would be evened-out to four of each, a proportion changed to all service-type when the state of training warranted. Further, the War Department was considering organization of a mechanized force as an integral part of the army in Fiscal Year 1931, and a British Carden-Lloyd light armored vehicle was being considered as an Infantry carrier. Each Guard observation squadron was getting one ground-to-air radio set. The solid tires on some of the aging wartime "Liberty" trucks were being changed to pneumatic.

Training objectives were being broadened; the 28th's and other Guard and Reserve division, brigade and regimental commanders and staffs joined in the first of what would be many Corps command post exercises at Fort George G. Meade, Maryland, in July. The occasion was used to launch a campaign for $1-a-year membership in a projected Society of the 28th Infantry Division, American Expeditionary Force, at General Shannon's Command Post. The wartime commanding general, Muir, was President; General Price, Vice-President; Lieutenant Colonel Frank A. Warner, Secretary and Treasurer, and Captain George W. Phillips, Assistant Secretary.

Nostalgic veterans responded enthusiastically, and in the Summer of 1931 approximately 2,000 of them reunited at Mount Gretna under canvas left standing for the purpose after field training and re-lived their "glory days" in more-or-less their old environment. The practice was continued for several years before evolving into get-togethers in more comfortable surroundings in various cities around Pennsylvania. The 28th still stood high in public favor. Initially it was proposed to designate

Highway #28 from Avella through Pittsburgh and New Bethlehem to Brockway the "Iron Division Highway." Much more representative of the statewide spread of the Division was the alternative selected: the choice of United States Route #322, zig-zagging from the Northwestern to Southeastern corners, passing through Boalsburg, site of the Division's shrine and museum on land donated by Colonel Theodore D. Boal; close to Indiantown Gap, and through Mount Gretna. Allegheny County dedicated a mammoth red keystone of blooming sage bushes—100 feet long and 70 feet at its greatest width—at South Park.

Paradoxically, at the very time that emphasis was being placed on the competitive aspect of combat range firing, the disappointing performance of Pennsylvania's team at Camp Perry, Ohio, in the known-distance National Rifle Matches (42nd among 108 teams) found unit commanders under pressure to concentrate on the basic principles of marksmanship training. The State team's poor performance at Camp Perry produced results in 1931: it placed seventh among all Guard teams and fourteenth among 113 from all Services.

That Fall General Shannon was elected Lieutenant Governor as Gifford Pinchot's running-mate, after the latter's four-year absence from the Chief Executive's office. The new Adjutant General was David J. Davis, who had been the 28th's Chief of Staff ever since the postwar reorganization. Economy was even more emphatically the word than before, with the Depression's pinch tightening, so Harrisburg-area units alone provided the Guard's inaugural parade representation.

Pay never had been the big attraction to Guard service, even in a day when an ordinary man could boast of "dragging-down" $25 a week. To the growing numbers of jobless, and those who were taking successive pay cuts, even the dollar per drill plus $22.50 for two weeks' field training (the State provided 50¢ a day) plus two weeks' wear on "G.I." shoe leather rather than one's own was most welcome. Further, a new Congressional pay bill proposed a nickle-a-day boost to $1.20 for privates first class!

Congress made its first, puny million-dollar appropriation for improvements at Guard training camps as a step towards alleviating unemployment; Pennsylvania got $8,610 for a warehouse at Tobyhanna. Much more was to come in later years.

It was not long before many were experiencing real suffering. The Guard's Motor Repair Section at Gettysburg picked, hauled and delivered peaches, pears and apples to sixteen needy families. Twenty-three jobless 107th Field Artillerymen in Pittsburgh, their families broken-up, lived in the Hunt Armory, working twenty-five hours a week on maintenance and reconditioning chores in return for their sleeping quarters and three meals daily, the wherewithal being provided from the organization's "take" for the rental of its privately-owned armory for riding, stabling and polo.

It helped some jobless a bit when the Legislature in a special "Work Relief" session late in 1931 allotted, among its programs, $300,000 to employ men on land-clearing, road-building and sewer line-digging at Indiantown Gap. That six and one-half by two mile area had been selected, rather than Mount Gretna, because it would provide room for artillery firing and for infantry and cavalry maneuvers.

Bad news, however, was the Secretary of War's directive to cut from the schedule five paid drills in Fiscal 1932. Although that order was rescinded, the next threat was a Budget Bureau recommendation—vigorously contested by Guard spokesmen at the national level—to pay only three-quarters of the Guard's authorized strength for drills and field training. Drill pay cuts were staved-off for a time, but ammunition expenditures were chopped in half. With the Depression deepening, although the law called for forty-eight drills, Guardsmen would be paid for only thirty-six! The State would make up the difference in reduced Federal pay for privates while at camp. General Shannon—the new Division Commander with General Price's retirement—called on all Guardsmen to respond to what today would be termed the "freebies." Division Headquarters was shifted from Philadelphia to Harrisburg. (Nostalgia note: the price to attend the testimonial banquet for the new commanding general was $2).

For decades there had been mutterings about "The Yellow Peril," and Japan's aggression in China didn't sit well with many Americans. It was seen as but a small cloud on most Americans' horizons, but it encouraged some small military advances which directly or indirectly affected the 28th.

Ten officers and men of the 103d Observation Squadron joined in the Air Corps' 1st Provisional Air Division maneuvers at Fairfield Air Depot, Dayton, Ohio, along with various other Guard air squadrons. It involved 672 aircraft of all types. The Keystoners' 02HA craft "stood up splendidly" on the flight from Philadelphia to Middletown, Pittsburgh and Dayton.

That summer also, a plane manned by a combination pilot-observer and a radio operator worked with the 3d Battalion, 108th Field Artillery, on an aircraft-directed "shoot"; the gunners' eighth round landed on target. Various 28th Division units were chosen to test new infantry drill regulations based on "Columns of Threes."

Senior artillery officers were sent to Fort Bragg for two days of instruction in the tactical use of tractor-drawn artillery; the 107th and 109th soon were to be motorized. Governor Pinchot and General Shannon and Davis fired the first three 75mm rounds from pieces supplied by Battery A of the 109th, from Nanticoke, to formalize opening of the new Indiantown Gap range. For economy, however, much artillery firing was done with 37mm sub-caliber.

Still, to a degree the Guard was living in "genteel poverty." The antiquity of the ambulances following the 108th Field Artillery in a parade drew comment from Philadelphia's Mayor J. Hampton Moore; moreover, the Regiment had been forced to borrow money to buy the gas, and General Shannon commented sadly about the requirement to issue some men reclaimed articles of uniform—hardly conducive to appearance and morale.

Strike duty near Brownsville pulled 325 officers and men of the 2nd Battalion, 112th Infantry, from field training in the summer of 1933, the first 100-man increment leaving by motor convoy six hours after Governor Pinchot's call to General Shannon.

The 28th's fliers suffered somewhat of a setback in 1934 when they had to surrender their two Douglas observation planes. President Roosevelt, cancelling air mail contracts, had turned the chore over to the Army, and it needed more aircraft.

The millions of dollars poured into work relief and other "recovery"-stimulating construction projects produced many badly-needed local armories—plain but functional as contrasted with the fortress-like structures of old. By the time America became embroiled in World War II the Indiantown Gap Reservation had been expanded; there were cinder block mess halls and bath houses, concrete floors for the traditional pyramidal squad tents. Acres of "temporary" frame barracks and other structures would come later.

It was the Governor's "cherished ambition" to see the entire Guard in field training at the same time, and he got his wish in 1934 although the 28th as well as non-divisional troops had to be split among "the Gap," Mount Gretna, Colebrook and Middletown, all converging at the new reservation for a review.

Considering the shortage and age of trucks, and lack of water carts, rolling kitchens, and gasoline tank and repair trucks, the move was a chore, accomplished partly by motor, partly by rail and partly by marching. A spectator was General-Lieutenant Franz von Boetticher, a member of the Field Marshal Paul von Hindenburg's staff in World War I and now Military Attaché at the German Embassy. Reports were that for the vast majority of troops—by now, non-veterans—it took some doing to get accustomed to the "wash-basin" steel helmets that normally were not part of their gear. It was the last review for the 107th and 109th Field Artillery in their horse-drawn mode.

The surest sign that "the old days" of relative sameness in the scale and scope of peacetime Guard training were gone forever came in the mid-'30's. For one thing, the War Department's training directive called for emphasis on operations against high-speed armored vehicles and aircraft, night operations, cooperation with friendly and defense against hostile aircraft, and the use of non-toxic chemicals and defense against chemical warfare. For another, it projected the largest troop concentration since "the War." Third Corps area troops—the 28th and the 29th (Maryland-Virginia-District of Columbia) Divisions and a Regular Army infantry brigade and field artillery brigade would be united at Indiantown Gap and Mount Gretna. As fate would have it, Virginia's and the District's troops (as well as a Virginia-based Regular Army engineer regiment and field artillery battalion) were excluded because of prevalence of poliomyelitis in Virginia

Field Artillery - 1936.

Cavalry on Parade - 1936.

Standing Retreat - 1936.

Heavy Artillery - 1936.

Camp at Indiantown Gap - 1936.

"New" 155mm Guns with pneumatic tires and drawn by trucks - 1936.

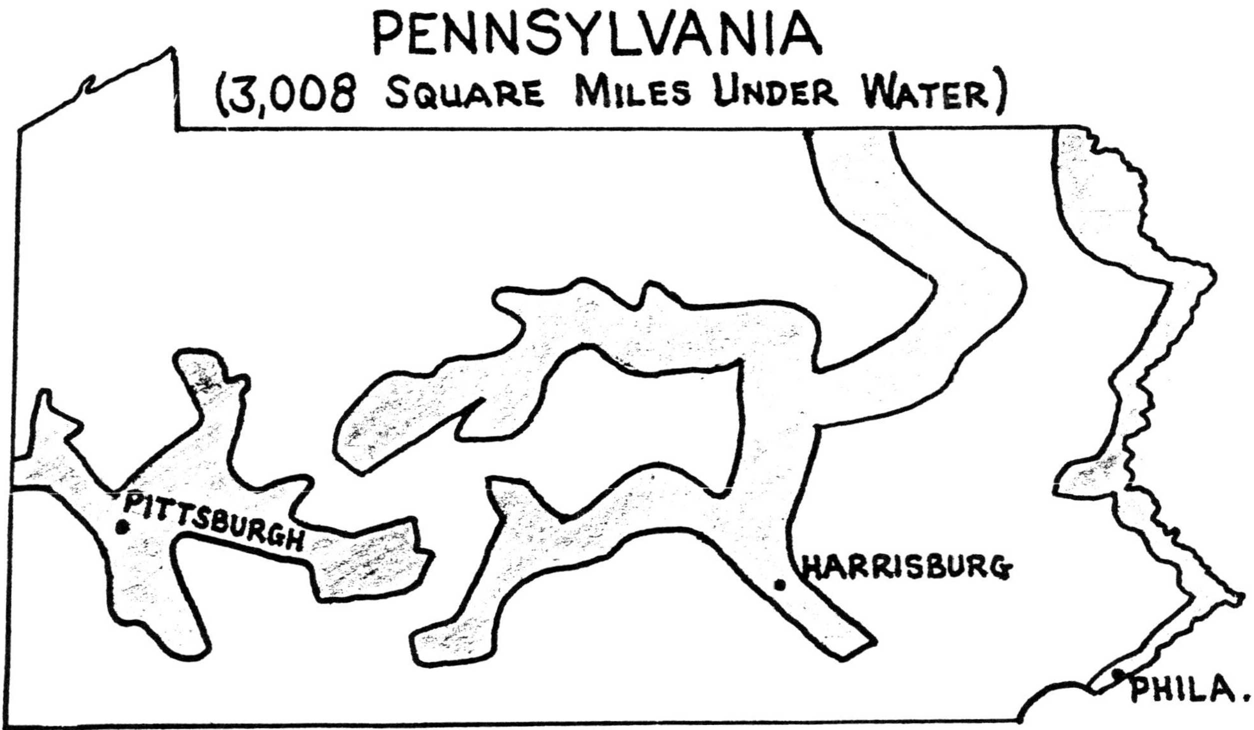

Extent of flooding in Pennsylvania, 1936.

and the District.

The approach of Spring in 1936 brought on rapid thaws and flooding that paralyzed much of the Commonwealth, affecting nearly 4,000,000 people in an area embracing 3,000 square miles. More than 6,000 Guardsmen were ordered to duty. Authorized to rent trucks or buses, and to buy medicines and food for the needy, sick and injured, officers found sleeping quarters for the victims and provided cots and blankets. Troops manned boats, removed people trapped in their homes, and carried out sick and injured.

As to military equipment, there was gradual improvement. For some years the 103d Motorcycle Company had been that virtually in name only; thanks to cooperation by the State Highway Patrol, it got essential equipment for one field training period; now, it had ten cycles of its own.

For a time, the Guard was glad to have inherited a fleet of cast-off light trucks from the Post Office Department. Sides and back of the body consisted of interwoven heavy-gauge wire; nevertheless, the non-TOE vehicles came in handy for hauling impedimenta to and from camp even if Guardsmen unlucky enough to have to ride inside provided an impression of offenders en route to jail.

However, even the 108th's Schneider 155mm howitzers now were riding on pneumatic-tired wheels and the cumbersome, gas-guzzling, slow old tractors that had drawn them were replaced by powerful trucks.

The equally slow and cumbersome World War I-vintage Renault light tanks creaked, groaned, backfired and emitted smoke from exhausts as they staggered the few hundred yards demanded of them in their last field training review in 1936, two of them coughing their final gasp just past the stand. The Division had become completely horseless for the first time in its history when the Tank Company drew two new model "lights" for training purposes a year or so later.

Staff and supporting personnel of the 28th joined in a First Army CPX at Fort Devens, Massachusetts, in August, their motor convoy halting at West Point for overnight billeting and messing. Other participants were the Guard's 26th and 29th Divisions and the Regulars' 1st, 8th and 9th.

In 1938, approximately 300 Guard officers and

men helped look after the needs of tottering veterans of the Blue and the Gray in the last reunion of those who had fought at Gettysburg seventy-five years earlier. The Guard had built 3,300 tent platforms, provided 6,000 cots and 18,000 blankets, laid sewer and electric lines and provided other facilities for the old warriors.

The Guard did some "innovating" of its own and leaped at the chance to try new developments. The 111th Infantry had the Mack Truck Company build into a bus-like body a self-sufficient mobile Command Post. In the rear lengthwise seats straddling a conference table provided working space for fourteen staff officers; up front there was a well-equipped kitchen with gas stove, hot and cold water, refrigerator, storage space for two weeks' food and room for two cooks. Air-minded 108th Field Artillerymen arranged with the Kellett Autogiro Company for two Regular Army pilots to try one of its "whirlybird" forerunners on a tactical reconnaissance for a battalion position and observation of fire.

The veteran Major General Edward Martin succeeded to the post of Adjutant General in January, 1939 and, with General Shannon's retirement in June, began serving concurrently as Division Commander. His old Regiment, the 110th Infantry, in which he had served in both the Spanish-American and "Great" wars, was chosen to be the first in the National Guard to be issued the new Garand M-1 rifles, based upon proficiency in the nationwide "Pershing Trophy" competition.

War talk and increasing distrust of "Red" subversion were in the air that spring. Secretary of War Harry H. Woodring was "vitalizing" the Regular officer corps and asked all Governors to apply "a high standard of physical fitness" to their Guard officers.

That summer's field training found the 28th and others at the Civil War battleground, Manassas, in 1st Army maneuvers under the eyes of observers from a dozen nations including both near-future military enemies and allies. As General Martin later wrote in *The Pennsylvania Guardsman*, they were "pitted against modern tanks, swiftly moving motorized infantry and artillery and skilfully handled horse and mechanized cavalry." For the first time in the 28th's history it employed a reinforced battalion—the 2nd of the 111th Infantry with Battery C, 107th Field Artillery—as a mobile reserve. Given mobility by twenty-three, one and one-half-ton trucks, the force repelled two cavalry squadrons that debouched from woods at a crossing of Occoquan Creek.

That Fall President Franklin D. Roosevelt moved to boost the Guard's strength, Pennsylvania's troop allotment going up to 14,500. By late Fall, four units rated "Honor Roll" listing by having gone to one hundred percent of the new "peacetime" figure—"war strength" had yet to be authorized. Two drills per week were authorized up to a maximum of twelve additional, and seven days of field training—not necessarily consecutive—were to be accomplished during the cold weather months up to March 1, 1940. Most units accomplished the latter by going as far as Cape May and Camp Dix, New Jersey, and Fort Meade and Fort Hoyle, Maryland, as well as Indiantown Gap. Many found outdoor sites near home. Those that took the seven days in one stretch reported better attendance than those that used "short takes." Many employers responded to a request that they let Guardsmen employes have the time off without loss of pay, seniority or vacation time.

The stimulus of even what scoffingly was called "the phony war" in Europe (before the fall of France) had its psychological effect on Guardsmen; published figures showed drill attendance up markedly from that of the mid-'30's.

The Regular Army was changing its infantry divisions from the old "square" to a new "triangular" configuration; the Guard's underwent a lesser TOE switch. Infantry battalion headquarters companies were reduced to eleven-man detachments; rifle companies got 60mm mortar and light machine gun sections; one company per battalion was switched to heavy weapons designation with both light and heavy machine guns and 81mm mortars; regimental headquarters companies received the new 37mm antitank guns (the old "one-pounder" howitzers went into history).

The 103d Observation Squadron's latest craft were too fast to use the parade ground "strip" at Indiantown Gap, so they got a 3,000-foot-long, 100-foot-wide shale runway.

A sartorial change was substitution of the "overseas" version for the visored "Pershing" cap for all

28th Division tanks pass in review, Manasses, Virginia, August 12, 1936.

"dress" formations. Of far greater significance and importance was the Army's decision to expand the 1940 field training period to three weeks at Field Army level employing all Regular and Guard units and with larger-than-usual Reserve officer augmentation.

While getting their basic training requirements out of the way in preparation for their stint in northern New York State that coming August, roughly 650 of the Pennsylvanians again found themselves on flood duty in early spring: the 109th Field Artillery's 1st Battalion Headquarters and Company A, 103d Medical Regiment, in the Wyoming Valley, and the 103d Cavalry's 2nd Squadron Headquarters, Medical Detachment and Troop E at Sunbury.

The "Blitzkreig" of May 1940 that had put all of western Europe in Hitler's hands had sharpened most Americans' sense of the importance of sound military training by the time the 28th and thousands of other Guardsmen entrained and entrucked for what was to be the largest peacetime troop concentration in America's history up to that time.

Spread over rock-encrusted terrain near the Canadian border, the Guardsmen progressed through successive stages of small unit to battalion-level, regiment-vs.-regiment and brigade-vs.-brigade exercises before the 28th and its neighbors of the 29th were pitted against each other. All culminated in a final maneuver engaging some 100,000 troops. The 28th's Tank Company earned a special citation from the Third Corps Commander for its role in a special Combat Team's seizure of river crossings.

Twice the 28th was host to President Roosevelt. Governor Arthur James and General Martin met the President, Canada's Prime Minister Mackenzie King and Secretary of War Henry L. Stimson on their arrival at Lisbon, New York and then again when the distinguished visitors attended Division memorial services.

The "boys" of the 1917-1918 "Iron Division" were aging but not forgotten, nor did they forget their own when, in September, their society dedicated a memorial altar to the late Colonel Boal at the site he had given the State in honor of his comrades. In 1916, he had raised and equipped a Cavalry Troop entirely through private funds, got ninety men to submit to military discipline without pay or sworn obligation, for two months, before their muster-in for Border Service. He was a recipient of the Distinguished Service Cross and the Croix de Guerre for wartime service with the 28th.

Now a new generation of "boys" and a sprinkling of the "old" were about to carry its colors into their nation's service for yet another great war.

All through their three weeks at what now is Fort Hugh Drum that late summer Guardsmen had been wondering whether their "hitch" might now carry into extended active duty. They hardly had returned home before the President called four Guard Divisions, not including the 28th, and other organizations into Federal service for what originally was scheduled to be one year. All others would be summoned eventually, but training camps were insufficient to take all the men in 1940.

Indiantown Gap facilities were expanded; hordes of workmen and heavy construction equipment toiled into the winter months, transforming open space into acres of hastily-erected structures, smelling of fresh gray paint outside, with raw unpainted wood inside.

On February 17, 1941, the 28th Infantry Division, Pennsylvania National Guard, disappeared; on February 17, 1941, the 28th Infantry Division, Army of the United States, was born.

It was about to write a new chapter in the history of the "Keystone Division."

CHAPTER **10**

World War II

September 1, 1939 - October 1, 1944

Robert Grant Crist, Camp Hill, Penna.

October 1, 1944 - December 13, 1945

Harold E. Myers, SP4, HHC, 2-109th Inf

World War II — September 1939-October 1944

By Robert G. Crist

BY 1939 A NEW senior generation was facing awesome decisions. In the world's capitals sat men of power who themselves had experienced a war that their own fathers had failed to prevent. That memory—of millions of dead and billions in destruction—hung over the chancellories and the continents.

For the United States there were no guns of August until 1943, but in the summer of 1939 tension built in every home. Into the living room came the voices of Kaltenborn and Agronsky and Murrow, as correspondents dispersed around the globe. Day by day, hour by hour, Americans shared foreboding and crisis.

By coincidence the top figures of the hour were four men who had been civilians in the Great War, facing three who had worn uniform. The lineup was the political prisoner Joseph Stalin, the civil servants Neville Chamberlain and Eduard Daladier plus an assistant secretary of the Navy, Franklin D. Roosevelt, versus Private Benito Mussolini, Corporal Adolph Hitler and Lieutenant Hideki Tojo.

Until Spring brought the fighting weather season of 1940 a world thought it all might just go away. True, Poland had disappeared the previous Fall, but this was the third such partition in recent memory—or was it the fourth? Not until the 1940 blitzkreig had unified Europe under the Swastika did the Axis powers face men of sterner stuff—Paul Reynaud, who had suffered front line duty in 1914; ex-brigade major Anthony Eden, who had won the Military Cross in 1915; and the incomparable First Lord of the Admiralty, whose war service stretched back to duty under the Queen in the century before this one.

For France the stiffening of the national backbone came too late; for Britain, just in time. For the other two great Allies Britain bought the time which was the *sine qua non* of later victories outside Moscow and Manila, at Stalingrad and Normandy.

In the United States War Department planning for the next war had been in process since the end of the last one. In charge after Munich was a native of Carlisle, Pennsylvania, Major General Stanley D. Embick, who was approaching retirement age. The Chief of Staff, General Malin Craig, whose own retirement was imminent, moved Embick up to the Deputy post in order to groom another Pennsylvanian, Brigadier General George C.

Marshall, in the War Plans Division for eventual succession to the top position.

The plans of Embick and Marshall in mid-1939 called for an immediate mobilization of twenty-seven divisions in the event of war. Nine would be Regular Army and eighteen National Guard. When German tanks rolled over the Polish border in September the Regular Army of the United States stood at 187,886 officers and men, the National Guard at 199,491 and the Reserve Corps (principally officers) at 119,773. If the nation had declared war it could not have fielded any divisions ready for combat.

Pennsylvania lay within the Third Corps Area of the First Army Command. The territory also covered Maryland, Virginia and the District of Columbia. Within the Corps jurisdiction were 14,468 Regulars. However, head counts do not a fighting division make, even though plans called for each of the Corps to form a division. Personnel were spread out over all kinds of posts, quarters and situations, from service as clerks in recruiting offices, to typists accounting for Spanish-American War caissons at ordnance depots, cooks at officers clubs, physicians learning military nomenclature at Carlisle Barracks, artificers at coast defense installations, ROTC instructors and honor guards at Arlington Cemetary. They had never gathered as platoons, much less as battalions, regiments or a division and would not have had the fighting skills if they had been assembled.

The Third Corps share of the eighteen National Guard Divisions, the 28th of Pennsylvania and the 29th of Maryland and Virginia, bore greater resemblance to what military men mean by an infantry division. The Keystone Division was short of full war-time complement, but it mustered 11,318 when orders came to mobilize early in 1941.

By one of the minor coincidents of history Marshall's first day as Chief of Staff in September 1939 was D-day for the German panzers. To him fell the task of implementing his own plans during the seventy-four months ahead until his own retirement following V-J Day in 1945.

Chief of these, insofar as the mission of the Nation Guard was concerned, was the series of "Rainbow Plans" that took United States neutrality in a European War as a premise and provided for hemispheric defense and the protection of both overseas possessions and the sea lanes leading to them. Accomplishing the defensive role called for the deployment beyond the continent of four divisions, or approximately 60,000 men. To permit the withdrawal of 60,000 from the total Regular roster in North America, immediate trained replacements were necessary. There was only one possible source, the National Guard.

However, in 1939 hope lingered that a declared war could evolve into a negotiated peace before land armies would clash or air armadas substitute explosives for the propaganda leaflets each was dropping on the other's homelands. There were, additionally, practical and political reasons why civilian soldiers could not immediately be wrenched from home to meet contingency war plans. Many people believed the United States could avoid the new war. Only a few Americans in 1939 foresaw a triumphant Hitler possessing the French and British fleets and conquering Latin America in preparation for an invasion of the United States itself. By summer of 1940 such a contingency seemed more likely: France had surrendered; amateur yachtsmen were taking the British Army survivors into tiny craft at the beaches of Dunkirk; London burned; fighter pilots—fewer than a battalion in numbers—battled the largest bombing fleet in history; His Majesty's Prime Minister spoke of fighting on the beaches, in the hills and, if necessary from the Empire beyond the seas.

With doubts increased and hopes decreased the Congress by Joint Resolution of August 27, 1940, authorized the Commander-in-Chief to order the National Guard into Federal service for twelve consectuive months. In companion legislation passed to achieve equity in the calls for personal sacrifice, the Congress authorized the first peacetime draft in the history of the nation.

With the Joint Resolution Marshall could now implement the Rainbow Plans. The 60,000 Regulars were replaced with 66,646 National Guardsmen, principally in the four divisions which were mobilized September 16, 1940: the 30th, 41st, 44th and 45th. Indicative of what the War Department conceived as its immediate problem was the identity of the other units called into Federal service: four observation groups and eighteen coast artillery sections.

A principal bottleneck delaying further callups was the shortage of facilities to handle the recruits who began to pour into the armed forces. Some 422 Selective Service Boards soon were supplying long lines of men, and reserve officers were moving out of civilian life. By logistical reckoning of the time an infantry division needed a training area of 40,000 to 100,000 acres and an armored division even more. There simply were not enough military reservations to handle the other fourteen National Guard divisions. Accordingly, camp construction or acquisition had to precede full mobilization. As part of the expansion program, on September 30, 1940, the War Department negotiated a lease with the Commonwealth of Pennsylvania for the Indiantown Gap facility and its thirty-three buildings. Construction people descended on the camp to provide 1,400 new structures.

The Guard was mobilized in twenty-two increments through June 23, 1941, when the last of 297,754 men were called to one year's active duty. Orders for the fourteenth increment came on February 1, 1941, by which the entire 28th Division and the 1,057 men of the 104th Cavalry Regiment were ordered to their home armories for induction and subsequent movement to Indiantown Gap. The 28th was the fourteenth of the Guard Divisions mobilized.

Over a ten day period following February 22, the units arrived at the Gap. Among the first was the Division Headquarters and the commander, Major General Edward Martin, who had been a sergeant in the War with Spain and had won a Distinguished Service Cross for valor fighting in France in the Great War as a battalion commander of the 110th Infantry Regiment. For three months under his eye the Division engaged in the most basic of training: manual of arms, care and cleaning of weapons, close order drill, marksmanship training and the like. Mid-way through the cycle First Army Commander Hugh A. Drum conducted his inspection and pronounced the work satisfactory.

At the end of the basic training period the 28th Division received its first group of draftees in a contingent arriving July 17 from Camp Wheeler, Georgia. It was part of what finally amounted to 18,000 "fillers" who would join the Division during the thirty-one months it served stateside.

Men of the 28th Division are activated and begin training.

Virtually all had to be processed through the basic elements of training concurrently while more seasoned members of the Division proceeded with advanced and specialized training, unit exercises and maneuvers.

While individuals were arriving, others were departing, so that on some date that passed unnoticed and subsequently cannot be delineated, the 28th Division, Pennsylvania National Guard evolved in fact into what it had become in law in February 1941, a division of a national army. Pennsylvanians were outnumbered unquestionably while the Division was still in the United States. The mix changed in various ways: 2,000 men transferred to the Air Corps, 2,500 to Officers Candidate Schools, and 400 joined the new paratroop force. A much larger group, 6,750 men, left the Division when the Army reorganized the 28th from a "square" to a "triangle" division. Detached was the entire 111th Infantry Regiment, its artillery and other supporting units. A still larger loss was the total number of trained men who periodically were bled off the Division and sent as cadres to form the nuclei of the reserve divisions, particularly in the series numbered 75-106. Outflow in all numbered more than 18,500 officers and men, more than the total authorized strength of an infantry division.

By July of 1941 the men of the 28th were adjudged sufficiently trained to learn how to act together as a divisional organization. That training had to take place in an area with more maneuver room than Indiantown Gap could offer. Accordingly, in August the Division was ordered to sprawling Camp A.P. Hill in Virginia. Some idea of the state of preparedness, so far as equipment is concerned, is to be found in the expedients necessary to get the men to their new station. Civilian automobiles supplemented by leased trucks and moving vans were pressed into service. For three weeks following August 25, 1941, the Division engaged in Second Corps training. After getting the experience of acting with other commands as a Corps, the Division then moved almost directly into First Army maneuvers in North and South Carolina to learn for the first time something about action with tanks, air support and other appurtenances of the largest mobile command in modern warfare.

The first Sunday of December 1941 found the Division in transit from Halifax, Virginia, to its home base at the Gap. Ten of the twelve months of Federal service had passed when news of Pearl Harbor reached the men. Now it was "for the duration," and the length of that period of time seemed prospectively to be endless. The enemy now boasted dominion over the littoral of Africa except Egypt, all of Europe save Portugal, Spain, Sweden and Switzerland, and the eastern borders and peninsulas of Asia and Oceania almost to Australia and Hawaii. It was a time of unrelieved, grim news as the Philippines and Singapore fell, burning oil tankers lit the night sky of Atlantic City, Britain starved, defeated Europeans were cowed by Quislings and millions died in titanic battles deep in the Soviet Union.

Although the victorious Axis powers racked up wins with divisions in the hundreds, the United States and its Western allies could muster only scores. These were largely underequipped and untrained for the new style of warfare against tanks, dive bombers and jungle guerillas. The United States Army consisted of 1.7 million uniformed men but only thirty-three divisions (two cavalry, five armored and twenty-six infantry), all untested in battle and most mere aggregations of men. Two-thirds of the force were quasi-reluctant products of the draft machinery.

Plans there were on December 7, 1941, but none to demobilize the 28th and send home its men. Instead, 182 more divisions were supposed to be formed by June of 1944, which would require a total army manpower count of a full 8,000,000 men and women.

One consequence of the actual start of war was to begin to remove from command those older men who were deemed too old for combat duty. Thus it was that the actuarial tables, if not the physical examinations, caught the commanding general. Edward Martin at sixty-four was at the age of compulsory retirement. A younger man had to be assigned to hone the skills of the Division and to lead it into battle. He retired February 27, 1942, after serving the Division in its (and his) third straight war. Beginning a new career nine months later as Pennsylvania's ninety-seventh Governor, he was asked by the Capitol press corps how they should address him. His answer was quick, and its substance indicative of his first love: "It took me a year to become governor and forty to earn a star; call me general."

As commander-in-chief of Pennsylvania's militia, among other duties, he supervised the new force brought into being as substitute for the National Guard units, all of which had been put under Federal jurisdiction, the Pennsylvania Reserves Defense Corps, whose authorized strength was 5,000. Governor Arthur H. James immediately after Pearl Harbor had called this blue-uniformed force to active duty. They served continuously, and without pay, for forty days following the declarations of war, guarding key bridges and other key places in the Commonwealth that were thought to be the targets of saboteurs.

The Guard and the Corps were only the first of nearly one million Pennsylvanians who saw service in the forty-four months of war that followed. A summary at war's end showed that about one of every eight persons in the Commonwealth had donned uniform. The count June 30, 1945, was 978,061 men and 22,730 women, exclusive of 179,000 who by then had been discharged, retired from duty or died. Thirty-five, more than the total in any other State, had won the Congressional

Medal of Honor. Two of the four Generals of the Army were Pennsylvanians by birth, three of the four-star generals, four lieutenant-generals, twenty-two of the major generals and seventy-four brigadiers.

Pennsylvania was to respond fully in the four years ahead, but in that December of 1941 the burden seemed to fall with particularly heavy weight on the men of the "Iron" Division who already had given ten months of service. Hardened by field training, they faced the likelihood that they would be the first now to face the ultimate test, combat to the death with a tough enemy. In the home armories were now fading records of the consequence of battle in an earlier war. The grim record of 135 days of combat in 1918 had been for the 28th Division 14,139 casualties, 2,874 of them deaths in action, including the only general officer in the American Expeditionary Force to die in battle, Brigadier General Edward Siegerfoos.

For those persons privy to the strategic principles that would govern deployment of troops in the war ahead, the prospects of personal survival were poor. The idea that developed was that 207 divisions would not be created but only about ninety. These would, unlike the pattern in other wars, not be withdrawn from fighting zones and substitutes put in their place. Instead, replacement training centers would send forward individuals who would be fitted into the existing units and the division kept in action or in position to move out of close reserve. Estimates were that a division would suffer 100 per cent casualties after ninety days of combat. In actuality many divisions counted 200 per cent casualties before peace came again.

As the 28th moved back into Indiantown Gap in December, the men found all leaves and passes canceled. The Division had to remain on the alert for contingencies no one could guess.

An early consequence of war was that the War Department ordered immediately the conversion of the eighteen National Guard Divisions from the square format into the triangular configuration that the commanders of the new Regular Army Divisions had found so much handier to deploy in the maneuvers that had been just completed. Now seventy-two National Guard Infantry Regiments of three battalions each and 107 battalions of light and medium field artillery could produce the cadres for six more full divisions. In January the 28th lost its 111th Regiment, which eventually was posted to the South Pacific.

How imminent, however, was the dispatch of the 28th to a theatre of operations overseas was anyone's guess. It was a time of confusion, of need for men in every quarter, and of rumor. One of the latter proved true, when advance elements of the 32nd, 34th and 37th Divisions (all National Guard) arrived in Ireland to prepare for the arrival of American forces to replace units of the English Army that was guarding Northern Ireland against an expected German attack but which now was being sent to North Africa as replacements for the British Eighth Army. The 34th, however, was the only one that went to Ireland in strength. Strategic considerations changed the destination of the 32nd and 37th. The Japanese Army had seized some of the outer Aleutian Islands and was simultaneously threatening Australia. General Douglas MacArthur, a commander without field forces, got the two divisions re-routed from the Boston and New York Ports of Embarkation and sent to him.

The destination of the 28th turned out to be Camp Livingston, Louisiana. By January 10th the first units left the Gap on a motor march south where they were to become a portion of the Fifth Corps of Lieutenant General Walter Kreuger's Third Army. The convoys were ten days on route, proceeding via Fort Meade, Fort Lee, Fort Bragg, Fort Jackson, Camp Wheeler, Fort Benning, Dinapolis (Alabama) and Vicksburg. For twelve months Camp Livingston was to be home for the 28th. It was here that General Martin relinquished command to Major General J. G. Ord, not a Pennsylvanian. He served until May of 1942 when the command was passed to Major General Omar N. Bradley. Under the latter the Division was to have a half year more of training in motor convoy practice, combat firing, maneuvers and, with it all, the absorbing of thousands of replacements sent it by the Selective Service system.

Bradley took the Division early in January 1943 to what was then known as Camp Carrabelle, near the Florida coast, to begin training in amphibious tactics, particularly shore-to-shore operations. The nature of the training suggested a future in Pacific island-hopping, but those familiar with the whole map of Europe found islands on the perimeters that could be used as staging areas. By February General Bradley was assigned command of the Second Corps then fighting in Tunisia. His replacement was the then Brigadier General, Lloyd D. Brown, who had trained the 101st Airborne Division and would remain in charge of the 28th until shortly after it entered combat seventeen months later.

Late in March 1943 amphibious training having been completed (with fourteen men of the 112th drowned during a storm-swept exercise), the Division was re-assigned, this time to Seventh Corps, Second Army, and its permanent station made Camp Gordon Johnson, which was simply Camp Carrabelle re-named. On June 3 another permanent change of station was ordered, this time to Camp Pickett, Virginia, where the 28th became part of the Twelfth Corps of the Second Army. Located within striking distance of both mountain and sea, the 28th sent units to both for special training.

With secrecy General Brown and three assistants flew to England on August 17 to confer with the Theatre Commander and prepare for assignment beyond the Atlantic. Two weeks later they were back in Pickett by courier plane via Iceland with knowledge of the assignment that had been given the 28th.

On September 20, 1943, the advance elements of the Division sailed from New York on the *Queen Mary*, Cunard Liner now pressed into wartime transport service. Six days later the remainder of the Division moved toward staging areas at Camp Miles Standish, Massachusetts. By the twenty-seventh the first elements of the Division were disembarking at Greenock, Scotland, and moving south to Tenby, Wales, to arrange for billets for 15,000 men.

Seven days into October the remainder of the Division began the thirty mile train ride to the Boston docks, and on the day following the men were under convoy on the U.S.A.T. *Cristobal* and other vessels. Five days out, the men of the Division heard their first sounds of war, as depth charges were dropped following the reports of enemy submarines in the vicinity of the convoy. With no other incident the ships sailed on through the Irish Sea, up the Bristol Channel and into Newport, Wales. In a ten day sea trip all had arrived safely and on time except for one ship that had been forced by mechanical troubles to return to Halifax, Nova Scotia, for repairs. The 28th Division men on her were two weeks late in arriving on the far side of the ocean.

In the European Theatre of Operations the 28th found itself part of Major General Leonard Gerow's Fifth Corps of the First Army. Commanding the latter was the former Division Commander, Omar Bradley, now wearing three stars and later to win a fourth and fifth.

Nineteenth months of stateside training behind it, the Division now found itself abroad but separated by half a continent from the Mediterranean area where the only combat was occurring. For the men of the 28th, scheduled to participate in the cross-Channel operations planned for 1944, the assignment to English bases meant still another round of training. Even if the operations in Northern Europe had been imminent, it would still have been necessary for the 28th to continue a full round of training for the simple reason that the arrival and departure of men, some the greenest of recruits, continued right down until sixty days before it entered combat. Additionally, as the war continued new knowledge had to be imparted to everyone, from the lowest rank enlisted man furnished a new piece of equipment to the highest ranking officers who needed to learn deployment tactics only recently conceived to meet the German techniques which also were continually changing. A Combined Chiefs of Staff organization, for example, permitted United States forces to share the lessons learned by British and Canadian armies that had for two years been fighting an experienced enemy.

One of the novel principles for staff to learn had emerged from the doctrine enforced by Lieutenant General Lesley J. McNair, commander of ground forces. It was that Divisions be small and shorn of many supporting combat and service units. Such groups, as for example independent tank battalions, anti-aircraft and tank destroyer groups, heavy artillery and specialty engineer forces were to be maintained as contingents in Corps or Army reserve. As needed they would be assigned out to division commanders where particular circumstances made such attachments profitable. Division commanders, accordingly, had to learn how to work with and then work without these specialty commands and, further, to engage in combined operations with the Navy and tactical air force elements. Some of the first training for the 28th, accordingly, was a session in Devonshire learning how to work with anti-aircraft and tank destroyer battalions.

Less sophisticated items were also on the training schedule: a 100-mile forced march through the Welsh mountains and serving as "the enemy" in a practice attack made by the 101st Airborne Division. Street fighting, demolition, using captured German weapons and similar items became part of the schedule, when it became clear that the 28th would probably be assigned the mission of capturing a European city in the years ahead.

Divisional records tell of events other than the grind of incessant training. A neighboring unit, the 101st Airborne, accepted a challenge to a football game which ended in a 6-6 tie but puzzled the Welsh. A Swansea newspaper, for example, reported seeing often "a conference between members of the side entitled to claim the ball ... grouped together with heads down as they prepared tactics... ." A "huddle" was not a word in the Welsh sports vocabulary. Familiar with soccer but anxious to be polite, the reporter found it all to be "a colourful adventure ... an acquired taste ... with numerous stoppages that do not lend themselves to speed and to constructive endeavor... ."

Another neighbor, the 15th Scottish Division, in January as suggested by higher authorities, engaged in an exchange of officers and enlisted men. The idea was to familiarize the two commands of two armies with the tactics, training, arms and equipment of the other. Christmas found the Division hosting several thousand British children, many of them evacuees from the cities, in Yuletide parties. The Division also accepted a challenge to a rifle marksmanship match from the 1st Pembrokeshire Home Guard Battalion of the Welsh Regiment at Tenby. The old gentlemen of the Home Guard won.

Training seemed to pay. By April 3, 1944, the Supreme Commander made his formal inspection. General Dwight D. Eisenhower wrote his formal opinion that the 28th Division was fit, efficient, serious and determined.

Throughout the entire period in Britain the greatest amount of stress was on amphibious kinds of training: assault over a beach, landing on a hostile shore, and ship-to-shore movements. When the 28th Division became especially proficient in this specialty, it was almost inevitable that the 28th Division would be in the mind of the Supreme Commander soon after his visit when particular attention was being given to forming a new Army for the European Campaign, the Third. As first constituted, it had no division at all which had gone through amphibious training. The 28th, accordingly, was added to its roster, as part of the Twentieth Corps. Bradley, whose First Army was to be the force assaulting the European beaches, wrote his regret at losting the 28th a second time in two years. The 28th had a new Army and a new commander, Lieutenant General George S. Patton, Jr.

The new assignment involved, among other matters, moving 160 miles from Wales to Swindin, Wiltshire, in southern England to make way for the 2nd Division which was transferred from Northern Ireland into the 28th's cantonments. Three shuttles took the entire "Iron Division" by truck to the new site. Patton immediately turned everyone out for Exercise "Eagle," a simulated assault on a hostile coast, and followed it all with intensive training in working in combat with tank units.

When D-Day, June 6, 1944, arrived, Patton's command remained behind. The 28th heard the fighter airplanes leave from nearby bases and saw the sky darken as squadron after squadron of bombers struck east. For an impatient Patton and his command six more weeks of training had to be endured.

The long wait ended July 18, 1944, when convoys of trucks at dawn began to pull out of the Wiltshire bivouac area bound for Weymouth and Southampton. Embarkation from the Channel ports began two days later. On the twenty-second the first men of the 28th Division stepped ashore on Omaha Beach at the base of the Cherbourg peninsula. The crossing had been uneventful on the A.T.S.S. *Cheshire*, the S.S. *Mechlenburg*, the U.S.A.T. *Jeremiah O'Brien* and other vessels. Troops climbed off the larger ships into smaller craft which then transferred them dry-shod to temporary piers.

To get the men clear of open areas that would be vulnerable to enemy air attack and to make room

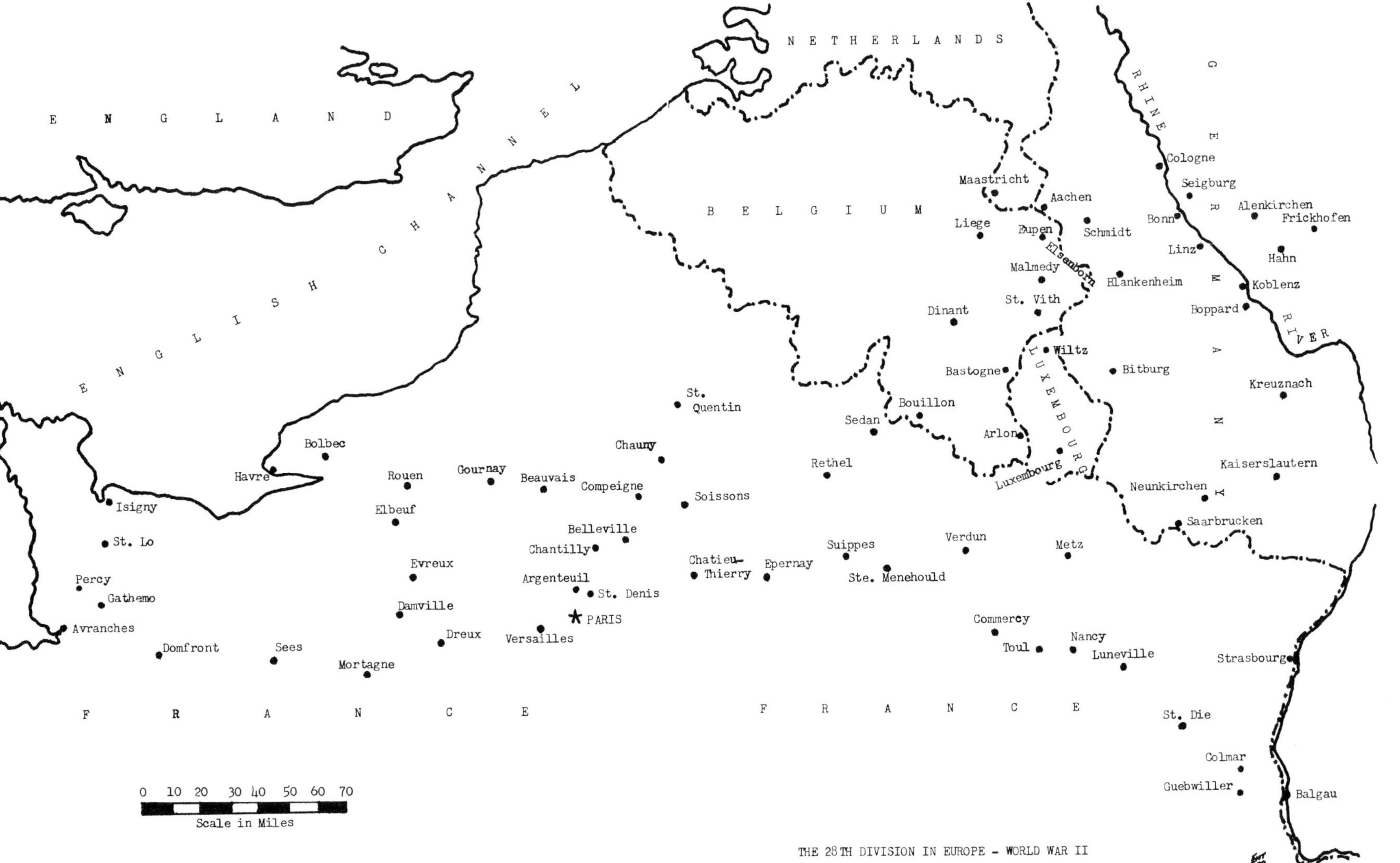

The 28th Division in Europe - World War II.

for additional units that were to follow, the Division was marched promptly one mile inland to a transit area near the French town of Colombiers. Occasional German airplanes circled overhead the first night, the first sign of active warfare for the men of the Division.

On July 23 the 28th received orders making it again part of the First Army, which on July 23 launched "Operation Cobra." The mission was to exploit the breakthrough of the German lines in western France that the 29th Division had accomplished on July 18 at St. Lo. Preparations included a saturation bombing of German defenses by 2,000 allied aircraft. Close after the air and artillery preparation came Bradley's Army in a great wheel south to the latitude of Paris and then a race east. Breakthrough and breakout, it was a furious, unrelenting rush involving only occasional delaying attempts by the Germans, whose own mission became that of retreat behind the West Wall in the Reich and defense of the homeland itself.

Although the news correspondents at highest command posts could see the front move on maps at dizzying pace and could write of the seeming ease with which Eisenhower's men seized whole Provinces, the riflemen and tankers at the front had another story. Very stubborn pockets of very determined Germans defended not vast provinces but tiny pillboxes, fortlike farmhouses, and picturesque hedgerows. It was the last of these, the standard features of pre-war postcards, that spelled death for the 28th as it pushed into Normandy. Packed solid during millenia of use, rock-filled walls of earth from four to six feet high, they were perfect cover and defense. Tanks were anticipated as the solution to burrowing out the German defenders, but tanks could not penetrate the rows. Instead, they had to climb pependicularly, exposing unprotected underbellies to German anti-tank fire that would have bounced off armor plate elsewhere. It was "Hedgerow Hell" to the 28th and the two divisions on the flanks in July, the 4th and 29th.

Movement into actual battle began July 27 as the 28th cleared the rear assembly area of the Nineteenth Corps sector. In the move the 28th suffered its first casualty, Captain Leo Metz, commanding the Anti-tank Company of the 112th Infantry, who was struck by German artillery fire while reconnoitering forward positions. His mission as an advance officer had been to prepare the way for the 28th to relieve the 35th Infantry Division on July 28. Actually, the substitution never occurred. Before orders could be executed German defenses crumbled in the St. Lo area, and the 28th Division was ordered to move south together with the 30th Infantry and part of the 2nd Armored Division, which comprised the Nineteenth Corps of Major General Charles H. Corlett. The breakout was in full rush.

Jumpoff from St. Lo came July 29. Elements of the Division rode south by truck through the smoking rubble of St. Lo to a point three miles south of the town. Here at Les Plains, an abandoned hamlet, soldiers of the 110th Regiment saw the first enemy casualties, four German dead at an abandoned field gun. Bloated bodies of draft horses and cows littered the area. After forty-one months of training, this was war: stench, decay, devastation, death.

The immediate target was the town of Percy; the next following was St. Sever Calvados, twenty-six miles to the south. At dawn on July 30 the Division found itself under attack for the first time, when enemy aircraft slipped through the usual allied air curtain and offered opposition. The first live enemy ground forces appeared opposite the Division at 10:00 in the morning—an anti-tank unit at Le Meanil Herman. In a sharp action in which men on both sides were killed the 110th Regiment took eighty prisoners and

In the hedgerows.

A German sniper in Gathemo, France, has just fired from behind the protecting wall of a battered church.

marched on to Maupertuis. With the morning of July 31 the entire Division was in place near Percy and prepared to launch its first major attack.

Heavy fighting ensued and lasted for three days as the Divisional units methodically pushed to the Corps objective. On August 4 the 110th Regiment advanced at a brisker rate, covering ten miles and earning congratulations from both the Division and Corps commanders. The cost was high: 600 casualties or twenty per cent of the Regiment. The 109th was put into its place and the 110th placed in reserve.

Facing the 109th as it moved toward Gathemo on August 7 was half of the German 84th Infantry Division, beefed up by an 88 mm. anti-tank battalion. At night on Augut 7, the 84th counterattacked with tanks bearing searchlights. It regained the few yards that the 109th had seized from it during daylight hours. On the eighth day the 109th countered with an attack that again gained the few yards, against bitter resistance. It was not until August 10 that the Division could enter Gathemo permanently, and that gain was made only after units of the 2nd Armored Division joined in the final assault. A measure of the intensity of the battle is to be found in the ammunition supply records of just one company of the 109th Regiment: it expended 4,000 rounds of 81 mm. mortar ammunition in three days of assault on Gathemo.

While the 109th and 110th were expending themselves against Gathemo, the 112th moved up to the Vire-Gathemo Highway amidst intense resistance that included ten machine gun nests in the sector of the 3rd Battalion alone. However, by August 12 the Division had accomplished its objective and was preparing for the swing east toward Paris and the advance to the German border. Just as the Division turned toward the heart of the continent General Brown left the Division. His successor on August 13 was Brigadier General James E. Wharton, who had been with the veteran 9th Infantry Division. On the same day, the assistant division commander since April 1942, Brigadier General Kenneth Buchanan, moved to the 9th as its divisional commander. General Wharton immediately went forward to one of the regimental Command Posts, where he became a sniper's target. Later in the day, his first in command, he died while being evacuated. In two world wars the 28th had lost a general officer in combat.

Appointed commander was Brigadier General Norman D. Cota, of Massachusetts, who had been assistant division commander of the 29th and was a veteran of the North African Campaign. He was to serve with the Division until its demobilization following V-J Day.

It was Cota's privilege to command during the most unforgettable and most publicized of the events in the history of the 28th Division, the celebrated march through Paris on August 29. In the two weeks preceding, the Division was assigned to the Corps of the First Army, which now was under the command of Lieutenant General Courtney H. Hodges. Bradley, meanwhile, had moved up to command of the Twelfth Army Group, consisting of the First and Third Armies.

The nearly 200 mile march to the French capital was no unrelieved shower of champagne, embraces by doughty mayors and buxom belles, and other aspects of the joy with which the French welcomed the liberators. Determined German rear guard elements had prepared their own reception. Part of the two weeks was occupied by a detour halfway across France when elements of the Division and companion units cleared Verneuil, Damville, Evreux, Bonneville and other cities. Veteran *panzertruppen* and 700 of the despised S.S. infantry were among the defenders who had to be bombed, strafed and finally rooted out of defensive positions.

It was, accordingly, a boost to morale when orders came through to the 28th notifying it that the

Division was to assemble at the famed town of Versailles and on August 29 have the honor of being the first American Division to parade in Paris.

After assembling in a driving rain and in darkness at the Bois de Boulogne at the southwest edge of the capital city, the Division stepped off at daylight twenty-four men abreast down the Avenue Foch by the Arc de Triomphe and down the Champs Elysees. Following the infantry, two regiments abreast, came the entire train of artillery, engineers, service units, tanks and tank destroyers. It was the official liberation of Paris, first allied capital to be reclaimed from the Axis. An elixir, an electric charge, a heavy dose of adrenaline to a tired allied world, it was plainly neither the beginning nor the end but certainly the beginning of the end.

Taking the salute at the Place de la Concorde were Bradley, Cota, Hodges and "Le Grand Charles" DeGaulle. By the hundreds of thousands Parisians flung fruit and flowers, cheers and cried. *"Boche Kaput; Vive les Americains!"* French musicians and the fifty-six piece 28th Division Band helped put spirit in the steps of the men at march. On hand were war correspondents from the newspapers and wire services of all allied nations. The Associated Press took a photograph that was printed in publications around the world, a shot destined to become with the flag-raising at Iwo Jima one of the great historic photographs of the war. It showed the 28th in full battle gear curb to curb, with the Napoleonic Arch of Triumph looming behind. It was chosen as the theme for a United States postage stamp issued a month later to honor the wartime army.

Seasoned men of the 28th knew on August 29 that the march through Paris was something more than a triumphal parade. Those on foot carried full combat equipment; behind rolled all the gear, supplies and ammunition needed for a fighting division to move directly into battle. The route had actually been chosen because it was the shortest distance between Normandy and the German border. Thus it was that the 28th, after a bivouac at the Parisian suburb of St. Denis, proceeded the next day to the Oise River, flushing out the Compeigne Forest, site of the signing of the Armistice in 1918 and of the surrender of France to Hitler twenty-two years later. On September 8, 1944, the Division passed Sedan, site of the French humiliation by Prussia in an even earlier war, and crossed into Belgium.

In pursuit of the Germans.

Majors Paul and Robert Gaynor, brothers, and both members of the 28th, each received the Distinguished Service Cross for heroism.

The entire 28th Infantry Division marches through Paris.

Without hesitation or serious impediment the Division crossed the corner of Belgium rapidly and in two days was on the Luxembourg border playing host to the Consort of the Grand Duchess of that small principality. Prince Felix, after four years of exile, accompanied the 28th to the edge of his wife's domain but was escorted into it by another command. The 28th at the last moment received orders to veer north toward Bastogne, just another of a hundred Lowland towns at this point in 1944.

For six weeks the fighting had been at a minimum and the main body of the German army out of contact. However, by September Germany itself was within artillery range of the 28th, which lobbed its first shell September 10, 1944, into the "Thousand Year Reich." Plainly, the war of fast movement was at an end. Resistance stiffened. Germans were at their last ditch.

Late on September 10 the first men of the 28th Division, a squad from the 109th Infantry stepped on German soil. Two nights later three entire battalions were digging in to stay, the first Allied units to cross the border in force. For the first time since the invasion of Bonaparte, an armed enemy was in Germany.

Pursuit ended with the initial breach of the Siegfried Line in mid-September. For the 28th and other units, a new kind of warfare began. Pressed against the border of their home country, the Germans fought desperately, not just the usual combat divisions that had survived the great retreat but the new units of former garrison troops, overage men and youth plummeted prematurely

A captured German soldier on his way to the POW cage.

Many combat medics lost their lives going to the aid of their comrades under enemy fire.

into uniform. For both sides, including the 28th Division, the casualty rate was fearful and defeat bitter. For both sides during the rest of 1944 there was just that—defeat in the campaign, with units on both sides decimated, overwhelmed and outfought.

At first the 109th and 110th Regiments had much their own way beyond the Wall. Quickly, however, German units counter-attacked and often took back the same pillboxes that the American forces had seized just hours earlier. It became policy for the 28th to do more than occupy such bunkers; they began methodically to detonate TNT in them, using up to 1,400 pounds per pillbox, in order to deny future use should the enemy succeed in counter-attacks.

In this see-saw battle at the West Wall the first member of the Division—and the only during World War II—won the coveted Congressional Medal of Honor. Staff Sergeant Francis J. Clark, of the 109th Regiment, in a period of five days became a virtual one-man army, taking over the command of two platoons when the higher-ranking leadership became casualties, rescuing one unit, single-handedly wiping out a machine gun nest and forcing an entire company of Germans to retreat. Wounded several times, he refused to be evacuated and continued in combat command. During this phase of action Lieutenant Colonel Carl L. Peterson took command of the 112th Infantry Regiment, the same unit which he had served continuously starting as a private in 1916. Moved into command of the 109th Infantry was another 28th Division veteran of World War I, Daniel B. Strickler, later a major general in charge of the 28th. For the eleven months beginning in December 1943, Theodore A. Seely commanded the 110th.

On September 29, 1944, the Division received orders to yield its positions to the 2nd and 8th Infantry Divisions and to move north. By October 5 the transfer was completed, with little German opposition. At Aachen, once Aix-la-Chappelle, it was placed on six-hour alert for movement with the 5th Armored Division, if a German attack were to develop. While near Aachen a soldier who had served as a lieutenant in a unit of the 28th Division in 1906-1907 paid it a visit: General George C.

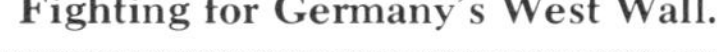

Fighting for Germany's West Wall.

Congressional Medal of Honor winner Technical Sergeant Francis J. Clark, Company K, 109th Infantry.

Marshall, chief of staff, who praised the Division "for its excellent work over here."

October 1, 1944—December 13, 1945 Operations

By Harold E. Myers

At the time of General Marshall's visit the Division was in Belgium in the Elseborn area for reorganization, reinforcement and training with tank battalions and other support groups. While it was so engaged other elements of the allied armies during October made additional breakthroughs in the German lines, particularly near Aachen and Rotgen.

Schmidt—October 26-November 16, 1944

First Army was designated to make the principal effort and, after crossing the Roer River, to secure crossings over the Rhine. Heading the drive was the Seventh Corps. Fifth Corps, of which the 28th Division was a part, was assigned the supporting mission of taking the small cross-road town of Schmidt.

The attack against Schmidt was to accomplish four things: open up additional supply routes for the Seventh Corp attack to the north, protect the ground for later attacks to seize the Roer Dams, and draw enemy reserves away from the Seventh Corps—preventing an attack against First Army's main effort. The Hurtgen Forest was a blackish-green mass of fir trees and undergrowth, the kind of place that would frighten the bravest of men. If it could speak, it would tell a story of death and suffering, of human courage and heroism. In this forest at a small crossroads lies the town of Schmidt, which in the early part of October 1944 the 9th Infantry Division was supposed to attack and secure. With German pillboxes hidden in the Forest and the Germans counter-attacking, the 9th Infantry lost 4,500 men and gained only 3,000 yards.

The Fifth Corps, under the command of Lieutenant General Gerow, designated the 28th Division to make a second attack on Schmidt. Under the West Point graduate, Major General Norman D. Cota, the 28th took over the 9th Infantry sector on October 26, 1944.

The terrain consisted mainly of valleys, ridges, and gorges, with the Kall River running diagonally across the 28th Division's zone of attack. There were also three distinct ridges: to the northeast was the Brandberg-Berstain ridge; in the center was the Germeter-Vossenack ridge; the third ran between the Kall and the Roer Rivers. The latter had the highest elevation west of the Roer River, and the 28th would be under the observation of the enemy on the way to Schmidt. This area was important to the Germans, who knew they had to keep the Americans bottled up in the forest. General Cota, knowing what his Division would be facing, did not like the overall situation and informed Corps that he did not like the terrain, or the fact that he would have split his Division for three different attacks.

The 109th Infantry Regiment, under the command of Colonel Strickler, was to attack northward toward the Hurtgen, then through the forest for about a mile. Then it was to advance on either side of the Germeter-Hurtgen highway overlooking Hurtgen and to block against counter-attacks. The 110th Infantry Regiment, under Colonel Seely, was to attack southward through the forest to secure the road from Schmidt to Strach.

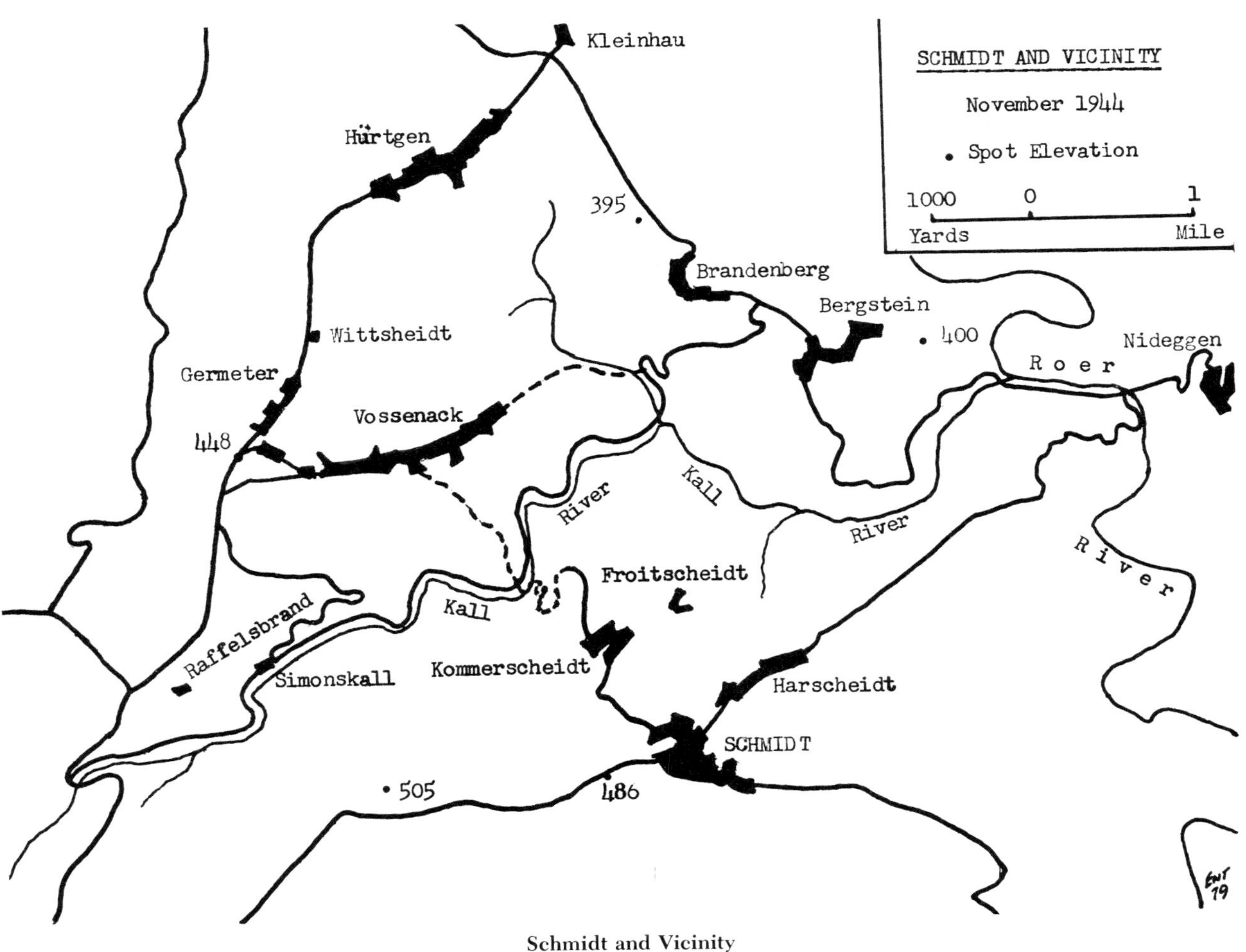

Schmidt and Vicinity

This strategy was for the planned attacks on the defenses of Monschau. The 1st Battalion of the 110th was the Division's only reserve. Colonel Peterson's 112th Infantry Regiment was ordered to make the Division's main effort in the center. The 112th was to attack and capture the towns of Kommerscheidt and Schmidt. The 2nd Battalion of the 112th was to secure Vossenack and Vossenack ridge while the rest of the 112th was to attack south of Vossenack, cross the Kall River and secure Kommerscheidt and Schmidt. The attack by the 28th Division had been scheduled to take place on the last day of October but was postponed because of bad weather. It began on November 2, with the Division artillery and other supporting artillery firing over 11,000 rounds into enemy positions.

THE ATTACK BY THE 109th: At 9 o'clock on the morning of November 2 the 109th Regiment, under LTC Daniel B. Strickler, jumped off and advanced toward the town of Hurtgen. The attack began with the 1st and 3rd Battalions abreast. The 1st Battalion made good progress and by 2:30 that afternoon had reached the wood line west of the Germeter-Hurtgen road. The Germans were everywhere and at times very close, but as the battalions advanced, the Germans took up their old positions in the rear of the passing Americans. The 3rd Battalion advancing along the road ran into anti-personnel mines, machine gun, and small arms fire, and was able to advance only 500 yards.

On November 3 the 3rd Battalion continued its attack. The plan was for 1st Battalion to hold its position, while the 2nd Battalion was to tie in with it. However, the 1st Battalion was hit with two counter-attacks, both of which it repulsed. The 3rd Battalion made very little if any progress in reaching its objective. On November 4, the 2nd Battalion was scheduled to pass through the 3rd Battalion in order to continue the attack of the 3rd Battalion. The 2nd Battalion met strong resistance, but by the day's end two platoons got across the Germeter-Hurtgen road. One company ran into mine fields and was pinned down with intense small arms fire. One platoon covering the road was completely wiped out by the mine fields. On November 5 the line companies spent the day improving their positions. For the rest of that day and the next the battalions in the woods came under intense tree bursts from the enemy and casualty lists began growing rapidly.

On November 6 the 109th was ordered to take aggressive action with the view of occupying the enemy ground and improving its position. On the left of the 109th the Germans were dug-in well in a draw along the Weisser Wen Creek. While advancing the battalions were to fire to the left in order to keep the Germans pinned down in the draw. The attack made very little progress, one company being hit with artillery and mortar fire, the other companies with machine gun, and mortar rounds. At noon that day the Regiment was informed that it was to be relieved that night by the 12th Infantry Regiment of the 4th Division.

THE ATTACK BY THE 110th: On November 2 the Regiment jumped off on its attack on schedule but ran into trouble immediately. The 2nd Battalion had been ordered to attack the pillboxes in the vicinity of Raffelsbrand, while the 3rd Battalion bore east through the woods toward the town of Simonskall. The forest floor was covered with fallen trees and branches from earlier artillery fire. Booby traps, wire, and entanglements were everywhere. As the battalions left the protection of their foxholes, they were hit with intense machine gun, and mortar fire. Both battalions were cut to pieces. Men became lost singly, in groups, and by platoons. Soldiers carrying explosive charges were hit by shell fire, detonating the charges and killing most of them. The 2nd Battalion did manage to get up to the pillboxes, but the battered enemy held. The battalion fell back trying to reach its own lines but was pinned down by machine gun and mortar fire. Hugging the forest floor, the men crawled and inched their way back to their own lines. For the 110th the day ended without gaining a single yard.

The 110th attack was renewed on November 3 with the 2nd Battalion making several attacks against the pillboxes in the Raffelsbrand area. Each time it was pushed back without gaining any ground while suffering heavy losses. When one company fell back, it had only forty-two men remaining. It was the most costly action of the day. Because of the events of the day in the 110th sector, General Cota released the Division's only reserves, the 1st Battalion of the 110th to the regimental commander.

On November 4 the 1st Battalion advanced south of Vossenack through the woods to Simonskall to try to turn the flank of the Germans at Raffelsbrand. The battalion met light resistance and by 9 that morning seized Simonskall. For the 110th on November 5 the situation was fairly stable, but it had made no progress in attacking the pillboxes between the battalions at Raffelsbrand and Simonskall. For the next few days the actions of the Regiment did not change substantially. Other than the 1st Battalion seizing Simonskall, the 2nd and the 3rd Battalions did little to accomplish the objective. The 110th had fought a hard, bloody and costly fight without gaining a yard. When it was relieved of its position, it was in no better shape than when it had begun on November 2. Decimated, the 110th was no longer an effective fighting force.

THE ATTACK BY THE 112th: On November 2, the 112th Infantry had an easier time of it than the other Regiments. The 2nd Battalion jumped off on schedule, advancing from Germeter east to Vossenack and Vossenack Ridge with a company of tanks. Meeting only light resistance, by noon it had seized Vossenack and the northeastern part of Vossenack Ridge. The battalion began to dig in immediately.

The 1st Battalion, making the Division's main effort, headed east through the woods south of Vossenack, across the Kall Gorge for Kommerscheidt and Schmidt. The lead company came under intense small arms fire and was unable to advance. The rest of the battalion was not committed. As nightfall approached, the 1st Battalion had gained only 250 yards.

Scenes typical of the Hurtgen.

At 0700 on November 3, the 112th continued its attack on the town of Kommerscheidt. Colonel Peterson decided to push his two battalions through Vossenack, then southeast through the Kall Gorge. The tanks with the 2nd Battalion on the ridge gave support to the attacking battalions by firing on the town of Kommerscheidt. The 3rd Battalion crossed the cold Kall River and up a hill. Meeting only light resistance in Kommerscheidt, it pushed on for Schmidt. By 2:30 that afternoon the 3rd Battalion had captured the town of Schmidt and began to dig in. Late on the night of November 3 the 1st Battalion moved into Kommerscheidt and also dug in.

On the morning of November 4 the situation for the 112th in Schmidt took a turn for the worse. At 7:30 in the morning the Germans hit the town with artillery fire. Thirty minutes later they attacked with infantry and two battalions of tanks. The 3rd Battalion was without tank support, and because of bad weather, air support was impossible. At the same time Division was having problems getting tanks across the Kall Gorge, which was traversed by a narrow and slippery trail impeded by overhanging rocks. Tanks trying to pass down the trail could not bypass them, and the engineers were having trouble removing them.

As the Germans attacked the 3rd Battalion in Schmidt, panic and confusion took hold. Those that could leave began to take off, leaving dead and wounded. Some 200 men ran in the wrong direction—southwest into German held territory. Of the 200 men, only sixty-seven lived to tell about it. Those men tho had stood fast to hold the attacking forces began now to evacuate in small groups towards Kommerscheidt. By late morning Schmidt was in German hands.

The 3rd Battalion had taken a beating. The Germans hit it with infantry, the artillery of six divisions, and tanks with mortar support. The 1st Battalion in Kommerscheidt had in support only one tank, with two others on the way. The Germans did not hit Kommerscheidt immediately but did use artillery and tank fire to harass the defenses of that position.

As the afternoon began to wind down, some 200 men of the 3rd Battalion joined the 1st Battalion in the town. During November 5, the situation was stable for the 112th in Kommerscheidt. The defenders held off three attacks by the enemy, and the tank force had increased to eight tanks. The next day the 1st and 3rd Battalions took heavy artillery and tank fire, but by using their own tanks and artillery, kept the Germans from making an attack. On the morning of November 7, the Germans opened with artillery and tank fire followed by infantry. For the 112th in Kommerscheidt the casualties began to grow, and the men remaining could not have numbered more than 100 in both battalions. Four of the eight tanks were knocked out of action, but by 0830 the Germans had infiltrated the town. By early afternoon the town was seized by the enemy forces. Orders were sent in the afternoon for all forces to withdraw east of the Kall River.

The 112th, with its 2nd Battalion in Vossenack and on the Vossenack Ridge, took its share of knocks. For days it had received a pounding from enemy artillery. Failing to take cover in the town, the battalion's toll of casualties was staggering. During the afternoon of November 7 some small arms fire broke out along the ridge. Men in their foxholes, nerves shattered, panic-stricken, groggy and tired, began running from their foxholes and the safety of their own lines. The company commanders tried to stop the mad flight of their men, but to no avail. The 2nd Battalion was on the run, gained none of the ground it had lost, and was destroyed as a fighting unit. When it was all over the Division had accomplished very little. Ground gained was Vossenack and Simonskall, only half of its objective in securing a line of departure for later attacks on the Hurtgen by another Division. The Division did draw enemy reserves from the Seventh Corps area, but this was only one of the four objectives of the operation. In the Hurtgen Forest the 28th Division suffered one of the worst defeats of the U.S. Army in World War II. The battle had cost the Division 6,184 casualties. Hit worst was the 112th Infantry Regiment, with 232 men captured, 431 missing, 719 wounded, 167 killed, and 544 non-battle casualties—a total of 2,093.

The 28th Division had completed its Hurtgen Forest operation by mid-November 1944. It then moved to the south, back to the point of its initial entry into Germany. The 28th Division was assigned to hold a twenty-five mile sector of the front line along the Our River, from the northeastern part of Luxembourg to the vicinity of Wallen-

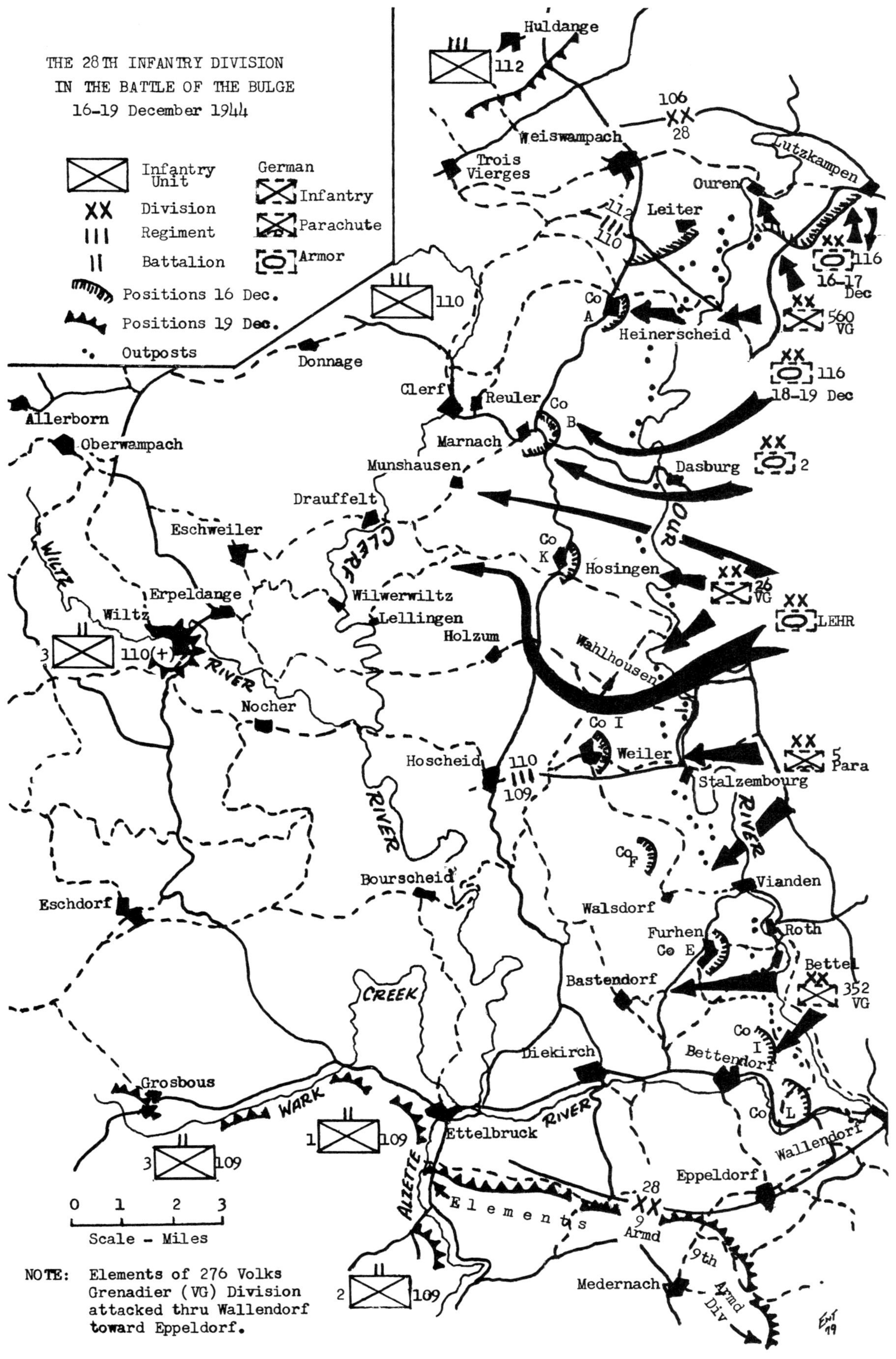
THE 28TH INFANTRY DIVISION
IN THE BATTLE OF THE BULGE
16-19 December 1944
Infantry Unit
Division
Regiment
Battalion
German
Infantry
Parachute
Armor
Positions 16 Dec.
Positions 19 Dec.
Outposts
Huldange
112
106
28
Weiswampach
Trois Vierges
Lutzkampen
Ouren
Leiter
112
110
116
16-17 Dec
560 VG
Co A
Heinerscheid
110
Donnage
116
18-19 Dec
Clerf
Reuler
Co B
Allerborn
Oberwampach
Marnach
Munshausen
Dasburg
2
Drauffelt
OUR
Eschweiler
CLERF
Co K
Hosingen
26 VG
WILTZ
Erpeldange
Wilwerwiltz
LEHR
Wiltz
Lellingen
Holzum
3
110(+)
RIVER
Wahlhousen
Nocher
Co I
Weiler
5 Para
Hoscheid
110
109
Stalzembourg
RIVER
RIVER
Co F
Vianden
Bourscheid
Eschdorf
Walsdorf
Furhen
Co E
Roth
Bettel
Bastendorf
352 VG
CREEK
Co I
Diekirch
Bettendorf
Grosbous
WARK
RIVER
Co L
1
109
Ettelbruck
Wallendorf
3
109
Eppeldorf
ALZETTE
28
Elements
9 Armd
0 1 2 3
Scale - Miles
9th Armd Div
Medernach
2
109
NOTE: Elements of 276 Volks Grenadier (VG) Division attacked thru Wallendorf toward Eppeldorf.

dorf. Here in this sector the full force of the German winter offensive would be thrown against the over-extended and thinly held line of the Division. The Germans hurled five crack divisions across the Our River the first day and four more in the next few days.

The Ardennes—December 16-20, 1945

THE ATTACK ON THE 112th INFANTRY REGIMENT: The attack by the Germans broke into the lines of the 28th Division on the very first day of the offensive. Enemy success was partly due to the fact that the 112th was separated from the rest of the Division. Under the command of Colonel Gustin Nelson, it was holding six and one half miles of the front line of the 28th Division zone, on the east, or German side of the Our River. The area around the position was covered with pine forest and good roadways. Some of the positions of the 112th were set up around German pillboxes on the German side of the River. The enemy was well aware of Colonel Nelson's position. Enemy artillery fire made it impossible for him to supply his battalions during daylight hours.

The 112th was opposed by the Fifty-eighth Panzer Corps, whose mission was to seize crossings over the Our River. In the early hours of December 16, the Germans tried to break through the lines of the 112th, using the bridges over the Our River. However, Colonel Nelson put a counter-attack plan into motion and restored his line in one battalion area. As dawn began to break, the Germans sent shock troops against the 112th. Caught in the open with machine gun and rifle fire the enemy made three attacks to try to break the lines, but the 112th pushed them back each time. All three attacks were made over the same ground, and the enemy had suffered a great many casualties. The 112th was holding its position and doing as much damage to the enemy as possible. On December 17, the Germans continued to hammer away at the lines of the 112th. However, the Germans found that they were opposing not only the 112th, but field artillery, tank destroyers, and American aircraft. The 229th Field Artillery Battalion was laying down fire on the enemy only 150 yards away from their position. The 112th Cannon Company, and Company C of the 630th Tank Destroyer Battalion, by direct fire, disabled eighteen enemy tanks. The 2nd Battalion of the 112th with the aid of the 103rd Engineers repeatedly counter-attacked enemy penetrations.

On the night of December 17, the kitchen areas of the 112th were overrun by the enemy. Kitchen personnel took up rifles and recovered their positions. Under constant small arms and artillery fire from the Germans along their entire front, the 112th asked for and received permission to pull back to high ground. However, the Regiment had to remain close to the river to keep the Germans from crossing. The 112th withdrew under the cover of darkness behind the river leaving much damage behind: at least 18 tanks, 186 men captured, and three times as many killed or wounded.

Late on the night of December 18, Colonel Nelson asked for instructions for his Regiment and learned that General Cota wanted the Regiment to support the defenses of Bastogne. It was ordered to fight a delaying action along Weiswampack, Trois Vierges, and on toward Bastogne. However, the Regiment did not receive this message until two hours later. Colonel Nelson had knowledge of enemy tanks in Trois Vierges, and under the cover of heavy fog moved his Regiment to Huldange. General Cota then ordered the 112th to hold the line vicinity Weiswampack-Beiler. Nelson now had two different orders and would jeopardize his Regiment if he followed either one.

He sent a message through the artillery communication channels asking permission to join forces with the 106th Infantry Division, a command that had just arrived overseas. Permission was granted, and the regiment, attached to the 106th, helped in the defenses around St. Vith. The 112th did heavy damage to the Germans through the effective use of its artillery, while their own casualties were moderate.

THE ATTACK ON THE 110th INFANTRY REGIMENT: The defensive position manned by the 110th west of the Our River was seven miles in length. The 110th Infantry Regiment was the 28th Division's center Regiment under the command of Colonel Hurley Fuller. Opposing the 110th was General Hasso-Eccard von Manteuffel and his Fifth Panzer Army. The task of the Regiment was to block entryways across the Our River and to keep open the "Skyline Drive." The 110th had the First and Third Battalions deployed along a ridgeline in small strongpoints. Its Second Bat-

The artillery did all that they could to help the infantry's defense.

talion was in Division reserve.

At 0615 December 16 word of the approaching enemy reached Colonel Fuller at his headquarters in Clervaux. As this message was being received, the Germans were already hitting his lines with artillery and tanks. In the southern part of his position some attackers had crossed the "Skyline Drive." The 110th was hit hard, and its situation was critical. Colonel Fuller, surveying his position, contacted Division headquarters and asked the Division Commander for his reserves. General Cota at first refused to give up the reserve because of the situation all along his front. However, Cota soon realized that his center Regiment was in a critical situation and released the 2nd Battalion to Fuller. The situation for the 110th Regiment improved during the day, but on December 17, any plans made by Colonel Fuller the day before were shattered. His 2nd Battalion was hit hard by German Infantry. Everything and everyone possible which could be found was put to use, but the Germans could not be halted. Time was running out for the 110th; supplies and ammunition were low, but it was holding, whenever or wherever it could. Although outnumbered four to one, it refused to give ground.

The 26th Volks Grenadier Division crossed the Our River but was unable to take control of the Clerf River crossing. The 110th was taking a beating with casualties piling up all over its sector. Men ran out of ammunition, were surrounded, but fought their way out. At times fighting was hand-to-hand. The overwhelming weight of the German forces was being felt all along the lines, but Colonel Fuller was still attempting to slow up the march of the Germans on Clervaux. Tanks were ordered by the Commanding General to support the 110th, but the tanks could not hold the attackers, and by early evening the town of Clervaux was in German hands. The battle was not all one-sided, however. The German 2nd Panzer Division paid heavily, their timetable was thrown off, and their race to Bastogne was lost. This was all because of the heroic actions of the 110th Infantry Regiment at the Clerf River crossings. On December 18, the remnants of the 110th withdrew to the west. Casualty losses were a staggering 2,750 officers and men.

THE ATTACK AGAINST THE 109th INFANTRY REGIMENT: The 109th Infantry Regiment, commanded by Lieutenant Colonel James E. Rudder, was in a defensive position in the southern part of the Division zone on the Our River. At 5:45 on the morning of December 16, the 2nd and 3rd Battalions reported a heavy concentration of enemy artillery fire on their positions. Orders were sent out to all battalions to be on the alert and to prevent the enemy from crossing the Our. The 1st Battalion, in reserve in the town of Diekirch, was put on alert. At 0830 the 2nd and 3rd Battalions reported strong enemy patrols on the west bank of the River. Both battalions, using small arms and defensive artillery fire, held firmly to their positions, and inflicted heavy casualties on the Germans. However, some German patrols did infiltrate the lines of the 109th and continued on to Walsdorf.

In the first day of fighting the 109th managed to hold its position. On the morning of December 17 at 5:45 the Germans laid down a heavy artillery barrage that lasted for an hour. Following this barrage, stiff fighting ensued, but the enemy was pushed back. Colonel Rudder and his Regiment had thrown everything they had at the Germans. Regimental casualties were climbing, but the Germans had to fight hard for every inch gained. The delaying tactics of the 109th completely exposed the left flank of the German drive at its base by almost destroying the entire 915th and 916th Volks Grenadier Regiments and a major part of the 914th of the 352nd Volks Grenadier Division. The reserve battalion was sent north to help in the defense of Hoscheid, but to no avail. By evening the town had fallen to the Germans. The 109th at the close of the day was holding against heavy attacks, but the reserves were committed, and the enemy was solidly fixed between the 109th and the 110th.

All during the night the Germans continued with artillery fire on the positions of the 109th, which was running low on supplies and ammunition. Tanks had to be used to remove the wounded. On December 18 the Germans hit the Regiment with more artillery fire. Heavily infiltrated by the Germans, the 3rd Battalion ran out of ammunition. Some companies fought hand to hand but were finally captured. Some were shot in their positions, helpless to fight off the enemy. By late afternoon the 109th was in a critical position. Colonel Rudder asked and received permission to pull his Regiment back to high ground at Diekirch. The 109th, however, was no longer able to hold a defensive position. The Regiment had lost 500 officers and men in the three days fighting. However, it did hold the enemy at the Ettelebruck crossing and prevented the 352nd Volks Grenadier Division from concentrating in the southern part of the Sauer. The 109th Infantry Regiment fought hard. In three days of fighting the 109th had used 280,000 rounds of small arms ammunition, 5,000 rounds of mortar, 3,000 grenades, and 300 bazooka rounds. The Germans hit the Regiment at Diekirch and forced it out. The Division Commander then ordered the remnants of the Regiment to pull back with the 9th Armored Division. The 109th Regiment was to take up positions on high ground south of the Wark Creek. On December 19 the withdrawal was successfully completed, and the 109th held excellent commanding terrain in this defensive position.

The 28th Division headquarters at Wiltz was having its problems with the enemy. On the night of December 18 the Division Commander rounded up every able-bodied man he could find, including bandsman, cooks, paymasters, and engineers, to defend Wiltz, but to no avail. Lieutenant Colonel Daniel Strickler and his band of men delayed the Germans, and the 28th Division headquarters transferred to Sibret. The 28th Division can be very proud of its efforts in holding the Germans from dashing across the Our River. The 28th Division held off eight crack German Divisions. It had inflicted over 11,700 casualties. 28th Division losses amounted to 3,850 men killed and wounded, and 2,000 captured, but this was less than half of the German losses. The 28th Division did its job; it gave the Allied Armies the time they needed. War correspondent Morley Cassidy once remarked, "For four days from 16 December to 20 December, the main roads were denied to the enemy. The type of resistance offered by the Keystone troops was one of the greatest feats ever performed in the history of the American Army."

Colmar—January 18-February 5, 1945

In the early part of January 1945 the 28th Division was assigned the defense of the Meuse River, an area stretching from Givet, Belgium to Verdun, France. Toward the end of January a move was made to the south to Alsace. Here the 28th Division had the experience of serving in the French First Army, in the reduction of the "Colmar Pocket" and had the honor of capturing Colmar. The 110th, 112th, and the 109th Infantry Regiments, with green troops filling their ranks, went back into the lines. The 28th Division was now on the Colmar plain. Here again the Germans were waiting for the men of the 28th "Bloody Bucket" Division, which they had faced in the Hurtgen Forest and the Ardennes.

Now on this field of battle, using leaflets fired by artillery into the Division lines, the Germans boasted that they would beat the Division again. The Germans were welcoming the Division with open arms. On February 1, 1945, the 109th Infantry Regiment with the 112th following, and the 110th in reserve, attacked the Germans in the Colmar Pocket. It was no picnic for the Germans this time. The Division hit the Germans hard. Lines began to bend; then they broke. This was not the Hurtgen Forest or the Ardennes. This time the 28th Division would not be denied, it would not be swept aside, and it wasn't. The Germans in Colmar had met their match. The 28th Division beat the Germans soundly on the Colmar plain.

During the month of February the 28th returned to the American First Army and took up positions on the Olef River. In the weeks to come the 28th then crossed the Rhine River and took up positions in the "Ruhr Pocket" to stop any German forces driving to the south. It was here on V-E Day.

Early in July 1945 the 28th Division started redeployment to the United States, arriving in the United States in August 1945. After V-J day the 28th Division reassembled at Camp Shelby, Mississippi and was deactivated on December 13, 1945.

Troops relax on board ship enroute home.

The 28th Division had spent 196 days in combat in Europe. It had fought against forty-five of the ninety German Divisions on the Western Front in Europe. The casualty figures for the 28th were staggering: an incredible 175 percent of the 28th Division's authorized strength; 9,157 wounded, 2,599 missing, 1,901 killed, 2,247 captured. At the time of V-J day only a small fraction remained of its original members. A fine officer and gentlemen should close this amazing story, for no one knows it better; General Omar N. Bradley. In 1944 General Bradley wrote of the 28th:

> It has always been my hope and desire to lead the 28th Division into combat, and it was with a feeling of great disappointment that I had to say goodbye to the Division when I was sent to Africa. You are soundly trained, your morale is the best, you are well equipped as well as any Division in history has ever been equipped for battle. And I am sure that you will make a glorious name for yourselves.

The 28th Division did not disappoint him.

CHAPTER 11

Post World War II - Korean War

Richard Seiverling, LTC, PAARNG (Ret)

THE END OF World War II for the 28th Infantry Division did not spell the end of the Division's service. Within six months after deactivation, plans had been completed to return the "Keystone" Division to its normal peacetime role as an integral part of the Pennsylvania National Guard.

On June 1, 1946, the Adjutant General of Pennsylvania authorized reorganization of Division Headquarters at Harrisburg. Immediately the wheels were put in motion to establish the entire unit across the Commonwealth.

Under the leadership of Major General Edward J. Stackpole, Division Commander, and Brigadier General Daniel B. Strickler, Assistant Division Commander, the new Division grew rapidly until November 1946, when Federal recognition was granted to the 28th Division after a Second Army inspection.

Most of the historic old units of the 28th were located again in their original areas of the state. The 109th Infantry Regiment, commanded by Colonel Thomas L. Hoban, was spread over northeast Pennsylvania with headquarters at Scranton. Colonel James Hawkins reorganized the 110th Infantry Regiment in southwest Pennsylvania with headquarters at Washington. In the northwestern section of the state, the 112th Infantry Regiment had its headquarters at Erie under the command of Colonel Kenneth C. Momeyer. Division artillery units were scattered at Philadelphia, Pittsburgh, Kingston, New Castle and Lancaster under the command of Brigadier General Brenton G. Wallace.

The Division's service units were spread throughout other areas of the state. The 103d Engineer Combat Battalion established headquarters in Philadelphia under Lieutenant Colonel Wilbur E. Duryea. At Lancaster Lieutenant Colonel James Appel organized the 103d Medical Battalion. The 28th Military Police Company was headquartered at Harrisburg, the 728th Ordnance Maintenance Company at Chambersburg, the 28th Quartermaster Company, the 28th Division Band at Altoona, and the 28th Reconnaissance Company at Philadelphia. The latter unit was the designation of the famed First Philadelphia City Troop, with a record of service since 1747.

During 1946 and the early part of 1947 the units organized themselves and began extensive recruiting drives. Training was conducted weekly at the home armories under a six-year training program established by the National Guard Bureau. This program was designed to build a Division

Organizational colors and standards of the 28th Infantry Division, called the "Bloody Bucket" Division by the Germans, are returned home during special ceremonies at the State Capitol, Harrisburg, on Armistice Day, November 11, 1946, after establishing a proud record of federal service during World War II. The Honorable Edward Martin (center), then Pennsylvania's Governor, participated in this memorable ceremony.

that could be put to immediate service at the completion of its training.

Not all the Guardsmen's time was devoted exclusively to training activities, however, for there were also the traditional manifold social functions: participation in hometown parades on National holidays, performance of various types of humanitarian, emergency and disaster-relief services in the local communities (fires, floods, blizzards, hurricanes, traffic control, etc.), and a variety of other state-related missions.

The customary two-week summer field training encampments were renewed in 1947 at Indiantown Gap Military Reservation (redesignated later as Fort Indiantown Gap) near Harrisburg. Thus, the "Gap" was the scene of the first postwar field training of the ground forces of the Pennsylvania National Guard (divisional and nondivisional units) August 9-23, 1947.

Under existing policies of the War Department only those units were authorized to participate in field training which had been granted Federal recognition by April 15 or which had qualified and filed application for Federal recognition prior to that date.

Annual Governor's Day Reviews and Parades were resumed at summer camp, usually well attended by visiting military and civilian dignitaries, and Guardsmen's parents, wives, relatives and friends from their hometown communities.

On June 1, 1947, General Stackpole retired as the Division Commander, and the command was given to Brigadier General Strickler, who was promoted to the rank of Major General on December 24, 1947. A Lancaster attorney, he was the only veteran in the 28th who had seen combat as a Division officer in both World Wars. Shortly thereafter Colonel Hoban of the 109th Infantry Regiment was named Assistant Division Commander, and he was succeeded at Scranton by Colonel Fred R. Evans.

In 1948 another reorganization of the 28th Division was initiated under a new Table of Organization authorized by the National Guard Bureau. Some of the major organizational changes included the addition of a heavy tank battalion and an antiaircraft artillery automatic weapons battalion, the latter being organic in Division Artillery. A replacement company was also added. An additional combat company was added to the Engineer Battalion. The Medical Battalion was reduced to one ambulance company and one clearing company. The Mechanized Reconnaissance Troop was reorganized as a Reconnaissance Company.

In the infantry regiments, the former cannon companies and antitank companies were eliminated. Each infantry regiment included, in lieu of those units, a heavy tank company and a heavy mortar company equipped with the 4.2-inch chemical mortar. The former medical detachments were reorganized as medical companies.

Added to the firepower and striking power of the Division were 144 medium and light tanks, twenty-four 4.2-inch chemical mortars and thirty-two, twin .40 mm anti-aircraft guns. Additionally, the firing batteries of the field artillery were increased to six pieces, thus increasing their firepower by fifty percent.

Full war strength of the new Division authorized 1,007 officers and 17,797 enlisted men. The ratio of combat troops to service troops was about eighty percent combat to 20 percent service. Authorized peacetime strength of National Guard Divisions was set at 926 officers and 12,882 enlisted men.

Training was conducted with relative ease, since thousands of seasoned World War II veterans of all ranks and grades—representing all branches of service—had joined the Division and performed as experienced cadre for the untrained recruits.

Long before the unification of the Armed Forces became a reality, the 28th had enlisted former officers and enlisted men from the Air Force, the Navy, the Marine Corps, and the Coast Guard. Former bomber pilots found themselves commanding rifle companies, and former seamen were now plotting the course of the Army's "tin cans," the heavy tank. While Congress debated and finally approved unification with cautious trepidation, the 28th Division made unification a fact and proved it could work well.

By 1950 the "Keystone" Division fielded a sound core of troops, adequately trained in organizational and operational procedures to function well as a unified team. They lacked only combat training in order to make them ready to meet any assignment. Although few believed that any major assignment was in the offing, the contrary was true.

Like the black cloud of a tornado, the war in Korea entered the scene in June 1950, and within weeks it became apparent that some mobilization of United States Reserve and/or National Guard units would be required—esecially since the Selective Service System had no pool of "draftees" available at this time to bolster the Armed Forces.

In July President Harry S Truman told the nation that he had authorized the call of four National Guard Divisions into active Federal service. This announcement was not unexpected, but until the units were actually identified, Guardsmen all over the country waited and hoped.

The 28th Division was conducting its annual summer field encampment at Indiantown Gap on July 29, 1950. No man in the Division could escape the feeling that his might be one of the outfits called into national service. The odds were, of course, that with more than thirty National Guard Divisions in existence across the country the 28th would not be mobilized. However, the Pentagon found the caliber of the 28th such as to make it a prime prospect. The President, himself a veteran of more than twenty years of duty as a citizen soldier, did not want just any six Divisions; he wanted the best. One of those was the 28th.

On August 2, 1950, the Army's alert order came to the Division. General Strickler immediately called a meeting of his staff and organization commanders to tell them the news. The encampment was terminated a week early to allow the men approximately thirty days at home before entering active military service. The "Keystone" Division was slated to go to Camp Atterbury, Indiana, for combat training.

On Sunday, August 6, the traditional Governor's Day Review was held, this one with a new meaning. Thousands of relatives and friends poured into camp to see their Guardsmen in final review before marching off to active duty. The presence of General Mark W. Clark, Chief of the Army Field Forces, on the reviewing stand with Governor James H. Duff and a host of other high-ranking dignitaries, served as a grave reminder that the gala reviews of previous years would be gone for a time.

The 28th Division broke camp early the next morning and started toward home. Uppermost in everyone's mind was the magnitude of personal problems to be solved. Countless details would have to be attended to before they could leave their families, homes and civilian jobs to embark on a new way of life. The date for induction was set as September 5, 1950. In four short weeks, the equipment and supplies for the entire Division, along with the men themselves, had to be ready.

To help speed up the process armory drills were increased from one to three a week to sharpen skills as rapidly as possible. Three major tasks had to be accomplished before induction. First, new physical examinations were required for every man. Then, new personnel records had to be completed. Finally, an exhaustive inventory of the supplies and equipment available to the entire Division had to be performed and turned over to Regular Army officials.

Mountains of paper work faced the Division at every turn. The aid of Regular Army instructors attached to the Division and others from the Pennsylvania Military District was procured to help the various headquarters units complete the job. Near the end of August ten percent of the officers and men of each unit were put on active duty status for the last ten days prior to the Division's activation.

On the morning of September 5, 1950, a formation of Division staff officers and Headquarters enlisted men at the Harrisburg armory listened to the reading of a brief order which automatically made them and all other Division troops in the State part of the United States Army.

A total of 8,856 enlisted men, 134 warrant officers, and 726 officers reported to their armories for duty that morning. They comprised the nucleus of a proud Division now entering its third tour of Federal service in half a century. Many units had changed commands in the years of reorganization, but the locations of units remained the same.

These were the units and their commanders when the 28th Division was inducted into Federal Service: Division Commander, Major General Daniel B. Strickler; Assistant Division Commander, Brigadier General Thomas L. Hoban; 109th Infantry Regiment, Colonel Fred R. Evans; 110th Infantry Regiment, Colonel Henry K. Fluck; 112th Infantry Regiment, Lieutenant Colonel Adam J. Dreibelbies; 28th Division Artillery, Colonel W. Willis Gilmore; 107th Field Artillery Battalion, Major Robert M. O'Donnell; 108th Field Artillery Battalion, Lieutenant Colonel Joseph L. Minter; 109th Field Artillery Battalion, Lieutenant Colonel Frank Townend; 229th Field Artillery Battalion, Major Alfred L. Barnes; 899th Antiaircraft Artillery Battalion, Lieutenant Colonel Harold E. Rochow; 103d Engineer Combat Battalion, Lieutenant Colonel Peter J. Broullire; 103d Medical Battalion, Lieutenant Colonel Gilbert N. Clime; 628th Heavy Tank Battalion, Lieutenant Colonel James A. Zimmerman; Headquarters Company, Captain Walter C. Moyer; 28th Military Police Company, Captain Donald E. Machamer; 728th Ordnance Maintenance Company, Captain Charles A. Linthurst; 28th Quartermaster Company, Captain Joseph L. Dumm; 28th Reconnaissance Company, Captain William S. Stokes, Jr.; 28th Signal Company, Captain Henry W. Cooper; and 28th Division Band, Chief Warrant Officer Lewis Lastort.

Guiding and directing the efforts of these units were members of the Division Headquarters staff, under the supervision of Colonel James G. Mackey, Chief of Staff; Lieutenant Colonel Adelbert A. Arter, G-1 (Personnel); Major Alvan Markle III, G-2 (Intelligence); Lieutenant Colonel Richard A. Dana, G-3 (Plans and training); Lieutenant Colonel George I. Macleod, G-4 (Supply); Lieutenant Colonel Albert G. Branyan, Adjutant General; Colonel William A. Boyson, Division Surgeon; Lieutenant Colonel Earl M. Honaman, Chaplain; Lieutenant Colonel William H. Hays, Inspector General; Lieutenant Colonel William M. Ruddock, Judge Advocate General; Major Alfred L. Beck, Chemical Officer; Major Carl F. Mauger, Light Aviation Officer; Captain Walter R. Ernst, Special Service Officer; Lieutenant Colonel Harry G. Swartz, Provost Marshal; First Lieutenant John F. Finley, Finance Officer; Lieutenant Colonel Albert J. Youndt, Quartermaster; Major Robert W. Cronenweth, Signal Officer; and Second Lieutenant Richard F. Seiverling, Public Information Officer.

The main body of the Division was scheduled to move out from Pennsylvania in two echelons. Troop trains of approximately 500 each were formed to transport the soldiers on September 10, 11 and 12. Simultaneously all wheeled vehicles of the units gathered at the four corners of the State to start their overland journey to Camp Atterbury in convoys.

Less than twelve hours after the first troop train left tragedy struck and shocked the Commonwealth into a state of mourning. Early in the dawn of Monday, September 11, a train carrying soldiers from the northeast Pennsylvania area stopped

HEADQUARTERS 28TH INFANTRY DIVISION

Camp Atterbury, Indiana

GENERAL ORDERS
NUMBER 2

19 September 1950

The death of the following members of the 109th Field Artillery Battalion, 28th Infantry Division which occurred near Coshocton, Ohio on 11 September 1950 is announced to the command with deep regret:

CAPTAIN ARTHUR J. THOMAS 012955459 Service Battery
WARRANT OFFICER JUNIOR GRADE JAMES F. McGINLEY ASN Unknown, Service Battery
WARRANT OFFICER JUNIOR GRADE WILLIAM M. WELLINGTON ASN Unknown, Battery B
Sergeant John W. Cox NG20316877 Battery B
Sergeant William C. Edwards NG33358869 Battery B
Sergeant Lester J. Kuehn NG23815200 Battery B
Sergeant Bernard S. Okrasinski NG23815813 Service Battery
Sergeant Gilbert B. Wharton NG23814614 Battery B
Corporal Carl W. Armbruster NG43043302 Service Battery
Corporal John L. Barna NG23815817 Service Battery
Corporal Joseph E. Fletcher NG23815870 Service Battery
Corporal Larry L. Luzenski NG23814667 Battery B
Corporal Thomas M. Ostrazewski NG23815869 Service Battery
Private first class Leonard Balonis NG23814690 Battery B
Private first class Edward W. Gallagher ER13354678 Service Battery
Private first class Harold Handlos NG23814703 Battery B
Private first class Clyde P. Harding NG57201273 Battery B
Private first class Martin Hornlein NG23814685 Battery B
Private first class Ronald J. Jackson NG23814695 Battery B
Private first class Raymond Pudlowski NG23814683 Battery B
Private first class Edmund Zabicki NG33108177 Battery B
Private first class Donald C. Zieker NG23814689 Battery B
Private William R. Disbrow NG23815881 Service Battery
Private Wallace R. Ludwig NG23815878 Service Battery
Private William F. Tierney NG23815872 Service Battery
Recruit Eugene Carr NG23814737 Battery B
Recruit William J. Dougherty NG23814736 Battery B
Recruit Hugh L. Fargus NG23814735 Battery B
Recruit Frank C. Martinez NG23814729 Battery B
Recruit Charles Norton NG23814739 Battery B
Recruit Richard A. Royer NG23814733 Battery B
Recruit William E. Sober NG23814732 Battery B
Recruit Thomas W. Wallace NG23815880 Service Battery

BY COMMAND OF MAJOR GENERAL STRICKLER:

OFFICIAL:

Albert G. Branyan.
ALBERT G. BRANYAN
Lieutenant Colonel, AGC
Adjutant General
DISTRIBUTION "C"

JAMES G. MACKEY
Chief of Staff
Colonel, GSC

General Orders Number 2, dated 19 September 1950, listing 33 Guardsmen killed in train accident at Coshocton, Ohio, on September 11th.

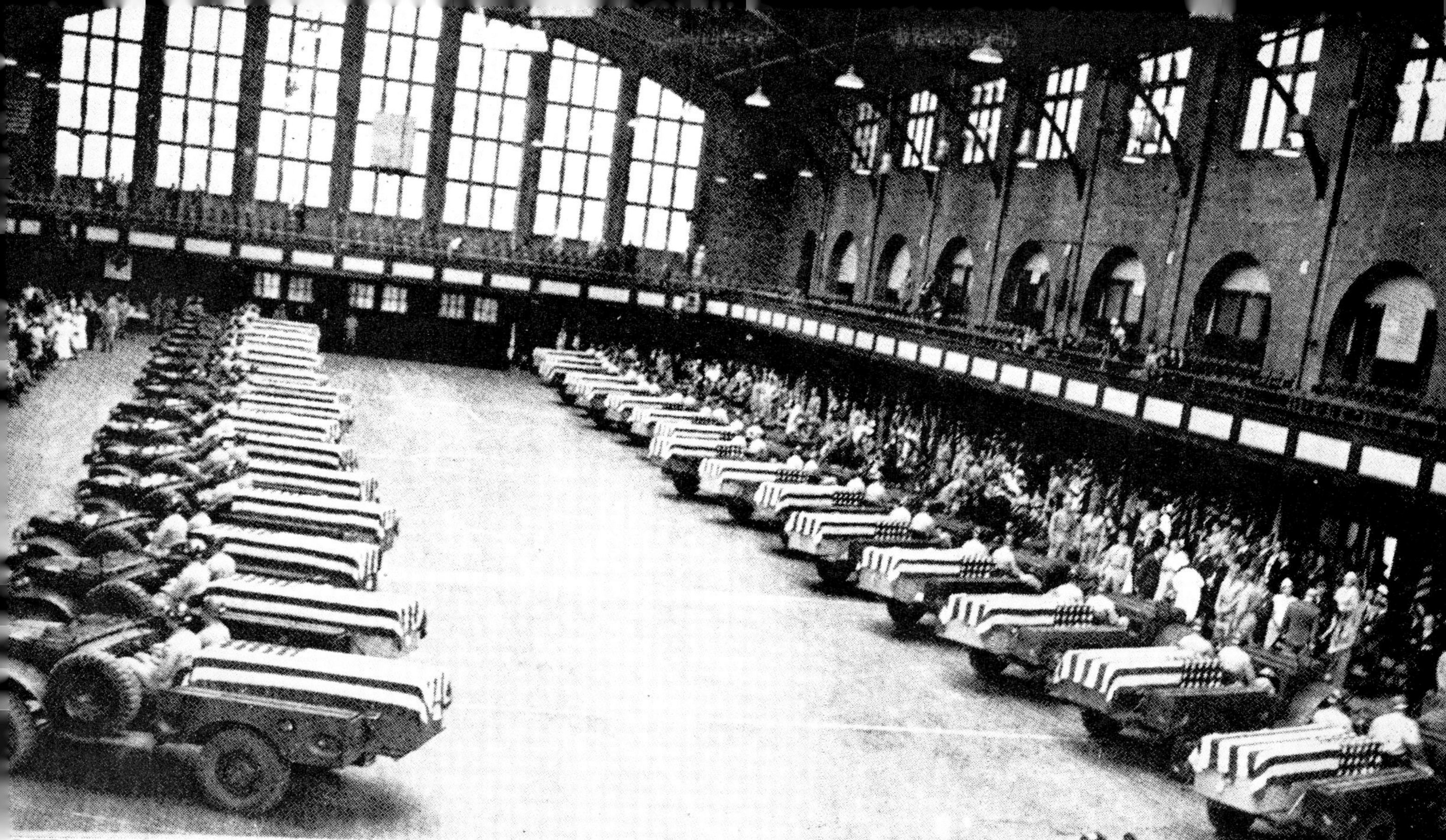

Flag-draped caskets of the 33 Guardsmen killed in the train accident pictured on the drill floor of the Kingston Armory. Adjutant General Frank A. Weber made available a 24-hour guard of honor to remain with each of the deceased until buried, as well as providing graveside firing squads and full military honors.

along the tracks near Coshocton, Ohio, while brakemen inspected a broken air hose. Suddenly, despite warning flares that had been posted along the tracks behind the last car, a passenger train crashed into the rear of the troop train.

Thirty-three Guardsmen were killed and hundreds injured. Hardly a man aboard the train was not bruised or cut by shattered glass. The dead were all members of Service and "B" Batteries of the 109th Field Artillery Battalion. At Wilkes-Barre and Kingston, home stations of the deceased, a gloom descended over the grief-stricken communities and lasted for ten days until the last of the soldiers were buried.

Five days later, on Saturday, September 16, General Strickler called an assembly of the entire Division on the Camp Atterbury parade ground. The roll of the dead was read as the stunned soldiers listened with heads bowed in tribute. Meanwhile, a thirty-six-man honor guard from the 109th Field Artillery Battalion accompanied the dead back for burial in Pennsylvania via a special funeral train.

Amidst the deeply-felt sorrow associated with the train accident, a six weeks' pre-cycle basic training period began promptly on Monday morning, September 18, while the Division received newly-inducted selectees to fill its ranks to full war strength. Several schools were established at Camp Atterbury, and leadership courses were given to all officers and non-commissioned officers. Hundreds of officers and enlisted men were also sent to numerous Army service schools at various installations for specialist training. The assignment of ten Regular Army teams to the Division during the pre-cycle training phase aided the 28th immensely.

The first "fillers" arrived on October 3, 1950, as the goal of bringing the Division to full strength was begun. Recruits arrived from induction processing centers in trainloads of several hundred a day. Most of them were from the East and Midwest, but, before the Division reached full strength about a month later, every state in the nation—and several United States possessions—were represented.

Army dignitaries began visiting the 28th Division to inspect its training activities, the first of whom was Major General Withers A. Burress, Commandant of the Infantry School at Fort Benning, Georgia. Next, Lieutenant General Stephen J. Chamberlin, Commanding General of Fifth Army (the higher headquarters of the Division), and several members of his staff, inspected the Division and camp facilities. The most distinguished visitor at this time was General J. Lawton Collins, Army Chief of Staff.

In late October the "Keystone" Division received its first major inspection by a seven-man team from Army Field Forces Headquarters. Led by Major General J. W. (Iron Mike) O'Daniel, the group observed all phases of training and operations. Before leaving, General O'Daniel remarked that the 28th "should be one of the best divisions in the U. S. Army" after completing its training.

Several significant changes had occurred on the Division staff. Lieutenant Colonel Oliver W. Robbins took over as G-1 and his predecessor, Lieutenant Colonel Arter, retired from active duty. Colonel Boyson, Division Surgeon, was transferred and he was succeeded by Lieutenant Colonel Albert J. Blair. Also, Lieutenant Colonel Irvin W. Gerth was assigned as Division Finance Officer, and Lieutenant Colonel Chester N. Rees was assigned as Ordnance Officer. The position of Headquarters Commandant was assumed by Major Edward A. Hitchin when the former commandant, Major E. I. Plaskow, was transferred to the 112th Infantry Regiment.

Combat training for the 28th Division soldiers was initiated on November 6, 1950. A twenty-eight-week cycle training program, it was specially planned by General Mark W. Clark, Chief of the Army Field Forces, for the training of National Guard Divisions.

The most intensive training cycle ever implemented to prepare full divisions for combat, the program comprised three phases. The first eleven weeks were devoted to individual training in combat skills, a program similar to basic training. The next thirteen weeks involved unit training at the company, battery and battalion levels, including coordination of auxiliary forces such as tanks, engineers, signal corps, artillery, and other forces in combined unit exercises. The final four weeks stressed field exercises by regimental combat teams and the entire Division.

Three days after the combat phase of training had commenced, General Clark paid the Division a visit while on an inspection tour of all newly-activated National Guard units. He had high praise for the "Keystone" Division, calling it "a worthy successor to the magnificent 28th Divisions that have gone on before it." General Clark particularly cited the many veterans of former combat service in the Division.

Despite one of the hardest winters in Indiana history which set records for low temperatures and snowfalls almost daily, Camp Atterbury's firing ranges were opened to full use for the first time since 1943. Beginning with small arms firing for qualification and familiarization, the sound of heavier weapons—machine guns, mortars, recoilless rifles and heavy artillery—was soon added to the thunder that echoed at all hours from the range areas. Meanwhile, Brigadier General Guy O. Kurtz, a veteran of thirty years' service in the Regular Army, was assigned as the 28th Division's Artillery Commander.

In January 1951 the Division established a system of proficiency tests to determine how well the soldiers had learned their "ABC's" during their basic training phase. Employing a battery of ninety testing stations, the men were quizzed orally on the basic subjects and demonstrated what they had learned during the previous eleven weeks. This system incorporated many original features in its testing procedures and drew favorable attention from Fifth Army officials and others. Results of these tests proved highly successful and indicated that the 28th Division soldiers had learned their basic skills well.

To mark the end of individual training and the start of unit training, a review of the 28th Division massed at full war-strength was held on Saturday, January 27, 1951. On the reviewing stand were seven general officers including the reviewing officer, Major General Albert C. Smith, Deputy Commander of Fifth Army. Many other military and civilian dignitaries, as well as several thousand visitors, witnessed the impressive review held on Camp Atterbury's airfield. Weather

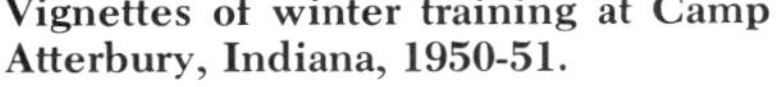

Vignettes of winter training at Camp Atterbury, Indiana, 1950-51.

conditions were ideal because, for the first time in weeks, the temperature rose into the fifties on that day.

As a highlight of the event, a hand-made American flag sewn by a Frenchwoman in Sarron, France—on the occasion of that town's liberation by elements of the 110th Infantry Regiment during World War II—was presented to a Regimental Commander, Colonel Henry K. Fluck, by M. Jean J. Viala, French Consulate-General from Chicago.

An aerial photograph taken of the 28th Division, because this review was the first held by a division at full war-strength in the United States since World War II, drew national and international attention and was reprinted in scores of major newspapers and magazines throughout the nation and numerous European countries as well.

News reports of the setbacks in Korea during weather conditions worse than those experienced at Camp Atterbury served to impress upon the "Keystone" Division soldiers the advantage of winter training.

The troops continued to make friends with the "Hoosiers" in Indiana and apparently were winning many friends in return. This relationship prompted the *Indianapolis Star* to print the following editorial about the men of the 28th:

> Members of Pennsylvania's Infantry have been in training at Camp Atterbury long enough now for the Indianapolis citizens to form some opinions of their new neighbors. We're happy to report that the opinions we hear are good.
>
> As a "soldiers' town" of long standing, Indianapolis tends to judge men in uniform by the way they behave away from Camp. The judgment of the men of the 28th is that they are soldierly gentlemen. On pass most of them appear to be alert, neat, courteous young men and not given to displays of "toughness" which some servicemen in training seem to think marks them as future Medal of Honor winners.
>
> As for real toughness, we're confident the 28th will be as tough an outfit as an enemy ever faced if it has to go into combat. We're proud to have Major General Daniel B. Strickler and his men as neighbors.

Christmas season arrived, and spirits of the 28th Division men picked up markedly. In order not to interrupt training half the Division got three-day

Impressive aerial view of the 28th Infantry Division, depicting the first full war-strength mass divisional review held in the United States after World War II. Photographed on Camp Atterbury's airfield January 27, 1951.

Muddy and tired, but still smiling, 28th Infantry Division soldiers march back to their barracks and a shower after completing the infiltration course at Camp Atterbury. Bursting charges, overhead live machine gun fire and inches of sticky mud made this one of the most realistic phase of the "Keystone" Division's training program.

passes over Christmas and the other half over New Year's. Not only was the prospect of a few days at home helpful, but also nearly every unit entered into Christmas projects and parties which helped take their minds momentarily off training and the weather. Many units staged Christmas parties for local underprivileged children, and others raised funds to help needy families in nearby Indiana towns. For the men remaining at Camp Atterbury over the weekends, there were special holiday dinners and numerous recreation activities planned, and work details were kept at a minimum.

Upon completion of the proficiency testing the scope of training turned swiftly to battle indoctrination, conducted under all types of weather situations including sub-zero temperatures, snow, rain and mud, designed to prepare the soldiers physically and mentally for combat by experiencing reproductions of actual battle conditions.

Battle indoctrination training was divided into four parts: close combat, infiltration course, combat village, and overhead artillery phases. Most of this training was conducted under live machine gun or artillery fire and incorporated such hazards as crawling through barbed wire, the wearing of full combat pack, and negotiating the courses in darkness.

On February 2, 1951, the Division was rocked by an unexpected levy involving several thousand of its trained fillers. The need for these fillers elsewhere resulted from the Army's recently-instituted policy of rotating soldiers after completing six months' combat service in Korea.

Before these soldiers could be reassigned, however, their personnel records had to be accurately updated, immunization shots had to be administered, special training had to be given them, and, prior to being shipped to embarkation points, they were all given seven days' leave. Hardly had this Herculean task been completed when a second levy was imposed in March, and fillers were again taken from all units of the Division.

The special efforts made by the Division in training and processing over 6,000 replacements for overseas shipment in record time and peak condition earned rewarding commendations from the Commanding Officer at Fort Lawton, Washington, and the Commanding Generals of Fifth Army, Sixth Army and Camp Atterbury. Although these levies had created an unexpected burden on the Division, training was conducted for those remaining practically without interruption after a re-arrangement of schedules.

Several additional training features were implemented as Spring progressed. "Aggressor" tactics and training in the "Maneuver Enemy's" techniques were taught by a special team of instructors from Fort Riley, Kansas. The team demonstrated work in psychological warfare, intelligence plan, aggressor-type equipment, and the umpire control system as applied to platoon level operations.

A provisional ranger company, organized from volunteers in the Division, was sent to Fort Benning, Georgia, for paratrooper and ranger training. The 899th Antiaircraft Artillery Battalion spent six weeks firing its guns on the ranges at Camp Oro Grande, New Mexico. Elements of the 628th Heavy Tank Battalion went to Fort Knox, Kentucky, for advanced armor training.

The 28th Military Police Company sent its men into Indianapolis and other Indiana towns to learn traffic control first-hand by working with regular policemen on city streets. The 103d Medical Battalion established a field clearing station and practiced working on "battle" casualties. The 28th Quartermaster Company conducted ration breakdown under field conditions.

During this period fillers arrived daily to replace the vacancies created by the levies. The new men were assigned to provisional training battalions to complete basic training, while veteran troops continued with the training cycle already in progress.

On April 12, 1951, the Division again reached full war strength as the last group of selectees arrived. The "Keystone" Division was again fast becoming a crack, fully-trained command 18,000 men strong.

On July 31, 1951, the *Indianapolis Times* published a special 28th Division Edition containing numerous pictures, news and feature stories about the "Keystone" soldiers' training. In addition to a front-page message from the 28th's Commanding General, letters from Indiana's Governor Henry F. Schricker and Indianapolis Mayor Phillip L. Bayt were reprinted.

Said Governor Schricker:

> It has been a proud and happy privilege to be the Host State to the 28th Division for the time of its extended training period at Camp Atterbury during the past year. A finer group of officers and men never set foot on Indiana soil, and their coming departure has brought a note of sorrow and sincere regret to a legion of Hoosier friends.
>
> General Daniel B. Strickler has every reason to feel proud of the men in his Division and especially of the excellent conduct record they maintained throughout their stay in Indiana. We shall miss them greatly when they are gone, but the best wishes of all Hoosier friends will accompany them throughout the days and months of their future service.

Mayor Bayt wrote:

> I would like to take this opportunity to wish you every success on your maneuvers in the southlands.
>
> In the year the Pennsylvania 28th Division has been at Atterbury, Indianapolis, as the largest host city in this locality, has been proud and honored to have been of service to you, your officers and men. I know personally that you have graciously taken time to participate in local civic functions, for which the City of Indianapolis is grateful.
>
> I wish to commend the troops of the 28th Division for the exceptionally gentlemanly manner in which they conducted themselves while on pass here.
>
> Especially do I want to commend your military police unit for the job they did directing downtown traffic. Their efficiency brought many favorable comments. Wishing you a speedy return to our Hoosier State... .

On October 26, 1951, the *Indianapolis Times* also published a 28th Division Souvenir Edition featuring a huge red Keystone imprinted on the center of the front page. This special edition was released just prior to the 28th Division's departure from Camp Atterbury.

In August 1951 Brigadier General John G. Van

Thousands of relatives and friends see the "Keystone" Division soldiers off on their voyage to Europe on November 12, 1951. In the background looms the 16,000-ton troopship *U.S.S. General Butner*, docked at the Philadelphia Navy Yard.

Houten, former Commandant of the Ranger Training Command at Fort Benning, Georgia, was assigned as the 28th's Assistant Division Commander, replacing Brigadier General Thomas L. Hoban. Also, Colonel Harry D. McHugh, former Director of the Reserve Officer Training Corps at Indiana University, was assigned Commanding Officer of the 109th Infantry Regiment, replacing Colonel Fred R. Evans.

Previously announced plans had called for the 28th Division to participate in Exercise "Southern Pine," largest joint Army-Air Force war maneuvers held in this country since World War II, scheduled for late June at Fort Bragg, North Carolina. Because the two levies had reduced the number of trained troops in the Division, however, "Southern Pine" maneuvers were postponed until August.

A fitting climax to the Division's individual and unit training was realized with its exemplary performance in Exercise "Southern Pine," officially conducted from August 13-27, 1951. Advance elements of the 28th had arrived at Fort Bragg as early as August 1 to prepare for the reception of personnel and equipment moving by both rail and motor convoys.

Joining the 43d Infantry Division (National Guard Division activated from the New England area) and other units in full-scale war maneuvers, this massive field exercise was designed to test the effectiveness of the participating units' organization, planning and battle worthiness.

Amidst temperatures soaring above the 100 mark, "Aggressor" troops attempted to crack the 28th Division's defenses for two exhausting weeks. At the end of the war maneuvers, the 28th and other defending units counterattacked and convincingly destroyed the "Aggressor" forces. In company with the Air Force and paratroopers of the 82nd Airborne Division, the National Guard Divisions—the 28th and 43d—had learned many valuable lessons should they be required to apply them to the battlefield.

At the completion of "Southern Pine," the "Keystone" Division had proven itself ready for any assignment that might be given it. High-ranking military and civilian observers gave the 28th a commendable rating in combat effectiveness.

The possibility for an overseas assignment of the 28th Division to join the North Atlantic Treaty Organization (NATO) command had been discussed for several months. By the time "Southern Pine" maneuvers were completed, this possibility had become a certainty.

After its return to Camp Atterbury from Exercise "Southern Pine," the Division began extensive preparations for overseas movement (POM). Mandatory training prerequisites still had to be fulfilled, including indoctrination lectures in military justice, chemical, biological and radiological warfare, and orientations concerning the European Command.

Upon completion of the various phases of POM, a full division review was conducted on October 26, 1951. On November 5 a division rally was held during which time the 28th's Commanding General addressed the troops concerning their movement to Europe and the mission assigned to the "Keystone" Division.

On November 12 about 3,000 soldiers of the 28th Division paraded in the city of Philadelphia before an estimated audience of 200,000 spectators. On the reviewing stand at Independence Hall were United States Senator James H. Duff, Philadelphia Mayor Bernard Samuel, Lieutenant General Edward H. Brooks, Commanding General of the Second Army, Major General Daniel B. Strickler, 28th Division Commander, and many other dignitaries including three former commanders of the 28th Division: Major Generals Norman D. Cota, William G. Price and Edward J. Stackpole, all retired.

At the conclusion of the parade, the first increment of 28th troops boarded the U. S. S. *General Butner*, docked at the Philadelphia Navy Yard. The ship sailed at 5:30 p.m., transporting the first units of the Division to their new assignments in Germany. Three additional increments sailed from Hampton Roads Port of Embarkation, Hampton Roads, Virginia, between November 13-26, 1951.

The first ship carrying the first increment of 28th Division soldiers, the U. S. S. *General Butner*, arrived at Bremerhaven, Germany, on November 21, 1951. After Thanksgiving dinner on the ship, the troops disembarked to be greeted by General of the Armies Dwight D. Eisenhower, Supreme Commander, Allied Powers in Europe (SHAPE), and other general officers at the scene.

After boarding trains 28th Division soldiers were transported to Camp Y-79 at Sandhofen, near Mannheim. This staging area was the site of the reunion of units with their equipment. Shipments from the United States were late in arriving, however, resulting in a much longer stay in the rustic and muddy tent city than had been anticipated.

Division Headquarters opened at Goeppingen on November 26, 1951, and Camp Y-79 had been completely cleared by December 17. *Kasernes*, former German Army barracks damaged during World War II, were completely renovated to house elements of the NATO Army, including units of the 28th in Southern Germany. The lavish *kaserne*, three and four stories high of steel and masonry construction, provided better living conditions and facilities than the drafty wooden barracks at Camp Atterbury.

Six area commands were formed by the 28th Division, each under the control of the major unit within the area. The 109th Infantry Regiment and its attached 109th Field Artillery Battalion comprised the Gablingen Area Command. The 110th Infantry Regiment, 107th Field Artillery Battalion and 728th Ordnance Maintenance Company were included in the Ulm Area Command. The Heilbronn Area Command consisted of the 112th Infantry Regiment and 229th Field Artillery Battalion. At Ellwangen were the 103d Engineer Combat Battalion and 103d Medical Battalion. The Leipheim Area Command housed the 628th Medium Tank Battalion and 28th Reconnaissance Company. The Geoppingen Area Command included 28th Division Headquarters and Headquarters Company, 28th Division Artillery Headquarters, 28th Signal Company, 28th Replacement Company, 28th Quartermaster Company, 28th Military Police Company, 28th CIC Detachment and 28th Division Band. The 108th Field Artillery Battalion and 899th Antiaircraft Artillery Automatic Weapons Battalion (SP) were also part of the Geoppingen Area Command, but were located in Schwabisch Gmeund and Nellingen, respectively.

The major problem facing the Division was that of becoming combat ready. Its primary mission

The first elements of the 28th Division arrive in Germany and line up on the dock at Bremerhaven on November 21, 1951.

Arrival at muddy Camp Y-79 at Sandhofen, near Manheim, Germany, on November 22, 1951.

MODERN MAP SHOWING AREA COMMANDS OF 28TH INFANTRY DIVISION IN SOUTHERN GERMANY DURING SERVICE WITH NATO (1951-1954)

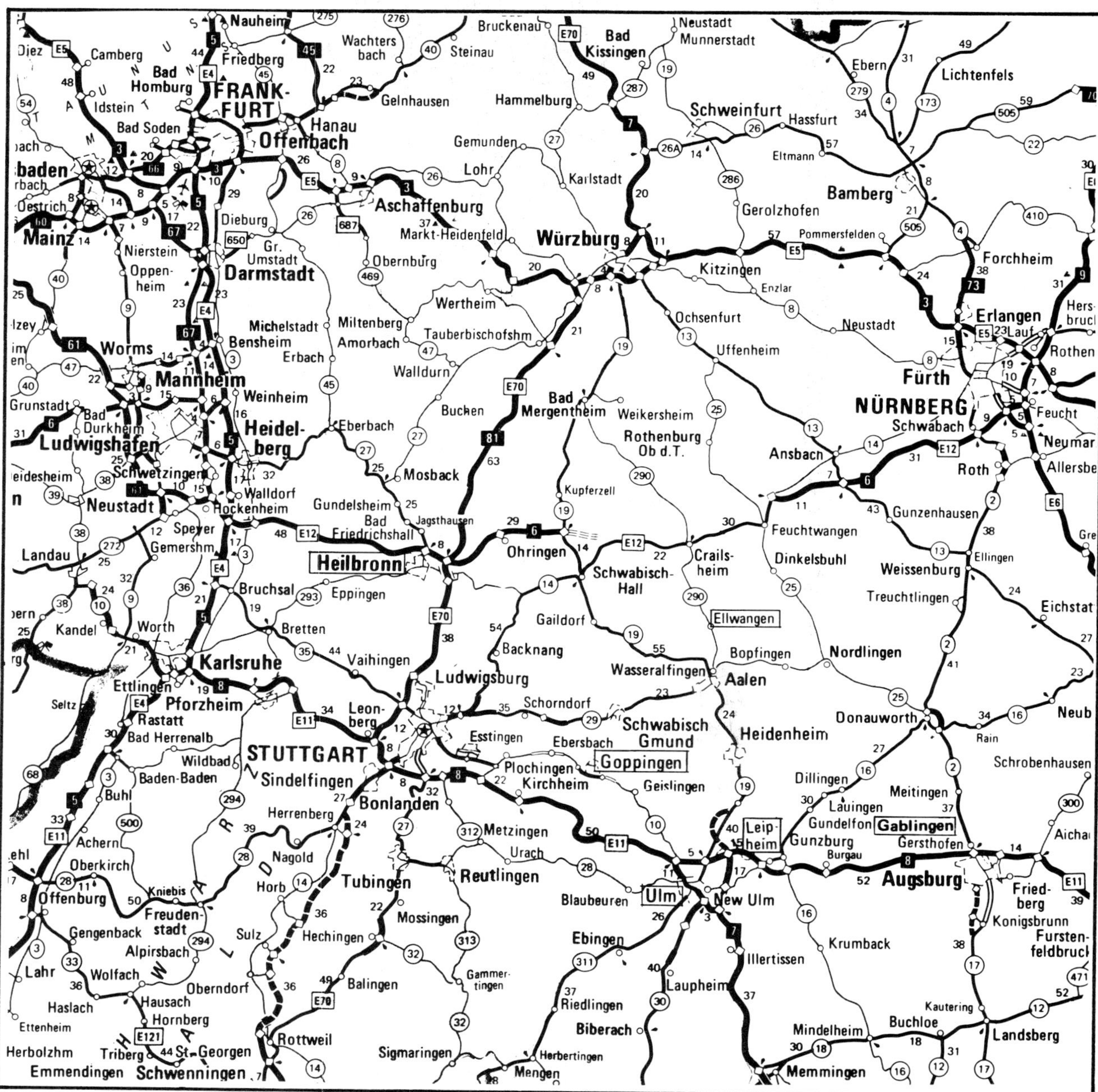

was to close with the enemy by fire and maneuver in order to destroy or capture him, or to repel an enemy assault by fire and close combat.

The specific mission of the Division was three fold: first, it was to engage in a training program designed to improve its capabilities to fulfill its primary mission; second, as part of the North Atlantic Treaty Army, it was to help safeguard the peace and freedom of the people of Western Europe and North America, and, if necessary, to defend that freedom against any aggressor; third, by demonstrating to the German people the high standard of American moral, cultural and material life, to inspire them with confidence in the democratic ideals.

In conjunction with these missions and the requisite of combat readiness, a forty weeks' intensive training program was launched, including

Combat-ready troops of the 28th in Germany board a plane during "Operation Airlift," a coordinated Army-Air Force maneuver. This and other kinds of maneuvers typified the frequently conducted large-scale field exercises designed to test the Division's mobility, communications and ability to sustain itself under "Aggressor" attack.

advanced individual and squad tactics through regimental combat team and Division exercises.

During the first several months in Germany much emphasis was placed upon the speed with which the 28th Division was able to load its equipment, ammunition, and personnel upon vehicles and move to a tactical assembly area prepared for combat. This training program was culminated in the Fall by three large-scale field exercises during which the Division was tested on its mobility and communications, the ability to sustain itself, and special techniques which had been practiced during the year, such as air-ground support, river crossings, night movement and delaying action.

During the first two of these exercises, the 28th acted in the role of Aggressor opposing the 43d and 1st Infantry Divisions, respectively. In the last, the "Keystone" Division was faced by elements of the 1st Infantry Division attacking from the East. In all phases of these tests, the 28th Division proved ready to fulfill its vital mission as a member of the North Atlantic Treaty Army.

By the end of 1952 the 28th Division had successfully completed more than a year of service in Germany and more than two years of service since its federalization in September, 1950. Most of the National Guardsmen, who had originally accompanied the 28th to Camp Atterbury, had since returned to their civilian endeavors in Pennsylvania. Many of the selectees, who had previously joined the "Keystone" Division at the Indiana camp, had also been rotated to the United States after completing their required two years of active duty.

On December 2, 1952, General Strickler was reassigned as Chief of the Military Assistance Advisory Group in Italy with headquarters at the American Embassy in Rome. For the next two months Brigadier General John G. Van Houten,

A machine gun team ready to fire at the approaching enemy.

Major General Cortlandt Van Renssalaer Schuyler, who assumed command of the 28th Infantry Division, United States Army, in Germany on February 5, 1953.

Assistant Division Commander, served briefly as the 28th's Commanding General. On February 5, 1953, Major General Cortlandt Van Rensselaer Schuyler, former Special Assistant to the Chief of Staff, Supreme Headquarters Allied Powers Europe (SHAPE), assumed command of the 28th Infantry Division, United States Army.

By this time most of the original Division staff members and organizational commanders, too, had either returned to the United States or, by virtue of their election to remain on active duty had been reassigned. The entire 28th Division was taking on a completely "new look" with regard to its personnel replacements, many of whom were career soldiers from the Regular Army. The once familiar nickname, "Keystone," which had long distinguished the 28th Infantry Division as a military unit primarily of Pennsylvania National Guard ancestry, was gradually becoming a fading memory.

Reactivation of the 28th Infantry Division, NGUS, in the Pennsylvania National Guard, was formally announced by Governor John S. Fine, Commander-in-Chief, on May 22, 1953, upon the appointment of the organization's three senior commanders. Major General Charles C. Curtis, from Allentown, was appointed Commanding General; Brigadier General Henry K. Fluck, from Somerset, Assistant Division Commander; and Brigadier General William S. Bailey, from Harrisburg, Division Artillery Commander.

The newly-reorganized 28th Infantry Division, NGUS, participated in its first summer encampment at Indiantown Gap from August 14-28, 1954, along with the non-divisional units of the Pennsylvania National Guard. Small unit training on the squad and platoon levels were stressed during the two weeks field encampment.

Under the leadership of Major General Charles C. Curtis, Division Commander at the time of reorganization a year previous (and since retired), and Major General Henry K. Fluck, who had subsequently been promoted to this rank while succeeding General Curtis as the 28th's Division Commander, 114 of the Division's 117 company and battery-sized units had been reorganized with a strength approaching 5,000 Guardsmen. Meanwhile Brigadier General Arthur D. Kemp, former Commanding Officer of the 111th Regimental Combat Team, was named Assistant Division Commander of the 28th Division.

The Division staff was composed of Colonel Herbert A. Vernet, Jr., Hershey, Chief of Staff; Lieutenant Colonel Oliver W. Robbins, Haverford, G-1; Major Joseph Head, Jr., Philadelphia, G-2; Lieutenant Colonel Nicholas P. Kafkalas, Harrisburg, G-3; and Lieutenant Colonel George I. MacLeod, Villanova, G-4.

Major units organized in the Division, their commanders and locations included in the 109th Infantry, Colonel Fred R. Evans, Scranton; 110th Infantry, Lieutenant Colonel William A. Boesman, Indiana; 112th Infantry, Lieutenant Colonel Donald D. Hays, Erie; 628th Tank Battalion, Lieutenant Colonel James F. Hogg, Johnstown; 103d Medical Battalion, Major James G. Trost, Lancaster; 103d Engineer Battalion, Major Jack C. Betson, Philadelphia; and 728th Ordnance Battalion, Captain William L. Greiner, Jr., Lock Haven.

On Sunday afternoon, August 22, 1954, an estimated crowd of 50,000 persons witnessed the Annual Governor's Day Review and Parade on Muir Field at Indiantown Gap as some 15,000 Pennsylvania Army and Air National Guardsmen paid tribute to their Commander-in-Chief, Governor John S. Fine.

The program was highlighted by the official return of the 28th Division's colors to the Commonwealth of Pennsylvania by Lieutenant General Manton S. Eddy, USA, retired, representing

The generals of the 28th Infantry Division, NGUS, appointed by the Governor on May 22, 1953, signifying reactivation of the "Keystone" Division. (Left to right): Lieutenant General Frank A. Weber, State Adjutant General; Governor John S. Fine; Major General Charles C. Curtis, Division Commander; Brigadier General Henry K. Fluck, Assistant Division Commander; and Brigadier General William S. Bailey, Division Artillery Commander.

Staff of the 28th Infantry Division, NGUS, on May 27, 1954. (Left to right): Lieutenant Colonel Oliver W. Robbins, G-1; Lieutenant Colonel Joseph Head, Jr., G-2; Lieutenant Colonel W. B. Salley, Jr., Inspector General, Second Army; Lieutenant Colonel Nicholas P. Kafkalas, G-3; Major General Henry K. Fluck, Division Commander; Colonel Herbert A. Vernet, Jr., Chief of Staff; and Lieutenant Colonel George I. Macleod, G-4.

the Department of the Army. After receiving the colors Governor Fine presented them to Lieutenant General Frank A. Weber, State Adjutant General, who, in turn, presented the colors to Major General Henry K. Fluck, Commanding General of the 28th Infantry Division.

The Department of the Army returned the numerical designation and colors of the 28th Division to the Commonwealth in conformity with Public Law 461, 82nd Congress, which required that National Guard designations and colors be returned to the States not more than five years after their call to active Federal service.

The 28th Infantry Division in Germany had already been redesignated the Regular Army's 9th Infantry Division on June 15, 1954. The famous 28th "Keystone" Division, along with its numerical designation, proud colors, battle streamers, historic lineage and glorious past, had resumed its familiar role once again as an integral part of the Pennsylvania National Guard.

Return of the 28th Infantry Division's numerical designation and colors to the Commonwealth of Pennsylvania, Governor's day Review and Parade, Sunday, August 22, 1954. (Left to right): Lieutenant General Frank A. Weber, State Adjutant General, Governor John S. Fine; Lieutenant General Manton Eddy, USA, retired (at microphone); and Major General Henry K. Fluck, Commanding General, 28th Infantry Division.

CHAPTER 12

1954 - 1979

Uzal W. Ent, Colonel, HQ, 28th Inf Division

THE LAST QUARTER century of the Division's history has been characterized by frequent call-ups of Guardsmen in service to the people of the Commonwealth during floods and other natural disasters.

For the 28th, 1954 was a year of recruiting and reorganization for the reconstituted Division. Starting in the Spring, non-commissioned officer and officer schools were held, four hours per month for officers and two hours per month for the non-commissioned officers. That year Regular Army support teams came to the annual encampment and assisted in the training of mortar and recoilless rifle crews and gave instruction to tank gunners, fire direction personnel, artillery crews, military policemen, mess teams and others. A three day recruit training program was also conducted at camp.

By January 1955 the Division numbered 6,399 officers and enlisted men. A program was opened by the Army whereby enlisted men with no prior service could volunteer for from eight weeks to six months training at Active Army training centers. After the training these men would return to their units. A twenty-four hour recruit training program was instituted in the Division, to be given in two hour increments over a period of twelve drills. (Twelve weeks, in those days.)

When the Division reported to Indiantown Gap on August 13 for its two weeks of annual encampment, strength stood at 7,490, or about one-half of that authorized for a division. The Regular Army again sent training teams to camp with the Division. Another three day recruit training program was also conducted that year.

On Thursday, August 18, 1955 very heavy rains fell throughout Northeastern United States as a result of Hurricanes Connie and Diane. In a thirty-six hour span on August 18-19 some ten inches of rain fell in the Stroudsburg area. On Friday the Division Commander, Major General Henry K. Fluck, alerted Lieutenant Colonel Jack C. Betson that his battalion, the 103d Engineer Battalion, would probably be going to Stroudsburg on short notice.

The Governor's Day Review, traditional in those days, was held on Saturday morning. As the Engineer Battalion marched off the field, their alert became a reality, and the battalion was ordered to Stroudsburg with the mission of erecting

Old span washed down stream.

Permanent bridge will be constructed just to the right of temporary site.

Bailey Bridges in the stricken area, in conjunction with plans of the Pennsylvania Department of Highways. The battalion arrived in Stroudsburg on Sunday morning, setting up its command post in the State Highway Maintenance Building along U. S. Highway 611, north of the town.

The scene of devestation was overwhelming to behold. Bridges weighing over 500 tons had been swept away, sections of highway washed out, and buildings completely disintegrated by rampaging water. East Stroudsburg was cut off from its sister community, Stroudsburg.

Some seventy-five people in and near Stroudsburg were killed in the flood, and another eleven lost their lives in nearby Wayne and Pike Counties. The high loss of life was attributable to the fact that people were lulled into a false sense of security when the rain ceased falling and the river and surrounding creeks did not appear to be rising. What they didn't realize was that the rain continued to pour down in the mountains to the north. When the local creeks and river did rise, the rate was twenty or more inches in five minutes. Many people were caught sleeping in bed.

More than half the deaths occurred at Camp Davis, a summer cottage retreat some five miles above Stroudsburg on Brodhead Creek. As water rose, the forty people in the camp rushed to the main building. The water swept around and through the cottage, rising all the time. The people sought refuge in the attic. By this time a wall of surging water battered the house, finally ripping it

Triple single 264 foot span and piling driven by civilian contractor.

Site of west Main Street bridge under construction.

Engineers of Philadelphia's 103d Engineer Battalion at the site of the ill-fated Camp Davis.

to pieces, spilling the refugees into the maelstrom of debris-filled flood. Only three people survived. Nothing remained of Camp Davis but the foundation of the main building.

Working with civilian contractors and the Highway Department, the 280 officers and men of the Engineer Battalion performed feats of engineering not normally to be expected of such a small and relatively inexperienced organization. Some 174 men of the command had had no previous military experience, yet they helped construct a Bailey Bridge 260 feet long between Stroudsburg and East Stroudsburg. This bridge continued to serve the area for several years before the Commonwealth replaced it with a permanent structure. The battalion also built another Bailey Bridge 100 feet long over Pocono Creek in order to provide a link with the community fire station which had been isolated by the destruction of a bridge over the creek.

The 260 foot span was completed in three days. Civilian contractors drove "I" beams into the creek bottom over which the engineers constructed the bridge. The 100 foot bridge was built in a day by a force of fifty men from the battalion.

The Engineer Battalion, in addition to accomplishing these two major projects, also built fords, culverts and small bridges, repaired roads, provided purified water and performed all demolition work for the Highway Department at quarries in Monroe County. Work went on 'round the clock, and personnel averaged some three hours sleep in a twenty-four hour period during the tour. On August 28 the battalion was relieved and re-

"Home, Sweet Home"?? Camp of the Engineers at Stroudsburg.

turned to their homes via Indiantown Gap on the 29th.

The Engineer Battalion was not the only organization from the Division on this flood duty, code named "Tempest Rapids." The entire Second Battalion, 109th Infantry, plus Company B from the First, a platoon of military police and men from the Quartermaster Company also saw duty. Individuals from many other Divisional units were attached to various of these commands for special tasks. In all, over 700 men of the Division were involved.

Flooding and damage was not confined to Stroudsburg, although it was most severe there, and martial law had to be declared. Many other areas of eastern Pennsylvania were also affected. In addition to units of the 28th Division, therefore, it was necessary to employ over 1,400 men from non-divisional commands. These included the 51st AAA Brigade, with the 337th and 213th AAA Battalions; six companies from the 111th Infantry Regiment; three units of the 165th Military Police Battalion; the 121st and 722d Transportation Companies and the 103d Ordnance Company. In all, over 2,100 men of the Pennsylvania Army National Guard were fielded in service to the Commonwealth in the wake of Hurricane Diane.

On September 19, 1955, Hurricane Ione roared into the eastern coast, and Pennsylvania was again threatened with floods. Lessons had been learned, however, and a flood relief plan was ready. The State had been divided into geographic areas and a command group and National Guard troops earmarked for each area. Fortunately, the danger passed by the 20th before the plan had to be implemented.

Another flood threat occurred on March 7, 1956, along the north branch of the Susquehanna River but failed to materialize. The Division encamped at Indiantown Gap August 12-26, with a strength of 9,189 troops. Training included small arms and crew served weapons training and small unit field training. Once more Regular Army teams assisted, and the three day recruit program was conducted. Two battalions of the Division went elsewhere for camp that year. The 628th Tank Battalion went to Fort Knox, Kentucky in July. The 899th Anti-Aircraft Artillery Battalion went to Camp Perry, Ohio.

Similar training took place when the Division

returned to Indiantown Gap for the annual camp August 3-17, 1957. Strength continued to increase, for by that time the 28th numbered 9,610 officers and enlisted men. Again, the 899th went elsewhere for camp, this time to Fort Miles, Delaware.

February and March 1958 brought heavy snowstorms to Pennsylvania. During a week in mid-February some 147 men, in Operation Snowfall, using trucks, half tracks, tractors, bulldozers, road graders, snowplows, wreckers and two tanks, were engaged throughout various parts of the Commonwealth in rescue operations, snow clearance, and delivery of medical and food supplies. Men driving a tank from Erie's Tank Company, 112th Infantry, were credited with getting a doctor to a stranded family in time to save the lives of a mother in labor and the baby she delivered. From March 17 to 24 Guardsmen were again performing similar missions after another snowstorm hit southeastern Pennsylvania.

The National Guard was authorized in 1958 to schedule a certain number of "multiple drills," consisting of two four-hour drill periods conducted on the same day (preferably on Sunday). Between September 10, 1958 and June 30, 1959, each unit of the Division was authorized three multiple drills and one week-end drill (two eight hour drill periods on a Saturday and Sunday).

By 1958 a new term sprang up—"Take Six." To "Take Six" was to volunteer for six months active duty. The program was going strong.

A major reorganization of the 28th Infantry Division occurred June 1, 1959. The regimental system was eliminated and the infantry reorganized into five battle groups. All combat service commands of the Division were collected under Headquarters, Division Trains. The 111th Infantry was returned to the Division. Each battle group contained a Headquarters Company, a Combat Support Company (containing heavy weapons of the group) and letter rifle companies A through E. The anti-aircraft battalion was taken from the Division, but two field artillery battalions were added. The Replacement Company was absorbed into the Administration Company. The chart below compares the "old" and "new" divisions. (Note: Certain battalions will be indicated by symbols such as "1-103 Armor," which represents "First Battalion, 103d Armor." Battle Group is abbreviated "BG")

OLD DIVISION		NEW DIVISION	
HHC, 28th Inf Div		HHC, 28th Inf Div	
Med Det, Hq 28th Div		28 Avn Co	
Band		Acft Maint Det	
MP Co		1st Recon Sqdn, 103 Armor	
Recon Co		2 - 103 Armor	
Signal Co		103 Engr Bn	
QM Co		28 Sig Bn	
Repl Co			
109 Inf Regt		1st BG, 109 Inf	
110 Inf Regt		1st BG, 110 Inf	
112 Inf Regt		1st BG, 111 Inf	
		2d BG, 111 Inf	
		1st BG, 112 Inf	
HHB, 28 DivArty		HHB, 28 DivArty	
107 FA Bn		1-107 FA	
108 FA Bn		1-108 FA (Honest John)	
109 FA Bn		1-109 FA	
229FA Bn		2-109 FA	
899 AAA/AW Bn		1-166 FA	
		1-229 FA	
103 Med Bn		HHD, 28 Div Tns	
103 Engr Bn		Band	
628 Tk Bn		28 Admin Co	
728 Ord Bn		28 QM Co	
		103 Med Bn	
		228 Trans Bn	
		728 Ord Bn	
Officers	697	Officers	800
WO	95	WO	105
EM	8,539	EM	9,503
Total:	9,331*	Total:	10,408*

* Strength figures and organization derived from Division Strength reports of May 29 and June 5, 1959.

This new Division was referred to as a "Pentomic" Division because of its five major groupings of infantry commands and its ability to fire artillery with atomic warheads. One battalion of artillery was an Honest John missile battalion. Operating with five elements instead of three, commanders had to learn new and very different tactical doctrine. The reorganization lasted four years, until 1963, when the 28th underwent another major reorganization.

The multiple drill and weekend drill concept became more firmly entrenched, as the numbers of each type drill authorized were increased in 1959. Units could schedule up to seven multiple drills during the period September 1, 1959 and June 30, 1960.

A Pennsylvania Guardsman on sentry duty on pier of the washed out State Bridge in Stroudsburg.

During the annual training period at Indiantown Gap in 1959 an experiment called "Exercise Ready Freddy," was conducted. Under the program some fifty officers and 950 enlisted men from the Ready Reserve were sent to camp with the 28th Infantry Division. These men were assigned to vacancies within the various units, performing duties commensurate with their military specialties and undergoing the same training as the rest of the men of the Division. The experiment was a success and led to a similar arrangement each succeeding year for many years to come. The name "Ready Freddy" was particularly odious to the Reservists involved and was dropped after 1959.

The "Take Six" program continued, with the Division being allocated monthly quotas for attendees. Coordination Officers were appointed to ensure that men taking part in the program got on the right train for their active duty station. Although the National Guard lost a few men who liked active service and stayed on active duty, the men returning to their units enhanced the training level of their companies.

In 1960 the Division marched on Muir Field during the Governor's Day Review, massed in a Division front, from a concealed position behind a hill across the field from the reviewing stand and spectators. At a signal, the massed Division marched over the hill and straight forward to a halt line which had been established for the commands. The normal Division review ceremony took place from that point on in the program.

By 1960 the assignment of Reservists to Divisional units for annual training was more formalized. Each command was required to submit requests, by Military Occupational Specialty (MOS), for Reservists, in a manner similar to that which would be required in combat. The requests were consolidated and forwarded for filling.

As in the previous few years a non-commissioned officers school was conducted at annual training by the Replacement Section (formerly Replacement Company) of the 28th Administration Company.

In September 1960 the Pennsylvania Army National Guard Officer Candidate School was opened. Twenty-eight candidates graduated in the summer of 1961, many of them from the 28th. The Berlin Crisis affected officer candidate school too. Three accelerated courses were conducted between August 1961 and the summer of 1963.

The United States and the Soviet Union confronted one another in the early fall of 1961, when the Soviets, along with the East German government blockaded all road and rail communication from West Germany to Berlin. The Regular Army, as well as elements of America's other Armed Forces, began to build strength in the event of war. Two non-divisional commands of the Pennsylvania Army National Guard were called to active duty. The 131st Transportation Company went on duty September 25 and the 165th Military Police Battalion on October 15.

The "Berlin Crisis," as it was known, led to orders for the National Guard and Army Reserve to increase drills. Referred to as the "Intensified Training Program," the plan called for every unit to schedule six drills during each month, commencing October 1961. Units could schedule two multiple drills and two regular drills, or a weekend drill consisting of a multiple drill and an eight hour drill, plus three regular single drills. The latter arrangement provided twenty-two hours of drill, the former but twenty. Training had to conform to Intensified Combat Training Programs (ICTP's) published by Headquarters, United States Continental Army Command (USCONARC).

In 1961, for the first time in history the Division underwent annual training out of Pennsylvania. From July 15 to 29 the Division assembled in Virginia for camp. The Division Artillery, Reconnaissance Squadron and Armor Battalion went to Camp Pickett and the remainder of the Division to Camp A. P. Hill. The entire Division was assembled at Parker Field, Richmond, Virginia on Saturday, July 22, for the Governor's Day Review. The plan was very unpopular with the men of the command, because the motor move to Richmond, the review, and the return to Pickett or A. P. Hill, consumed an entire day. The round trip between Richmond and Pickett was about five hours and between Richmond and Hill about four hours.

By June 30, 1961, about sixty-five per cent of the members of the Pennsylvania National Guard had completed a six month tour under the Reserve Forces Act of 1955, which had established the "Take Six" program. The remaining thirty-five per cent were either obligors who had completed a longer tour of active duty, were veterans, or had belonged to the Guard prior to 1955.

Strange as it may seem, for a time in 1962 the Guard were not permitted to enlist men with no prior service. However, commands were encouraged to concentrate on building strength by enlisting prior service men. As of June 30, the Division numbered 789 officers and 9,780 enlisted men. Some 576 officers and 1,235 enlisted men were listed as veterans. Seventy-three officers and 5,162 enlisted men had completed the six months training program. Another 1,122 men were then in that program. Obligors included fifty officers and 427 enlisted men.

In February the Intensified Training Program was terminated. By direction of the Continental Army Command all men awaiting their six months of active duty were required to undergo a sixteen hour recruit training program which was administered by Division personnel.

The Pennsylvania Army National Guard experienced a strength loss between 1962 and 1963. The reenlistement rate dropped from over seventy-five per cent to fifty-one per cent. A recruiting program and reorganization on April 1, 1963, offset the loss. The Division stood at 9,455 personnel on June 30, 1963.

As a result of the reorganization—code named ROAD (Reorganization of Army Division) the battle group was dropped, and the infantry battalions were grouped into three brigades of two battalions. The Ordnance Battalion was expanded and redesignated "Maintenance Battalion." The Aviation Company was expanded into a battalion of two companies. Company E (Aircraft Maintenance), 728th Maintenance Battalion, was attached to the Aviation Battalion. The Division Trains was redesignated Division Support Command. Its headquarters detachment became a company and included the Division Band. The Quartermaster Company was absorbed into a new battalion, called the Supply and Transport Battalion, or S&T Battalion. The Quartermaster Company became Company A of the new battalion. The Second Battalion, 109th Artillery was eliminated.

The chart below shows a comparison between the old and new. Commands not shown were not affected by the reorganization:

OLD DIVISION	NEW DIVISION
28 Avn Co	28 Avn Bn
1-103 Armor	1-223 Cav
2-103 Armor	1-103 Armor
	2-103 Armor
1st BG, 109 Inf	1st Inf Bde: 1-111 Inf
	2-111 Inf
1st BG, 110 Inf	2d Inf Bde: 1-110 Inf
	1-112 Inf
1st BG, 111 Inf	3d Inf Bde: 1-109 Inf
2d BG, 111 Inf	2-109 Inf
1st BG, 112 Inf	
1-107 FA	1-107FA
1-108 FA (HJ)	1-108 FA (HJ)
1-109 FA	1-109 FA (155mm/8 in)
2-109 FA	1-166 FA
1-166 FA	1-229 FA
1-229 FA	
HHD, 28 Div Tns	HHC & Band, 28 DISCOM
Band	228 S&T Bn
28 QM Co	728 Maint Bn
228 Trans Bn	
728 Ord Bn	

The "Take Six" program was replaced on August 11, 1963, by the Reserve Enlisted Program of 1963, known as REP-63. Under REP-63, the term of active duty required of every new enlistee into the Army National Guard was based on the specialist training which the new enlistee would require for the position in which he enlisted. However, the minimum active duty period, established at eighteen weeks, could be as much as a full year, depending on the training required.

Two innovations marked the annual training of the Division in 1964. The first of these was a command post exercise (CPX), called Exercise PICK-HILL, which was conducted for commanders and staffs during the annual training period. The other innovation was airlifting 255 men of the First Battalion, 112th Infantry, from Pittsburgh to Byrd Field, Richmond, and then returning them home by air at the end of the camp period. By 1964 the Division was hosting over 1,100 individual Army Reservists at annual training.

With 1965 came a marked increase in the United States' involvement in Viet Nam. The Division underwent another reorganization in November of that year. The 28th Infantry Division was split among three states—Pennsylvania, Maryland and Ohio, and became part of the Selected Reserve Force. Certain commands of the Reserve Components were selected for stepped up training. These commands got priority for Regular School quotas, equipment and money. They became what was known as the Selected Reserve Force (SRF). SRF commands could recruit to 100 per cent strength. Everything accrued to an SRF unit; little to a non-SRF unit. The non-SRF unit was restricted to fifty per cent strength level and had no priority for schools, equipment nor money. All SRF commands were required to conduct twenty-four additional training assemblies each year, plus annual training. Non-SRF commands were restricted to the normal forty-eight assemblies, plus annual training. Pennsylvania elements of the 28th were divided between SRF and non-SRF. The Maryland and Ohio elements were part of the SRF Division. The non-SRF commands of the Division were 2-111 Infantry; 1-110 Infantry; 1-112 Infantry; 1-103 Armor and 2-103 Armor. All of these commands were later designated "SRF" as certain organizations were relieved of that assignment and were replaced by other commands. This second generation SRF was called "SRF-2."

The Maryland Brigade was the 3d Brigade, 29th Infantry Division, which included 1-115 Infantry; 1-175 Infantry; 2-175 Infantry and Trp D (air), 1-183 Cavalry. The Ohio Brigade was the 3d Brigade, 37th Infantry Division, which included 2-145 Infantry; 1-147 Infantry; 1-166 Infantry and Co A (Airmobile), 137 Aviation Battalion.

By June 30, 1966, the Pennsylvania elements of the 28th Infantry Division (SRF) numbered 9,752 personnel, or 195 less than the strength authorized. Conversely, the five non-SRF battalions were still well over the fifty per cent strength ceiling imposed by the reorganization in 1965. They contained 1,966 men, 565 more than authorized. By the end of 1966 the 28th Infantry Division (SRF), including the Maryland and Ohio elements, stood at 15,324 men. Non-SRF commands in Pennsylvania added almost 2,500 more. According to the Division strength report for December 30, 1966, the 28th, including Maryland and Ohio brigades, contained 17,820 men. From the mid-1960's until the early 1970's all Reserve Components, including the National Guard, had no recruiting problems. The draft law, enacted for the Viet Nam war, contained provisions for young men to enlist in one of the Reserve Component forces for six years in lieu of being drafted for two years. Many young men took advantage of this provision of law to enlist in the 28th Infantry Division. Their service, on the whole, was honest, diligent and faithful.

Because of increased drills and demands for further training progress, SRF commands had to meet more rigid standards of training. As a result, a series of more stringent Army Training Tests were administered to selected companies and battalions of the Divison during the annual training period. Non-SRF battalions supported the tests and provided the "enemy." By so doing, these commands also derived training from the exercise.

Recognizing the importance of good communications, the 28th Infantry Division (SRF) conducted a communications exercise at Indiantown Gap in June 1967 for commanders, selected staff members of the various subordinate echelons, and communications personnel. Civil disturbance training also became an important part of training schedules throughout the Division. Special thirty-two hour schools were run for officers. Another thirty-two hour course in riot control training was also given to each unit.

Annual training in 1967 was scheduled for Camps A. P. Hill and Pickett, in Virginia. Advance detachments were already at these sites, when orders came directing that certain commands of the Division from Pennsylvania would remain in the State and undergo their annual training at Indiantown Gap. Many civilian leaders in Pennsylvania feared that racist elements in the State would take advantage of the 28th Division's absence from the Commonwealth to create trouble, even rioting. Accordingly, the 3d Brigade, with 1-109 Inf (Mech), 2-109 Inf and 1-111 Inf, elements of 1-166 FA, the Cavalry Squadron, Engineer Battalion, Aviation Battalion and the bulk of the Support Command were directed to "the Gap." Elements of the Division Headquarters, Support Command Headquarters and some support units were sent to Pickett and Hill to support Maryland and Ohio elements of the Division at those two sites. A feature of the training that year at camp was an eight hour riot control exercise for all units training at Indiantown Gap, plus brigade field training exercises.

The year 1968 was one of considerable activity for the National Guard and the 28th Infantry Division. The Division was reorganized again, and in April, in the wake of Dr. Martin Luther King's assassination, elements of the 28th were placed on riot duty.

The reorganization of February 1968 removed the Ohio Brigade from the Division and added the 116th Brigade (The Stonewall Brigade), from Virginia. However, at the same time the 56th Brigade was created in Pennsylvania, consisting of 1-110 Infantry, 2-111 Infantry, 1-112 Infantry, 1-107 Field Artillery, plus administrative, forward supply and truck elements, military police and other support units. This brigade became part of the 42d Infantry Division, headquartered in New York. The 28th was relieved of its SRF duties and the 42d, along with other commands, became part of the new SRF structure—SRF II.

The reorganized 28th Infantry Division was a more complicated organization than ever before. Subordinate elements of almost every combat support and combat service support organization of the Division were fragmented among the three states. The most striking facet of the configuration and operation of the "new" 28th was how well all parts of the Division worked together. The rapport between the leaders from the top to bottom in this tri-state Division was outstanding and a matter of great pride among those who served in this organization. Very close official and personal relationships soon developed. Many of these personal relationships remained after the Division was reorganized again into a single state entity in 1975.

Under the 1968 reorganization the Division Headquarters remained in Pennsylvania. However, the Division acquired one brigadier general, Assistant Division Commander in each state, for a total of three. One platoon of the Military Police Company was in Maryland and another in Virginia. A company of the Aviation Battalion was also in Virginia. The Engineer Battalion also had one company in Maryland and Virginia. A company of the Signal Battalion was in Maryland. The Cavalry Squadron had a troop in Virginia and two more in Maryland. Division Artillery lost First Battalion of the 107th Artillery to the 56th Brigade and the First Battalion of the 166th Artillery was deactivated. However, the First Battalion of the 110th Artillery of Maryland and Virginia's First Battalion of the 111th Artillery were added to the Division's artillery. The Second Battalion of the 103d Armor was deactivated.

The Support Command was also fragmented. The Administration Company had a Brigade Administration Section in both Maryland and Virginia. Each of these states also contained one medical company and a maintenance company from the parent battalion in Pennsylvania. A forward supply section and truck platoon from the Supply & Transport Battalion were also in the other two states.

The infantry brigades were configured as follows:

3d Brigade (Maryland)	**55th Brigade (Pennsylvania)**	**116th Brigade (Virginia)**
1-115 Inf	**1-109 Inf (Mech)**	**1-116 Inf**
1-175 Inf	**2-109 Inf**	**2-116 Inf**
2-175 Inf	**1-111 Inf**	**3-116 Inf**

Unlike the previous tri-state configuration, all

elements of the Division were known as part of the 28th Infantry Division. In the old configuration the Maryland and Ohio brigades had retained their old divisional affiliations in their titles: (Ohio—3d Bde, 37th Inf Div; Maryland—3d Bde, 29th Inf Div.). Further, regardless of state affiliation, everyone in those commands wore the red keystone patch of the 28th Infantry Division. The new 28th numbered 906 officers, 104 warrant officers and 14,283 enlisted men. Although each state had its own regulations, most were based on Army or National Guard Bureau directives so that there were strong common policies on important matters, facilitating command, control and supervision and also aiding in staff visits and inspections.

On Thursday, April 4, 1968, following the shooting of Dr. Martin Luther King, a prominent black civil rights leader, serious rioting broke out, accompanied by wanton destruction of property, looting and intimidation of the people in many cities throughout Pennsylvania. Pittsburgh, Pennsylvania was one of those cities.

On Friday, April 5, the Mayor of Philadelphia declared a state of emergency and closed all city taprooms. The Adjutant General, Major General Richard Snyder, after consulting with Governor Raymond P. Shafer, alerted key National Guard officers.

By Saturday the situation in Philadelphia was stabilized but had deteriorated in Pittsburgh. Some twenty people had been arrested by the police, but it was obvious that large numbers of police and Guardsmen would be needed. Accordingly, at 6:30 that evening Pennsylvania Army and Air National Guard officers were placed on duty. The Pittsburgh-based 107th Artillery, 42d Division, was alerted at 7:00 P.M. Pennsylvania Air National Guard cargo aircraft were readied to transport troops into the area, if necessary. They could carry 700 troops in one haul. The Joint Operations Center (JOC) earmarked for Pittsburgh in accordance with Civil Disturbance plans, was the Headquarters of the 28th Division Artillery. Under the command of Lieutenant Colonel Fletcher C. Booker, Jr., it departed for Pittsburgh at 7:35 P.M. April 7. Small arms ammunition and chemical munitions were also flown to that city. By midnight rioting, violence and disorder were beyond the capability of local authorities to control. In addition to the 107th Artillery, the First Battalion of the 110th Infantry, (a unit also of the 56th Brigade, 42d Division) and First Battalion of the 229th Artillery were alerted for duty. All three battalions were drilling that week-end, facilitating their assembly and call to duty.

At one in the afternoon of April 7, local authorities asked the Governor for intervention by the State Police and National Guard. The 107th moved to the Civil Center, setting up headquarters there. Troops were deployed to the Hill District about 4:30 P.M. The First Battalion, 110th Infantry joined the 107th at the Civic Center and deployed troops to clear Hill District streets at about 6:45 that evening. The First Battalion, 229th Field Artillery arrived in Pittsburgh about 6:00 P.M. and headquartered at Pill Stadium. By 9 that night streets were cleared. All three battalions then established patrols in the area.

Four more batallions—28th Signal, 218th Military Police, 876th Maintenance, Second Squadron, 104th Cavalry and the 405th Maintenance Company—were alerted for duty. The Hill District was in complete disorder. A crowd of 3,000 blacks were preparing to march into the city. Looting went unchecked, and a sixteen block area was enveloped in flames. Guardsmen in riot formation cleared Centre Street in the District. Mayor Joseph M. Barr imposed a curfew of the city for five days, prohibiting citizens from the streets between 7 in the evening and 5 the next morning. The Governor proclaimed a state of emergency, closed all liquor stores and prohibited the possession of firearms by all citizens except law enforcement personnel. By midnight 2,752 National Guardsmen were on riot duty in Pittsburgh. On that day alone, April 7, there were twenty-three gunfire incidents (none by Guardsmen), forty-two fires and fifty-one known cases of looting.

A report early on the morning of April 9 that the two city reservoirs were being poisoned was found to be untrue. Television and radio stations "killed" the rumors at once. With daylight came the mobs once more. Troops again cleared Centre street, then Kirkpatrick, in the Hill District. Military Policemen were ordered to protect the Spring Hill and McNaughton reservoirs in the wake of new rumors. At 5:30 that evening reserve Guard troops were quickly sent to the Homewood-Brushton section, to prevent a potential holocaust

Scenes from Pittsburgh, April, 1968.

of fires before they could be well started. Many suspected arsonists were arrested by the police.

At 6 that evening, the Governor, Adjutant General, and their staffs visited the Joint Operations Center. Sniper and fire incidents were now being reported from the vicinity of Chateau Plaza. Several men of the 218th Military Police Battalion were shot at but not hit.

At 11 the night of April 8 in the Hill District an automobile ran through three road blocks and attempted to run down several Guardsmen on duty. The white driver and his female companion were captured and jailed by the police. By the end of the day more fires, shooting incidents and looting were reported. Over 4,000 Guardsmen were on duty, deployed as follows:

HHB, 28 DivArty	**Hunt Armory (JOC)**
1-107 FA	**Hill District**
28 Sig Bn*	**Civic Center**
1-110 Inf	**Hill District**
1-229 FA	**Pitt Stadium**
218 MP Bn	**Critical utilities**
Hq & Co A, 876 Maint Bn	**Homewood-Brushton**
405 Maint Co.	**Homewood-Brushton**

* The Signal Battalion was employed as "Street Troops," in riot control duty and not in their traditional role.

On Tuesday, April 9, 3,500 more Pennsylvania Army National Guardsmen were activated. Every headquarters was placed on full duty in the event that the entire 18,532 man Pennsylvania Army National Guard would be required. Civil disturbances slackened during the day in Pittsburgh, where now 4,389 Guardsmen were on duty.

The curfew was lifted on Wednesday, and Guard patrols and security posts were reduced by fifty percent in the Hill District. In the afternoon of April 10 a peaceful memorial march for Dr. King by some 3,000 people became unruly as the crowd returned from Point Park to the Hill District. Disorders threatened. Troops in riot formation cleared Centre Street once more. Reserve troops blocked attempts by hostile people to attack the rear of these riot formations. At 10:30 that night a thirty percent troop reduction was ordered for the following day. As of midnight on the 10th, troops, numbering 4,389 men, were deployed in Pittsburgh as shown below:

HHB, 28 DivArty	**Hunt Armory (JOC)**
1-110 Inf	**Hill District**
1-229 FA	**In reserve**
1-103 Armor	**BN (-) Homewood-Brushton. Liquor stores. Hq Co - Reserve**
218 MP Bn	**Critical utilities**
876 Maint Bn	**In reserve**
1-107 FA	**In reserve**
28 Sig Bn	**Bn (- a plat in reserve) Hill District**
2-104 Cav	**Sqdn (- 200 men in reserve) Homewood-Brushton**
167 Sup & Svc Bn with 722 Sup & Svc Co atch	**Manchester area**

By Thursday, April 11, local authorities, together with responsible citizens in the troubled areas of the city, were beginning to restore order. Violence, met by resolute Guardsmen and police, was ended. The intensive riot control training which the Guard had undergone paid off. Not one incident of gunfire or undue reaction or violence by a Guardsman occurred, although militant blacks and other rioters repeatedly baited, taunted, cursed and defied the resolute soldiers. The 3,500 men of the various headquarters were all released from duty. By Friday, only the 107th remained on duty. That battalion was dismissed on Saturday, April 13.

Although many of the commands serving at Pittsburgh were part of the 42d Division and not the 28th, everyone worked in complete unanimity of purpose under the direction of the 28th Division Artillery commander, who was the officer in charge of Guard operations in the area.

The Division training program in 1969 stressed basic and advanced unit training of platoons and companies, aiming for battalion exercises during the annual training period. As in 1968, the Division was split for annual training in 1969. In 1968 the Division was split among Camp Drum, New York, and Camps A. P. Hill and Pickett, Virginia. Elements of the Division attended training in different time frames from the beginning of June until July 20. In 1969 the scene was the Pickett-Hill complex, but the time frame was awkward—June 7 through July 26. From 1969 through 1979 the Division has been split up for annual training, and the time frame has been expanded. In some recent years elements of the Division were in annual training from May through July or August.

Under such conditions command and control become difficult for the Division Commander and his staff. Further, this fragmentation of an organization which was designed to fight as an entity, undermined the divisional concept and the soldier's sense of belonging to a Division.

Although the Pittsburgh riots were the most trying and difficult for the Commonwealth and the National Guard, they were not the only major civil disorders. York, in July 1969, was the scene of lesser, but serious, racial disorders. Racial problems in the city during April and May led to the imposition of a curfew which was lifted the third week in May. The community returned to normal.

However, on July 17 a racial gunfight resulted in the shooting of a black youth. Scattered disorder resulted. Over the week-end of July 19-20 the situation deteriorated. Sniper fire, along with arsonist-set fires, were reported. Violence became so great that outside assistance for civil authority was necessary. More police were deployed.

At 1:30 on the morning of July 22, Major General Richard Snyder, the Adjutant General, ordered Lieutenant Colonel Donald E. Enders, Sr., commander of the 3d Battalion, 103d Armor, to assemble the members of his headquarters company for duty. The same day Governor Shafer proclaimed a state of emergency in York and declared an 8 p.m.-6 a.m. curfew to remain in effect until the emergency was concluded.

Four patrol teams were formed, each consisting of one armored personnel carrier (APC) containing four Guardsmen, a State Policeman and two city policemen, along with a State Police car and four State Policemen, plus a ¼-ton truck containing two Guardsmen and two city policemen. Three patrols were deployed at 7:30 that evening. The other patrol remained on standby at the armory.

Because the Third Battalion was scheduled for annual training on the 25th, the Headquarters and Headquarters Company, along with Company C, Second Battalion of the 109th Infantry were brought into York on July 23, relieving the Armor Battalion. At the time that both the 109th and 103d were both on duty, some 618 Guardsmen were deployed in York.

Utilizing the men of the 109th, patrols were continued, but York was now relatively quiet; violence had ended. Guardsmen, however, did participate in two raids. One on July 23 and another the next day. In both raids Guardsmen were used to seal off the area, while civilian police entered the buildings being raided. The Guard also provided security for fire stations and firemen from July 24 through 26 and a security force at York General Hospital from July 23 to July 26.

Between July 17 and 28 York police reported that twenty-eight persons suffered gunshot wounds, two of them fatal. Twelve more people were injured from other causes, and ninety-two persons were arrested for firearms violations or breaking curfew. The Guard never had to resort to gunfire. Once the National Guard, along with their armored personnel carriers, were deployed with the police, violence and disorder came to an end. Once again the Guard demonstrated good discipline, bearing and professionalism.

During 1970 Division strength rose from 14,330 to 14,923. Training that year continued to stress the unit, and Army Training Tests were again a feature of the program. Annual training was highlighted by a command post exercise which featured actual live fire of artillery and mortars in support of simulated combat action by the infantry.

The 28th received the new M16A1 rifle in 1970. This was a 5.56 mm gas operated rifle capable of semi-automatic or full automatic fire. It replaced the M-1 Garand rifle. A special training program was instituted in order to insure that all Guardsmen—officers as well as enlisted men—became proficient in the new weapon as quickly as possible.

The Division Headquarters and Headquarters Company, Military Police Company, the Headquarters and Headquarters Battery, Division Artillery, plus the Brigade and Division Support Command Headquarters and Headquarters Companies, the Headquarters and Headquarters Company (or Battery) and one lettered company (or battery) of every Divisional battalion, plus the Administration Company all took part in a mobilization exercise named LAB CHECK II. The exercise required that an extensive series of premobilization and mobilization actions be tested by the participating units. Because of the great

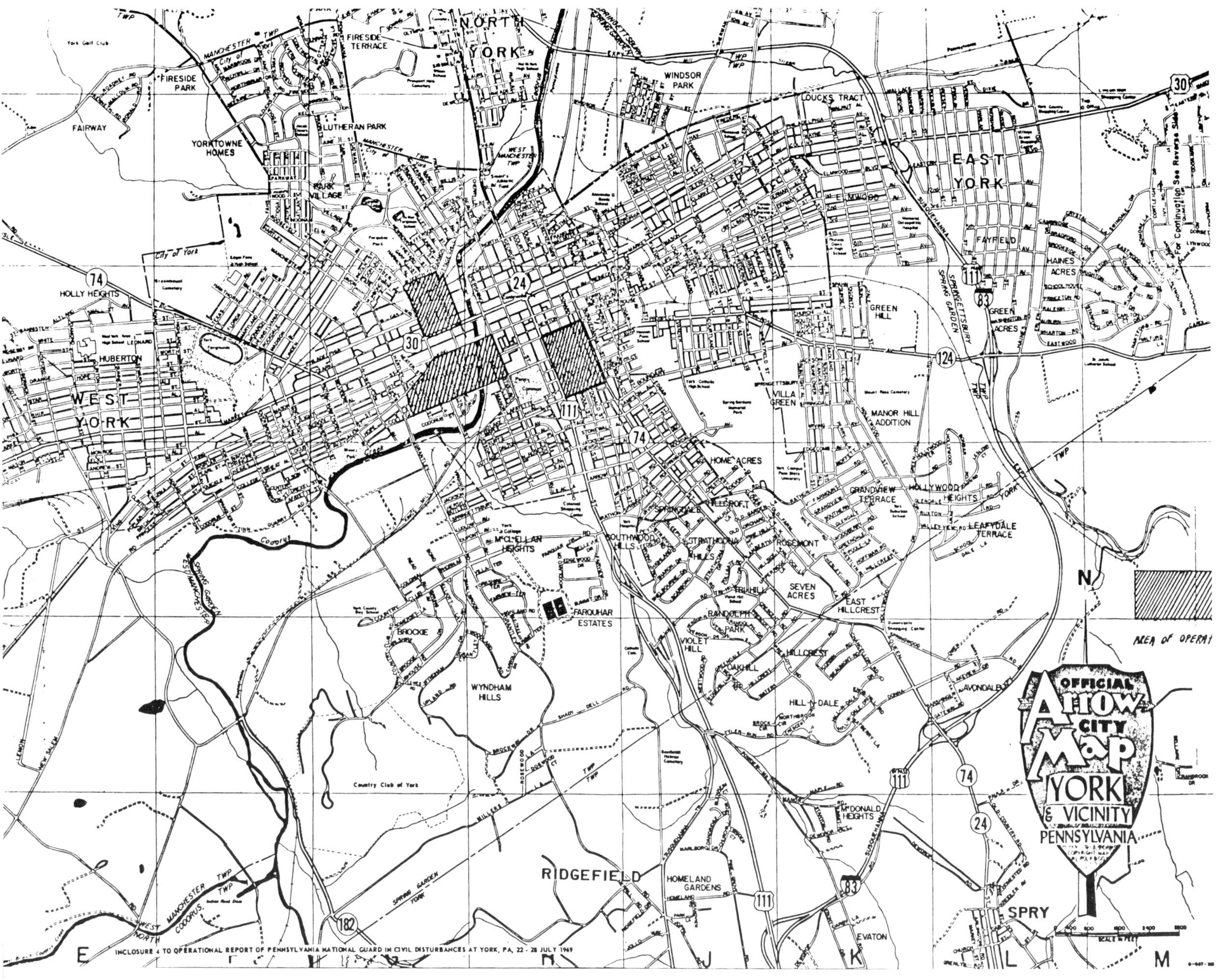

INCLOSURE 4 TO OPERATIONAL REPORT OF PENNSYLVANIA NATIONAL GUARD IN CIVIL DISTURBANCES AT YORK, PA, 22 - 28 JULY 1969

size and complexity of the exercise, it was conducted by people in a drill status or on special active duty days between October 1 and December 12, 1971.

The Viet Nam War continued, and the pool of Individual Reinforcing Reservists (IRR) available to National Guard units attending annual training continued to grow. Some 198 officers and 2,000 enlisted men were ordered to camp with the Division in 1971.

Operation Chester! Although most of the National Guardsmen employed in and near Chester in the wake of floods in that locale in September 1971 were not part of the 28th Division, the unit and unit members belonged to the Division, or are now members.

Heavy rains between September 12 and 15, 1971, caused Chester Creek to overflow, inundating much of Chester and surrounding communities. On Saturday, September 18, Company E of the 103 Engineer Battalion and Company D of the 102d Engineer Battalion, both scheduled for drill that day, were sent into Chester instead to help clean up debris and erect a 100 foot, light tactical raft bridge over the creek. Company B, Second Battalion, 111th Infantry, went to the flood-damaged Chester armory to help with clean up operations.

The next day the remainder of the Second Battalion was sent to Chester to assist. By Monday, September 20, 625 Guardsmen, under the command of Lieutenant Colonel Harold J. Lavell were on duty in the city, remaining until Friday. Their work included clean up of streets, schools, homes, debris removal, pumping water from basements and assisting the restoration of telephone circuits. Selected men from the following units were on duty:

HHD, PAARNG	**Co A, B and E, 103 Engr Bn**
HHC, 56 Bde, 42 Div	**Co D, 102 Engr Bn (42 Div)**
1st Adm Sec, 42 Admin Co	**144 Engr Co (Non-div)**
3d Lt Truck Plat, 42 S&T Bn	**HHC, 2-111 Inf (56 Bde)**
HHC, 103 Engr Bn	**Co A, B and C, 2-111 Inf (56 Bde)**

When units were relieved from duty, certain men were retained in service in order to clean and maintain unit equipment.

Said to be the worst natural disaster to strike Pennsylvania in historical times was Tropical Storm Agnes, which treated the State in a most unladylike manner. As a result of the extensive damage caused by this storm and flood, the Pennsylvania National Guard was engaged in relief operations in 122 communities in thirty-five of the Commonwealth's sixty-seven counties. The flood destroyed some 50,000 homes, 126 bridges and nearly $35 million worth of farm crops. Thousands of miles of roads were closed, and 200,000 telephones went out of service. Forty-nine deaths were attributed to this flood, and property damage was estimated at $3 billion.

The storm moved into Pennsylvania on Wednesday, June 21, 1972, without heavy force, but with drenching rains. Previous rains had saturated the earth, causing the rain to run off into rivers and streams which soon overflowed their banks.

National Guard troops, reacting to local conditions, started disaster relief actions before daylight on June 22. From that date through August 6, 12,036 Army National Guardsmen and 644 members of the Air National Guard served in disaster relief operations. Between June 24 and July 10 over 5,000 men were on duty each day. Because the emergency stretched over such a long time span, commands were rotated on duty. Further, since the disaster occurred during the summer when the Guard normally attends annual training, most men were credited with annual training, or a portion of it for their flood duty. Of 152,117 man-days expended on this duty between June 22 and August 6, some 135,486 man-days were credited to annual training. At the outset, however, the status of personnel was undecided. Some commands had already performed their annual training. The First United States Army Commander finally ruled that disaster relief duty, up to a maximum of fifteen days per man, could be credited as annual training.

In addition to the National Guard, the Regular Army, Army Reserve, Navy Reserve, Marine Corps Reserve, Coast Guard Reserve, State and local police, Civil Defense, Red Cross, Salvation Army and a host of other civilian agencies were all engaged in providing relief to the disaster areas. Virtually every unit of the 28th Infantry Division had at least some of its men on duty. Of 16,666 men assigned to the Pennsylvania Army National Guard, 12,036, or 72.2 per cent, were on

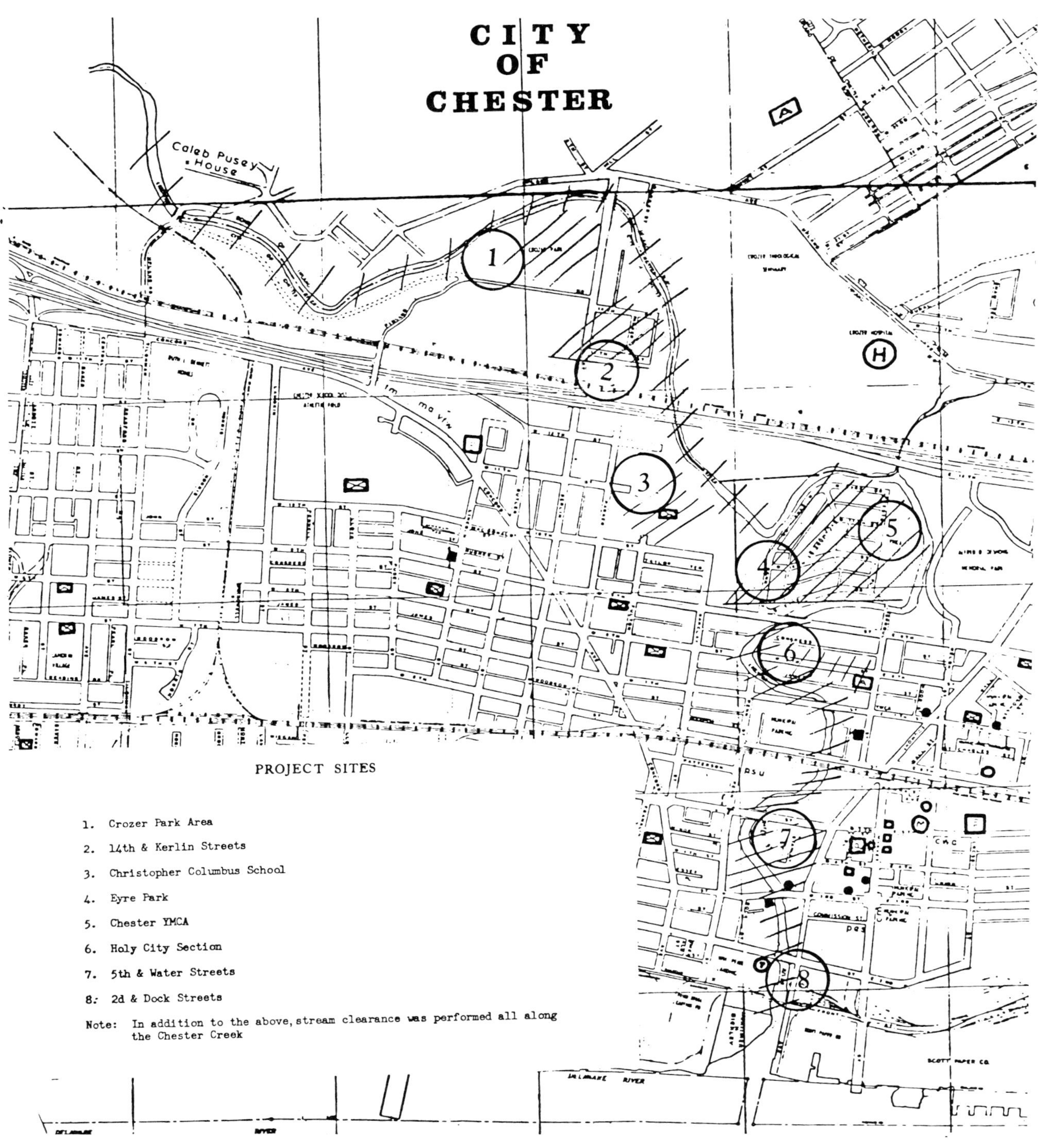

emergency duty during the time of the disaster. Emergency duties included evacuation, rescue, debris clearance, security, traffic control, medical service, providing transportation for water and food, food and material distribution, aviation support and the recovery of disinterred bodies and caskets. (Water broke through a dike and cascaded through a cemetery in Forty Fort, washing away caskets, bodies and the skeletal remains of people who had been buried there. United States Army Graves Registration units were brought in to help in recovering and identifying bodies. The ceme-

GEOGRAPHIC LOCATIONS WHERE PENNSYLVANIA NATIONAL GUARDSMEN SERVED ON DUTY DURING AGNES' FLOOD DISASTER FROM 22 JUNE–6 AUGUST 1972

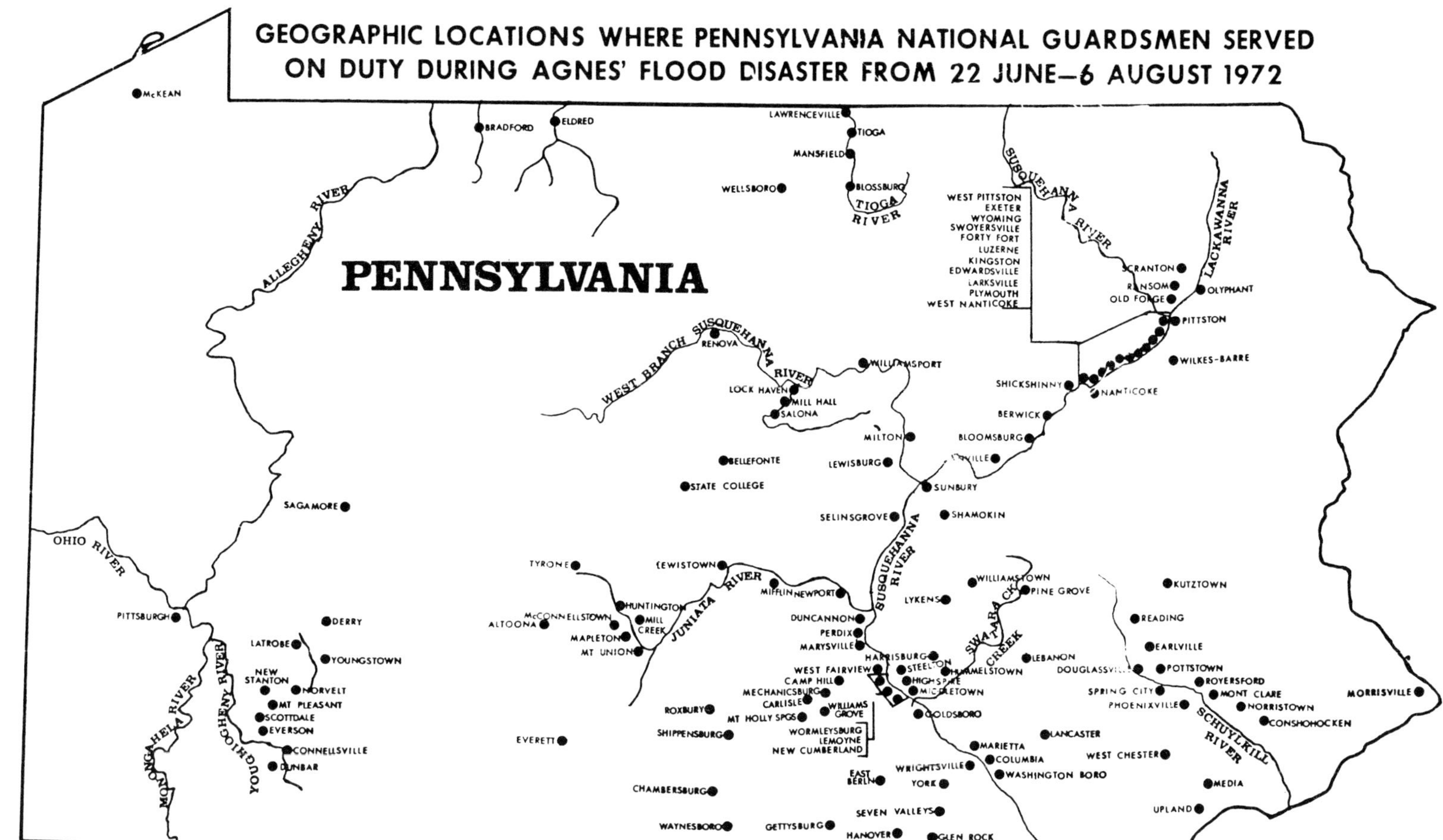

June 22, 1972 is a day which will remain in the minds of hundreds of thousands of Pennsylvanians who were caught in the worst storm to hit the Commonwealth in history.

The fury of Tropical Storm Agnes resulted in billions of dollars in damage, the evacuation of thousands, and a rebuilding effort which will take months.

Over 10,000 Pennsylvania National Guardsman were ordered to state active duty at one time or another by Governor Milton J. Shapp.

There were three primary areas where the Guard played a major role: Wilkes-Barre-Scranton, by far the hardest hit area in the nation; the Harrisburg-York-Lancaster areas; and the region covered by Milton, Williamsport, and Lock Haven.

The Wilkes-Barre, Forty Fort, and Plymouth areas suffered the heaviest damage when the Susquehanna River shot tons of water into the low lying regions along the river. The primary work of Guardsman took place in these areas.

Units from all over the state were called to help. Engineer units from Philadelphia moved north and put their sophisticated training to good use; medical and supply units saw that the human needs were taken care of; and infantrymen provided the back-breaking labor which became a necessity as residents went back to their homes only to find them in ruins.

The Pennsylvania National Guard did an admirable job—a job that showed to the people of Pennsylvania that the Guard belongs.

Grateful citizens receive emergency supplies from a Guard helicopter.

tery had to be fenced off and placed under guard in order to keep ghoulishly inclined people from looting and further desecrating the washed out graves.)

The Wilkes-Barre area was the hardest hit area of the State. (Of the 411 men which the First Battalion of the 109th Field Artillery, from that city, had on duty, the homes of 202 had suffered flood damage, and the homes of twenty-seven others had been destroyed.) A special task force was organized for the area, consisting of both civilian and military organizations. As the water receded, engineer, transportation, quartermaster and maintenance units were required to support both military and civilian needs in the area.

Engineers, including the Division's 103d Engineer Battalion, started to repair bridges and clear away debris from streets and highways. Because almost all food in bulk storage facilities in and near the city had been destroyed, massive amounts of food had to be flown in. The Pennsylvania Air National Guard's 171st Aeromedical Airlift Wing flew in 76,000 pounds of food. In all, a record 1.45 million pounds of food and supplies were eventually delivered to the Wilkes-Barre airport by military and civilian aircraft. A food distribution point for the 68,000 people of the area was established at Pocono Downs Raceway. On July 1 the 228th S&T Battalion, with the 721st Supply & Service Company attached, was placed on duty to assist in the distribution system. On July 13 the Red Cross assumed the food distribution task, when the Guard units were relieved of the duty.

Additional medical support was provided to the troops on duty as well as to the people of the Wilkes-Barre area by doctors, nurses and medical technicians from the 103d Medical Battalion, the 911th Medical Company and Company D of the 102d Medical Battalion, along with medical personnel from the Air National Guard and Reserve and Active Army units. Mass care centers, aid stations and dispensaries were established by these commands, effectively augmenting what remained of civilian hospitals. Two hospitals in the area had been evacuated, and eighty percent of the practicing physicians in Wilkes-Barre had lost their equipment in the flood.

Army National Guard aviators performed heroically too. Using thirty-six helicopters, they performed 893 missions, involving 1,715 flying hours. In carrying out their missions they transported 1,919 personnel, plus 119,950 pounds of cargo without a single accident or incident. The air crews saved many lives. The rotor blades of one aircraft broke tree branches in order to complete a rescue mission. Other missions included providing searchlights in support of anti-looting security tasks, aeromedical evacuation of injured Guardsmen, airlifting medical personnel from one point to another in order to help in a mass care center, plus numerous reconnaissance missions for governmental and military agencies.

The scope and magnitude of National Guard involvement in the Agnes disaster was so great that it is impossible to cover adequately the subject in a history of this size. Observers state that the Army National Guard and 28th Infantry Division never proved their value to the people of the Commonwealth any better than they did in the

Four year old Ronald Bromsburg of West Wyoming expresses his feeling about the removal of three stitches from his injured knee by a Pennsylvania Army National Guard medical officer.

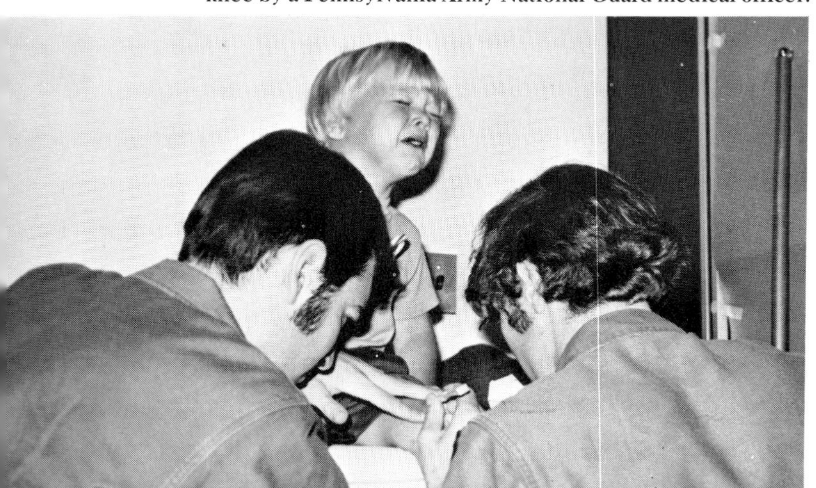

A daring rescue.

wake of Tropical Storm Agnes. Morale and discipline of the men on duty was summarized in the report of the operation by the Adjutant General: "The presence of well disciplined National Guard troops fulfilled a psychological need of reassurance for the general populace... ."

The Guard suffered some casualties. One man died of a heart attack, sixteen required hospitalization, and eighty-five others were treated for minor diseases and illnesses. For 152,117 cumulative days of duty this is a remarkably low injury/illness rate.

The original plan for annual training in 1972 was to assemble the entire Division at Fort Drum, N.Y. This was to have been the first time for such an assembly in twelve years.

In spite of the fact that Agnes had called most men to prolonged duty, the 28th, a tri-state command with over half the organization stationed in Maryland and Virginia, still had to engage in its annual training. Accordingly, a small command group from the division headquarters, under command of Major General Nicholas P. Kafkalas, with Lieutenant Colonel Edward Laufenberg as his acting Chief of Staff, led Maryland's 58th Brigade (since redesignated from 3d Brigade) and the 116th Brigade from Virginia to annual training at Fort Drum, New York. Lieutenant Colonel Uzal W. Ent, Executive Officer of the 28th Division Support Command, lead a small provisional headquarters from that command in order to supervise service support.

Through 1972 unit strength was not a problem. Young men of draft age could enlist in one of the Reserve Components and fulfill their military duty. As a result, many men enlisted in the Army

National Guard. However, by 1973 recruiting and retention became a problem for the National Guard. Compulsory military service was discontinued as the Viet Nam war closed, and most draft-motivated enlistees declined to reenlist. Over the next five years the recruitment and retention of personnel became *the* most important task of commanders. Some commanders who could not master these very difficult facets of command were relieved from command.

Initially, recruiting demanded the primary attention of everyone, but as the years passed the emphasis shifted to what was termed "Quality Recruiting," that is, attempting to enlist people who were better than the norm. Dependable soldiers were needed to strengthen each command. The theory was that high quality soldiers would attract more high quality soldiers. Recruiting and retention in the National Guard are highly decentralized activities, being concentrated in the locales occupied by each unit. Some units, for a variety of reasons (good salesmen, good training, positive recruiting attitudes in unit members, fair and just treatment by unit leadership) were able to maintain a reasonable strength. Others were barely able to exist. The recruiting and retention problem became so acute that many commanders had to spend so much time and attention on these two activities that training and other vital matters suffered.

A reorganization in 1975, returning the entire 28th Division to Pennsylvania, had an adverse effect on recruitment in some units which were converted from service missions to infantry companies. It was not until late 1978 that any real reversal of the downward trend in strength was reversed. The 28th Infantry Division sank to a strength level of some 72 percent. Numerous recruiting drives were launched: (Operation Keystone Plus - 1973; Operation Exploitation - 1973; Operations Full Strength, Keystone Second Effort and Opportunity - 1974; Operation Minuteman - 1975; Operations Keystone Muster '76 and Fill the Keystone - 1976). People were put on special tours of recruiting duty, and a Division Recruiting Command was created, composed of men on recruiting duty full time. The efforts only temporarily stemmed the loss. Selected people were left behind to recruit when the Division went to annual training in 1974, 1975, 1976 and 1977. Retention seminars were conducted, and special orientations for men whose term of service was about to expire were held for them during the annual training period. Between November 1978 and June 1979 retention improved, and the Division increased in strength by somewhat over two percent.

The increase in strength was attributable to two important factors. The first was the formation in 1978 of a small force of personnel on tours of active duty under the direction of then-Captain William E. Woodman, at the State Staff Headquarters. This force of less than sixty men, called the Full Time Recruiting Force, working with full-time employees and unit commanders throughout the State successfully recruited more and better personnel than the units, themselves, had been able to do. The rapport between the full-time recruiter and the unit full-time technician helped in this endeavor.

The second factor in this modest upward trend in recruiting and retention of personnel was the end of the very unpopular Viet Nam War. As this war progressed, anti-war feeling among the people of the United States became more and more pronounced and more violent, leading at times, to marches, riots and other civil disorder. Although these anti-war activities never required the use of National Guard troops in Pennsylvania, violence and disorder in many other parts of the United States did require the employment of large numbers of National Guard soldiers, the Regular Army and police to suppress them. The anti-war feeling often manifested itself in an anti-military feeling. Many people believed, wrongly, that "the military" was perpetuating the war. These people ridiculed and disparaged everything of a military nature. Shortly after the war ended much of the antipathy slowly disappeared. By 1978 it was no longer a factor inhibiting the recruitment of personnel.

In 1973 the enlistment of women was authorized in the National Guard for the first time in history. Major General Nicholas P. Kafkalas, in a letter dated August 13, 1973, implemented instructions from the National Guard Bureau and announced that enlistment of women into the 28th Infantry Division was permitted. Women were excluded, however, from units classified as "Category One." Category One units were combat units—infantry, armor, cannon artillery and combat engineer. By May 31, 1979, the 28th Infantry Division contained eight female officers and 246 enlisted women.

On February 1, 1974, the Pennsylvania National Guard was called upon to aid civil authorities during a massive nationwide strike of independent truckers protesting the increase in fuel costs, the fuel shortage and the reduction of speed limits to 55 miles per hour. In a two week span between February 1 and 13, fifty-nine percent of the Pennsylvania Army National Guard was employed in this duty, which was code named "Operation Safeway." Major tasks included aerial and vehicular highway patrols (794 such patrols), providing guards on over 100 overpasses, escorting a total of 2,523 trucks, hauling critical items, transporting 174,000 gallons of gasoline and assisting the Pennsylvania State Police in quelling disorders. Over 590 Guardsmen worked with the State Police in this last mission. Citizen's Band (CB) operators called the policemen "Smokey the Bear," and the Guardsmen "The Little Green Men."

Once again the Guardsmen performed their duties quietly and efficiently. The civil disturbance training which the men had undergone contributed to the professional manner with which they accomplished every mission. Although a number of men suffered minor frostbite, colds and other respiratory diseases, only one man was injured. He was struck in the face by a rock thrown by a trucker, as Guardsmen and State Police broke up a group of protestors.

Nineteen seventy-five was a year noted for two significant events—the 28th Infantry Division was "brought back" to Pennsylvania and Tropical Storm Eloise once more caused extensive flooding in the state.

The return of the 28th Infantry Division to Pennsylvania on April 1, 1975, caused a major adjustment of certain commands in the state. The entire 56th Brigade of the 42d Infantry Division was incorporated into the 28th, but there remained one full brigade, plus a number of other units to be formed, either as new units, or from existing organizations. Since the 55th and 56th Brigades contained troops primarily from the eastern half of the state, it was decided to form the additional brigade in the west. Western Pennsyl-

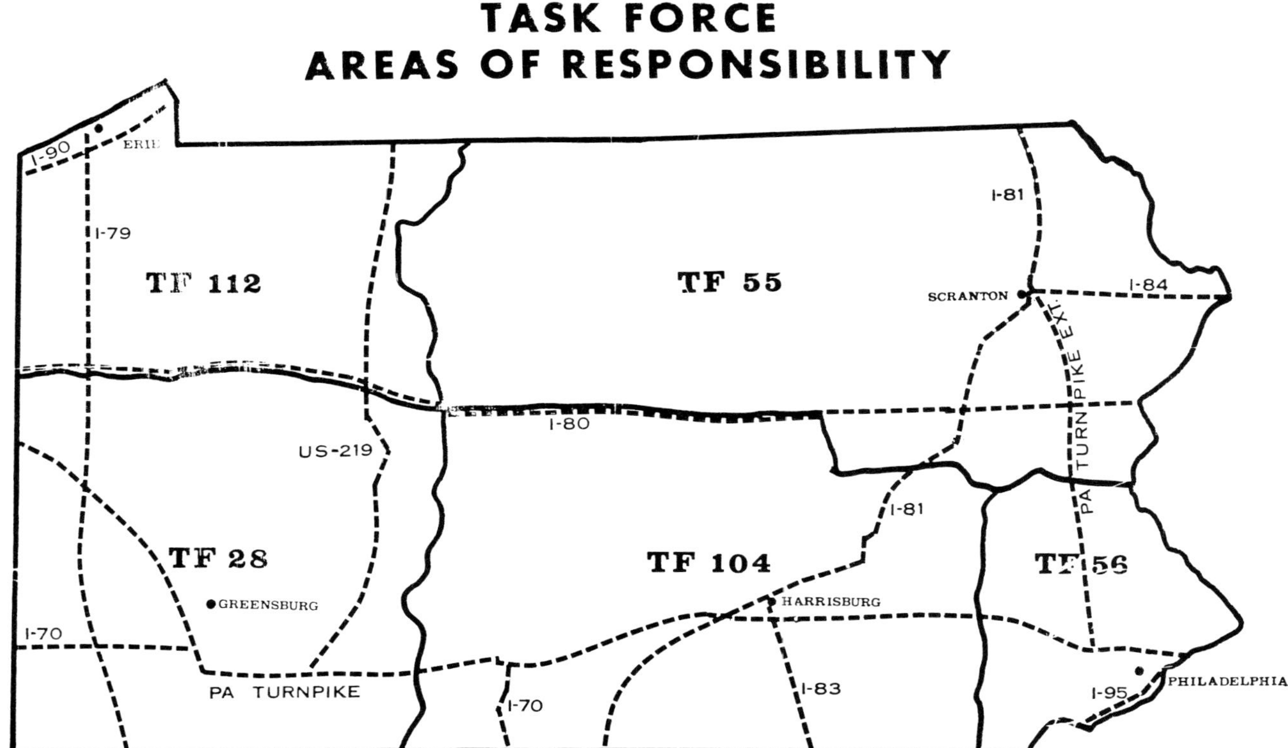

The Pennsylvania Army National Guard performed many duties during "Operation Safeway." These pictures show some of them.

vania contained a large number of non-divisional service support units. However, many of these units, many years ago had been infantry companies. Indeed, some were still housed in the same armories. Certain supply and service, maintenance and military police units thus reverted to their origins as infantry companies. The conversion of these commands from their service support roles to the infantry dealt a severe blow to recruiting and retention. It was a calculated risk, but little else could be done if the entire 28th Division were to become again an exclusively Pennsylvania command. It was hoped that leadership and a positive attitude among the non-commissioned officers of these units would overcome the adverse psychological effect upon the men. Such was not the case. For almost four years these units steadily lost strength. Only in 1979 is the downward trend finally being reversed.

The following compilation shows how the 28th Infantry Division was reorganized in Pennsylvania. A notation "NC" means that there was no change in the status of the unit. Other notations indicate what units were reorganized to form the new structure:

HHC, 28 Inf Div - NC
28 MP Co - Split between Johnstown/Greensburg. Greensburg element from HHD, 218 MP Bn.
HHC, A, B and E, 103 Engr - NC
Co C, 103 Engr - Newly formed
Co D, 103 Engr - From Co D, 102 Engr Bn
HHC, 28 Avn Bn - NC
Spt Co, 28 Avn Bn - Newly formed
Co A, 28 Avn Bn - From Trp N (Air), 104 Armd Cav (Note: The 104 Armd Cav Regt was disbanded at this time. The number 104 was preserved by redesignating the 223 Cav.)
228 Avn Co (Assault Spt Helicopter) - Formed Jan 1, 1976 and attached to 28 Inf Div.
Entire 28 Sig Bn and 1-103 Armor - NC
HHT, 1-104 Cav - Was HHT, 1-223 Cav, 28 Div
Trp A, 1-104 Cav - Was Trp A, 1-223 Cav
Trp B, 1-104 Cav - NC
Trp C, 1-104 Cav - Was Trp C, 1-101 Cav, 42 Inf Div.
Trp D (Air), 1-104 Cav - Was Co B, 28 Avn Bn
HHC, 2d Bde - (Number of new brigade) was 408 Gen Spt Co.
Entire 1-110 Inf - NC, except that Co C was moved from Monessen/Monongahela to a former missile site on a hill in the country outside Finleyville, Pa.
HHC, 2-110 Inf - Formerly HHC, 167 S&S Bn and 708 MP Co
Spt Co, 2-110 Inf - Was part of 722 S&S Co
Co A, 2-110 Inf - Was part of 722 S&S Co
Co B, 2-110 Inf - Was 645 Maint Co
Co C, 2-110 Inf - Was 724 MP Co
Entire 1-112 Inf - NC, except that Co B was split between Meadville and Corry and all of Co C was reorganized in Erie. The elements of Co B at Franklin were moved to Corry.
HHC, 55 Bde, plus entire 1-109 Inf (Mech) and 2-109 Inf - NC
HHC, 3-109 - From HHD, 164 Svc Bn and 1067 MP Co
Spt Co, 3-109 Inf - Was 911 Med Co
Co A, 3-109 Inf - Was Trp E, 2-104 Cav
Co B, 3-109 Inf - Was Trp F, 2-104 Cav
Co C, 3-109 Inf - Was Trp C, 3-103 Armor (3-103 Armor disbanded Jan. 1, 1976)
HHC, 56 Bde - Was HHC, 56 Bde, 42 Inf Div. The entire 1-111 Inf and 2-111 Inf came with the Bde. The only change was that Co C, 1-111 Inf was split between Doylestown and Kutztown, gaining the men of the former Co B, 3-103 Armor.
HHC, 2-112 Inf - Was HHT, 2-104 Cav
Spt Co, 2-112 Inf - Was How Btry, 2-104 Cav
Co A, 2-112 Inf - Was part of Trp C, 1-104 Cav and How Btry, 2-104 Cav
Co B, 2-112 Inf - Was Trp H, 2-104 Cav
Co C, 2-112 Inf - Was Trp G, 2-104 Cav
HHB, 28 DivArty - Entire btry moved to Hershey.
Entire 1-107 FA, 1-109 FA and L-229 FA - NC
HHC, 1-166 FA* - Was HHT, 1-104 Cav
Svc Btry, 1-166 FA - Was part of Trp C, 1-104 Cav
Btry A, 1-166 FA - Was part of Trp A, 1-104 Cav
Btry B, 1-166 FA - Was part of Trp A, 1-104 Cav
Btry C, 1-166 FA - Was Spt Co, 3-103 Armor
*** The 1-166 FA was redesignated 1-108 FA Oct. 1, 1975**

HHC, 28 Div Spt Comd (DISCOM), 28 Fin Co and 28 Adjutant General Co (formerly Administration Co) - NC, except AG Co split among Harrisburg, Altoona (Band), Washington and Philadelphia.
HHC and Co B, 103 Med Bn - NC
Co C, 103 Med Bn - Organized in Allentown.
Co D, 103 Med Bn - Was Co D, 102 Med Bn, 42 Inf Div
Entire 228 S&T Bn - NC, except that Co B (Truck Co) was split between Pine Grove, Butler and Phila.
Hq & Co A, B and F, 728 Maint Bn - NC
Co C, 728 Maint Bn - Was 405 Maint Co
Co D, 728 Maint Bn - Was Co D, 42 Maint Bn, 42 Inf Div
Co E, 728 Maint Bn - Moved from New Cumberland to Indiantown Gap.

The reorganized 28th Infantry Division went to annual training in increments in 1975, between May 31 and August 23. The Finance Company went to Fort Benjamin Harrison, Indiana; the Automatic Data Processing Section to Fort Pickett, Virginia. Most of the rest of the Division, except 2-112 Infantry, the 2d Brigade (then designated 57 Bde) went to Fort Drum. The 2-112 Infantry went to Indiantown Gap, as did the 57 Brigade, with the 2-109 Infantry and 3-109 Infantry attached. Company A (-), 1-112 Infantry trained in Allegheny State Forest, supporting training of other military forces.

Heavy rains fell on Pennsylvania from September 18 to 25, 1975, saturating the ground and causing extensive flooding in thirty counties of the Commonwealth. Although the rains diminished by September 27, rivers did not crest for another two days.

The Governor declared the affected counties a disaster area and ordered elements of the National Guard to duty on September 26. The order formalized and expanded the numerous instances where Guard units had already volunteered assistance to communities near their armories.

For the purpose of National Guard support to

PENNSYLVANIA

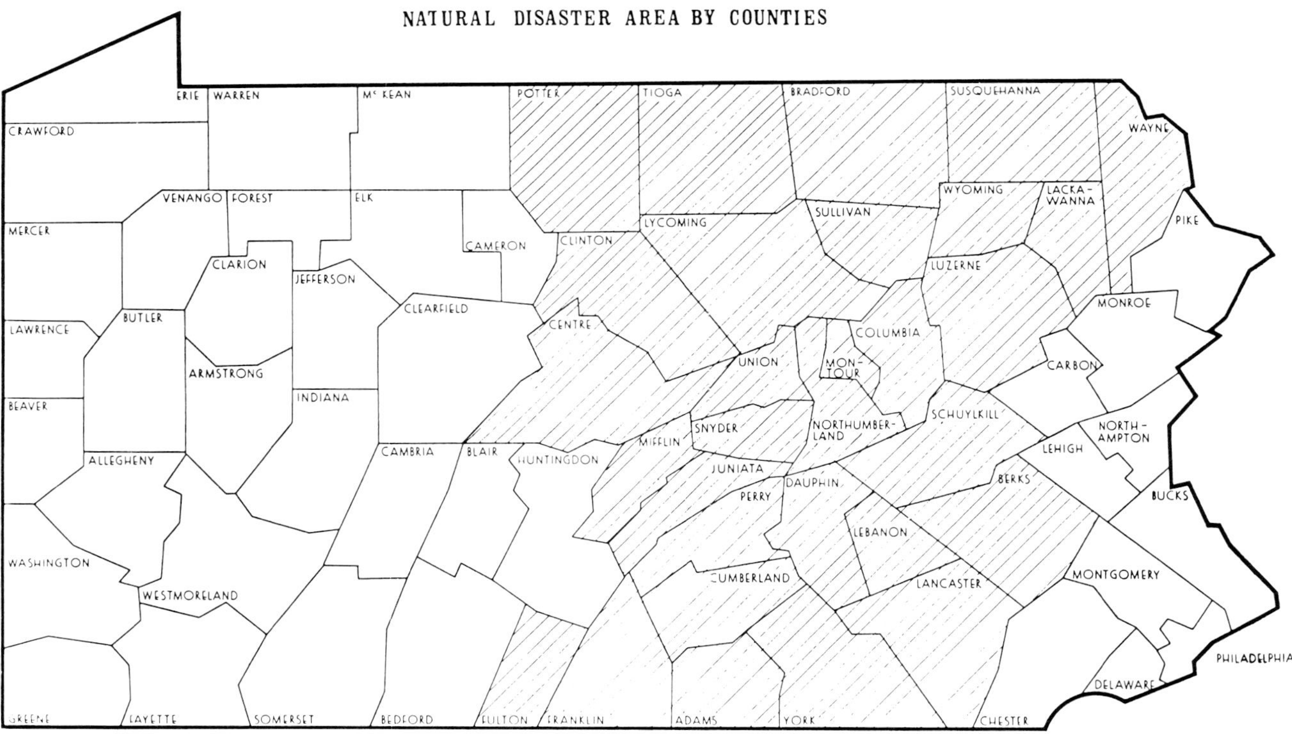

Pennsylvania - Natural Disaster Area, By Counties - Operation Eloise.

The community of Towanda lies partially submerged by flood waters.

civil authorities the flooded area was divided into a northcentral and southcentral sectors. The Commander of the 55th Brigade, Colonel Thomas A. Sharpe, had operational control of all Guardsmen in the northcentral sector, while the 28th Division Artillery Commander, Colonel Glenn E. McCarl, had operational control of Guardsmen in the southcentral sector. As in previous floods, the Guard rescued stranded people (including some rescued by helicopter), transported emergency supplies and equipment, operated and maintained water pumps and generators, provided security, patrolled dikes and reservoirs, helped sandbag, performed traffic control and medical support duties and many other tasks.

As flood waters receded in one locality equipment and support were shifted elsewhere. In the north on September 28, as the situation improved, units were relieved from duty and sent home. On the other hand, the entire 2-111 Infantry, which was on week-end training at Fort Indiantown Gap, was alerted for emergency flood duty in Harrisburg.

By September 29 Guardsmen were helping civil authorities by providing security, traffic control and pumping operations. However, waters had receded in many areas, and some troops were released from duty.

On September 30 all Guardsmen, except a few kept on active duty to clean and maintain unit equipment, were released from duty. A total of 3,662 Pennsylvania Guardsmen and women (for the first time) served during this emergency. On September 28, the peak day of support, 3,357 Guardsmen were on duty. Again morale and discipline were outstanding. The troops performed every duty in a creditable fashion, demonstrating once more the value of the National Guard.

In 1976 the 28th was introduced to a new acronym—ARTEP—Army Training and Evaluation Program. This was, and is, a comprehensive and very involved program to train individuals,

Infantrymen of the 28th Infantry train hard and realistically to improve their combat effectiveness.

The armor and both light and heavy artillery sharpen their military skills in the field.

A CH-47 "Chinook" helicopter of the 228th Aviation Company, Assault Support Helicopter.

squads, crews, platoons and higher echelons of the army and to measure the progress of that training.

Military authorities recognized that service support units should be in their support roles during periods of state active duty in service to civil authorities. Accordingly, headquarters and service support type units were gradually exempted from civil disturbance training. This training was retained in the programs of all other troops, however. All of these other troops had an "on the street" mission in the event of a civil disturbance, meaning that they would have to be prepared to deploy onto the streets of towns or cities which were experiencing civil disturbances. Service support troops would perform their traditional support roles.

While the bulk of the 28th Adjutant General Company (formerly 28th Administration Company) attended annual training at Fort Benjamin Harrison, Indiana, in the latter half of May that year, the remainder of the Division went to annual training in increments between June 12 and August 21. The Armor Battalion and Cavalry Squadron went to Fort Drum, New York, in June, and the Infantry, Artillery, Support Command and newly formed 228th Aviation Company (ASH) performed annual training at Forts Indiantown Gap, A. P. Hill and Pickett.

In 1976 another dimension was added to the training and exercise of certain staffs. For the first time the staffs of the division headquarters, as well as the headquarters of the various elements of the Division Support Command, attended a logistical exercise at Fort Pickett, Virginia. This was one of a series sponsored by the Regular Army for staffs of selected Reserve Component commands. Originally conceived as a logistical exercise for Quartermaster student officers earlier, this series of exercises, code named "LOGEX" were dramatically refined and expanded to include logistical commands of all sorts, divisions, as well as Air Force, Naval, Marine and allied representation. It was and remains the largest, most complex exercise of its kind conducted in the United States. Participants spend the bulk of two weeks in the preparation and conduct of the operation. Because of the time span, commanders and staffs are better enabled to train. Time is available to examine problems, explore various solutions—even make mistakes, and correct them.

Unfortunately, in 1976 and again in 1978, when

the 28th was again invited to participate in LOGEX, the command could not muster many of the principal staff members to attend. Therefore, substitute officers and enlisted personnel were sent. In spite of this serious shortcoming, the Division acquitted itself most creditably, according to a letter from the Exercise Director. The exercise was utilized to train the substitutes, and in the long run proved beneficial to the command and individuals concerned. The 28th Infantry Division accepted an invitation again to participate in LOGEX in 1980.

Command post exercises and the like were not new. Since 1953, Division staff, as well as the staffs of subordinate commands, took part in, or conducted their own, command post and field training exercises. Although there have been many such exercises, the fundamental objectives of each have been to train the individual staff officers, non-commissioned officers, and staff sections in performing their individual and collective staff functions. Generally, at least one such exercise was completed each year. Subordinate staffs might have conducted additional ones for themselves. Each exercise normally "ran" two or three days during a week-end.

All of these had code names of one sort or another. In the mid-1950's commanders and staffs journeyed to Fort Meade, Maryland, for a series of command post exercises (CPX) named "Tobacco Leaf." By 1958 the CPX name had changed to "Trap Line." "Trap Line II" was conducted that year. The chart below shows a representation of the various CPX's conducted by the Division since 1958. In the early years everyone travelled to Fort Meade, First Army Headquarters. Later these exercises took place either at the Harrisburg Military Post or Fort Indiantown Gap. Since 1976 the exercises have been conducted from field command posts, adding a touch of realism to the operation.

DATE	CODE NAME	REMARKS
1958-1963	Series named "Trap Line"	Series ran to Trap Line IV
1964	First "Bear Trap" CPX	Conducted in buildings at Indiantown Gap. Scene of exercise: USSR.
1966	High Card I	Conducted at Harrisburg Military Post. Scene of exercise: Viet Nam.
1966	Bear Trap III	Buildings, Indiantown Gap. Scene of exercise: USSR.
1967	Bear Trap IV	Same as Bear Trap III.
1971	Gobi Bear II	Buildings, Indiantown Gap. Scene: Germany.
1973	Gobi Bear III	Same as Gobi Bear II.
1975	Three Fires	Harrisburg Military Post. Scene: Harrisburg-Indiantown Gap area.
1976	Orbiting FIG	Field CP's, Indiantown Gap. Scene: Germany.
1977	Cabalistic Knight I	Field CP's, Indiantown Gap. Scene: Same as Three Fires.
1978	Orbit Rampart	Same as Orbiting FIG.
1979	The exercise is not yet named but is scheduled to take place from field CP's at Indiantown Gap in October 1979. Scene: Germany.	

Command post exercises or field training exercises were also run during some annual training periods, notably in 1970, 1971 and 1972.

Two natural disasters marked 1977. The first of these was an energy crisis brought on by heavy winter storms throughout the Commonwealth between January 15 and February 8. Heavy snow, subzero temperatures and high winds created serious fuel shortages in parts of twenty-eight counties. Some 800 Pennsylvania Army National Guardsmen, including personnel from the 103d Armor, both battalions of the 110th Infantry, both battalions of the 112th Infantry, First Battalion, 107th FA, First Battalion, 229th FA, 728th Maint Bn, Trp D of the 1-104th Cav, 228th Avn Co, Third Battalion, 109th Infantry, Division Headquarters, Hq, 2d Brigade, 228th Supply & Trans Bn and First Battalion, 109th FA were called into State service. The troops helped in snow removal, delivery of emergency fuel oil to

PENNSYLVANIA

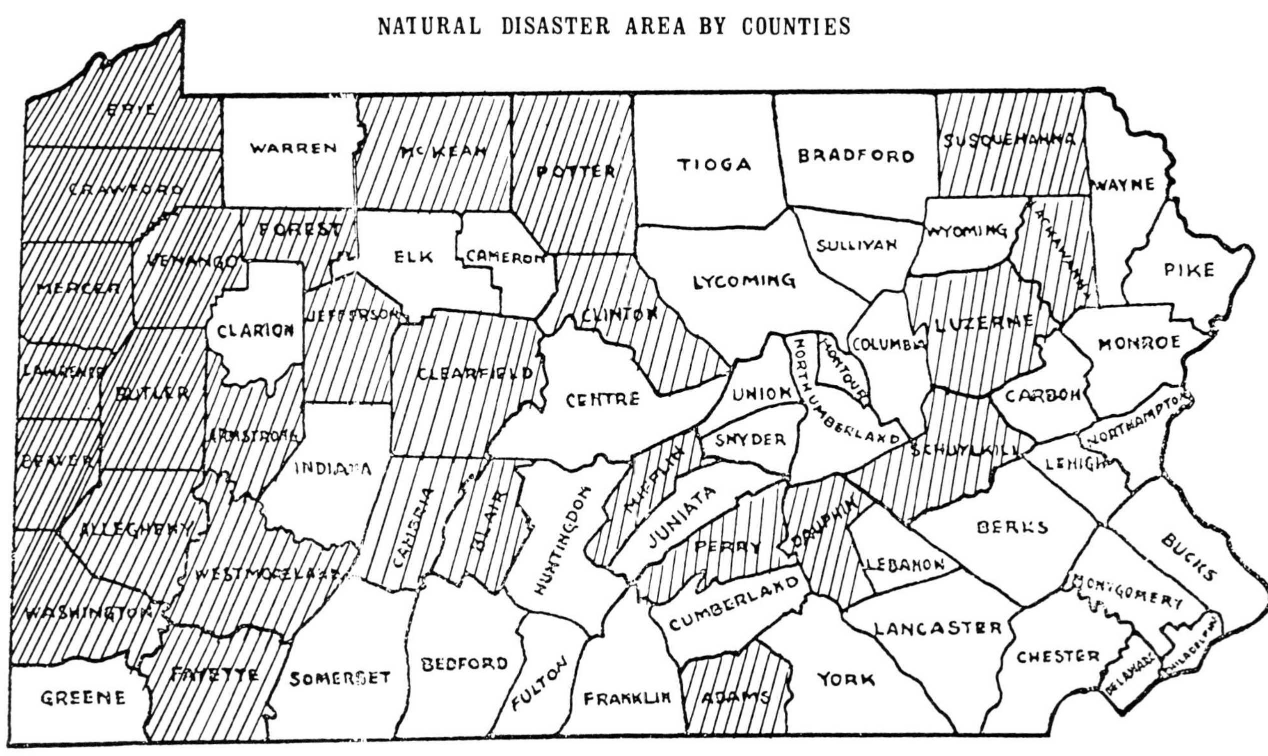

Pennsylvania - Natural Disaster Area by Counties - Operation "Energy Crisis."

citizens, rescue operations and the delivery of pipe and pumps to various communities which had lost water supplies because of frozen and damaged pipes. In one instance four CH-47 "Chinook" helicopters of the 228th Aviation Company airlifted 700 feet of eight inch pipe and four, 6000 pound water pumps to Curwensville, Clearfield County. Because of their bulk, the pumps had to be sling loaded.

Serious as the earlier crisis was, a flood which struck the Johnstown area on July 20 resulted in far more deaths. Confirmed deaths numbered seventy-two, while an additional twenty-eight persons were listed as missing. Some 140 communities in an eight county area suffered from the flood. The large number of deaths and missing persons were the results of the sudden collapse of the Laurel Run Dam. When the dam burst, over 101 million gallons of water was sent rushing down the valley toward Johnstown. In all, some 3,476 Guard members were placed on duty between July 20 and August 18. During five of these days more than 2,000 were on duty; during seven other days 1,500 were on duty.

Brigadier General Robert M. Carroll, Deputy Adjutant General, was sent to Johnstown on the morning of July 20. He and a small staff assumed operational control of National Guard units in the area until July 22, when Headquarters, 28th Division Artillery, under Colonel Robert Armstrong, was appointed Task Force Headquarters. General Carroll remained in over-all command. Headquarters, 55th Brigade relieved Division Artillery on August 1. Major General Fletcher C. Booker, Jr. relieved General Carroll the same day.

As in all past tours of flood duty, the Guard assisted in rescuing people. In one day 212 people were evacuated by Guardsmen. Helicopters were often used in these missions. The most harrowing—and most heroic—was the unselfish act of SP5 Barry P. Pope, Trp D, 1-104 Cavalry. SP5 Pope volunteered to rappel one hundred and

Many rural areas were isolated by the fierce snowstorms.

Some areas could be reached only by armored personnel carriers.

Guardsmen helped cut away broken tree branches in order to free power lines. They delivered emergency fuel to isolated homes. Chinooks delivered water pumps to Curwensville, Pennsylvania.

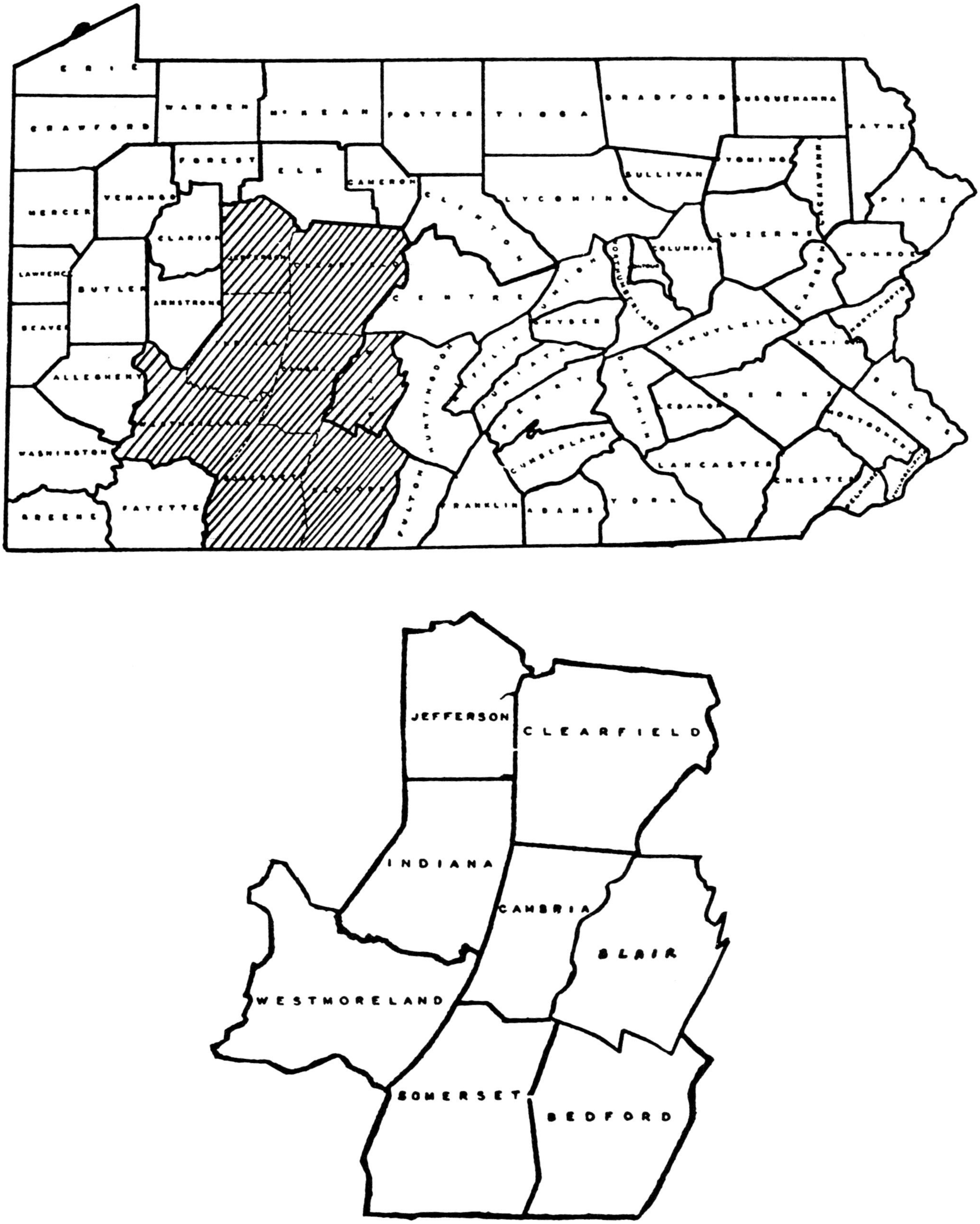

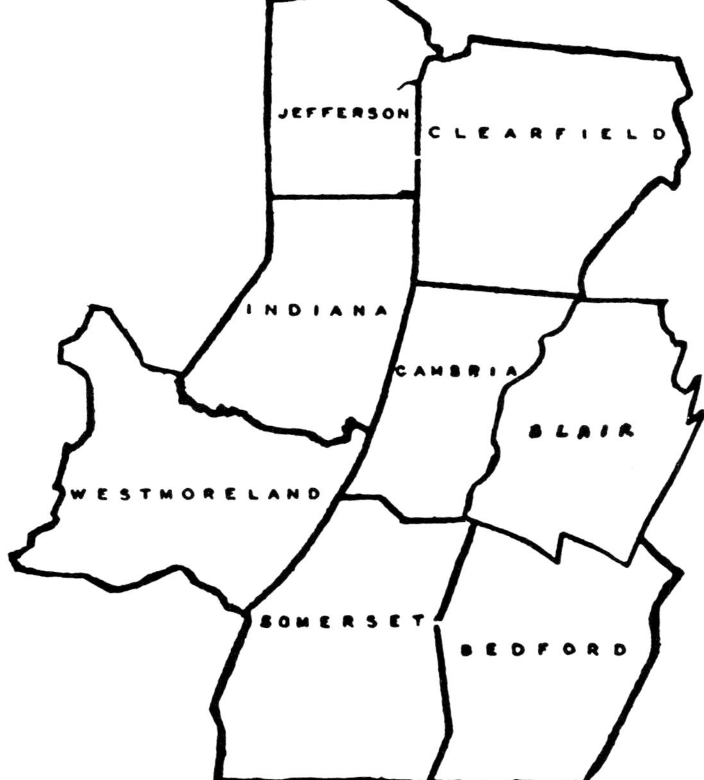

The diaster area, Operation Johnstown, July 1977.

twenty-five feet from a hovering aircraft to a flood-covered bridge in order to rescue a diabetic man trapped there. Pope avoided electrically charged wires and debris and successfully assisted the victim into the helicopter. Knowing that only one person could leave the bridge at a time and that low ceiling and bad flying conditions might prevent the aircraft's return, he waved the helicopter away so that the rescued man could be delivered to safety. He clung to the bridge, waist deep in rushing flood waters for twenty minutes until the helicopter could return for him. SP5 Pope was awarded the Pennsylvania Cross of Valor for his deed, as well as the Valley Forge Cross of Valor from the National Guard Association of the United States.

The crew of a CH-47 Chinook helicopter found three persons clinging to a bridge. Using the rescue winch of the aircraft the crew saved the people and transported them to a first aid center. Then working with the pilot of a small civilian helicopter, piloted by a National Guardsman who was flying it for his civilian employer, the crew went in search of other stranded people. The pilot of the small aircraft acted as spotter. When he located stranded people, the Chinook crew would maneuver the back end of their helicopter into the water and lower the exit ramp. The flight engineer and Civil Defense personnel on the craft would then wade into the water, often up to their waists, and haul the people into the helicopter. One woman, suffering from a ruptured spleen, and other people suffering many other injuries, were rescued in this manner and transported directly to Latrobe Hospital.

Meanwhile, the 28th Military Police Company, from Johnstown and Greensburg, was on annual training at Fort A. P. Hill. Most of the men from Johnstown had no knowledge of what may have happened to their families or homes. Communications with the Johnstown area were completely severed. In spite of their obvious and justifiable concern for the safety of their loved ones, the members of the company did not desert their posts with the Division but awaited authority to return to Johnstown. The forebearance displayed by these men speaks highly of their esprit and sense of duty. The unit was subsequently employed on flood duty in Johnstown from July 24 to August 5, when it was relieved by the 165th Military Police

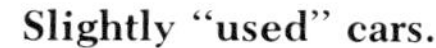

Slightly "used" cars.

Battalion.

On July 25 an explosion collapsed a large section of the Royal Plate Glass Company building in downtown Johnstown. Eleven people were injured. Seconds after the blast Sergeant Galizia R. Delpaine, 28th Military Police Company, being on duty nearby, squeezed through a narrow hole in the debris leading to the basement, which was filled with smoke and toxious fumes. Disregarding the possibility of further explosions or death by asphyxiation, Sergeant Delpaine thoroughly searched the ruins for victims. None were found. He also earned the Pennsylvania Cross of Valor.

For most Guardsmen, however, flood duty consisted of the more mundane, but equally important tasks of distributing water, controlling traffic, security missions, delivery of critical supplies, distribution of food and assisting in clean-up operations. The 876th Engineer Battalion, a nondivisional command, constructed a 127-foot Bailey Bridge across Two Lick Creek. The 28th Signal Battalion provided personnel to establish communications between the State Emergency Operations Center at the 876th Armory and the Military Task Force Headquarters at the Greater Johnstown Vocational-Technical School.

During the period of the flood and its aftermath over 2,400 members of the 28th Infantry Division were employed on State Active Duty. Members of the following units were on duty:

HHC, 28 Inf Div	28 Sig Bn
HHC, 55 Bde	Co B, 228 S&T Bn
HHB, 28 DivArty	Hq & Co A, 728 Maint Bn
28 MP Co	Co E (Acft Maint), 728 Maint Bn
HHC, 28 Avn Bn	Co F, 728 Maint Bn
Co A, 28 Avn Bn	1-103 Armor
Trp D (Air), 1-104 Cav	3-109 Inf
228 Avn Co. (ASH)	1-107 FA
28 Materiel Management Co*	1-108 FA
	1-109 FA

* This unit was organized in January 1977.

In April 1977 the 28th changed commanders. Major General Fletcher C. Booker, Jr. replaced Major General Nicholas P. Kafkalas, who became the Adjutant General of Pennsylvania.

Again, portions of the Division trained at Forts Indiantown Gap, A. P. Hill, Pickett and Drum. This year, however, Battery C, 1-229 Field Artillery attended training at Camp Santiago, Puerto Rico. A battery of the Puerto Rican National Guard trained with the 229th, replacing the missing Battery. A large contingent of personnel from the 28th Signal Battalion went to Fort Gordon, Georgia, for their annual training, where they under-

Members of the 28th Military Police Company discuss the situation.

Major General Fletcher C. Booker, Jr., Commander, 28th Infantry Division, congratulates Lieutenant Colonel Irvin, commander of the 876th Engineer Battalion, on the completion of a 127-foot Bailey Bridge over Two Lick Creek.

went occupational specialty schooling. As in recent years, the 28th Finance Company was on what was known as "Year Round Training." Under this system, individuals from the unit trained at different times during the year with the Army Finance Office at Indiantown Gap. During 1977 several small mutations occurred in the 28th. On September 1, Company D of the 103d Engineer Battalion was consolidated with Company B. On the same date pathfinder teams were attached to the 28th Aviation Battalion. Battery F (Target Acquisition), 109 Field Artillery was organized in York, becoming part of the 28th Division Artillery.

A severe snow storm on January 19 and another on February 6, 1978, required the activation of selected members of the Pennsylvania Army National Guard. Between January 19 and 23 Guardsmen volunteered their services to various locales in the Commonwealth, assisting in transporting doctors, nurses and other emergency personnel, towing stalled vehicles and rescuing stranded motorists. On January 19 these volunteers from the Division numbered thirty-four; seventy-one by the next day, sixty-one on each of the succeeding two days and nineteen on January 23.

The storm on February 6 was more severe than the earlier one and could not be handled by a few volunteers. Accordingly, several hundred Guardsmen and women were placed on State Active Duty. Again, services rendered included transportation of medical personnel, emergency transportation of sick and injured people to hospitals, wrecker service, rescuing stranded motorists, transportation of food and fuel, back-up emergency service for local and State Police, snow removal and temporary housing and feeding for stranded motorists.

Personnel from the following Division commands served during these storms. Units marked with one asterisk (*) provided volunteers in January and personnel on State Active Duty in February. Those marked with two asterisks (**) provided volunteers in January but were not called up in February. The commands not annotated were on State Active Duty in February. Some 260 people from the Division were on duty during February.

HHC, 28 Div*
HHC, 55 Bde**
HHB, 28 DivArty*
HHC, 56 Bde
103 Engr Bn
28 Avn Bn
228 Avn Co
Co D, 728 Maint Bn
Co E (Acft Maint), 728 Maint Bn
Co A, 228 S&T Bn**
Co B, 228 S&T Bn
Co C, 1-109 Inf (Mech)
HHC, 2-109 Inf
Co B, 2-109 Inf**
1-111 Inf
Co C, 2-112 Inf
1-108 FA
1-109 FA
HHC, 28 DISCOM
28 Mat Mgmt Co
28 Adj Gen Co
Hq & Co A, 103 Med Bn*
Co C, 103 Med Bn
Co D, 103 Med Bn**
Hq & Co A, 728 Maint Bn**

Annual training for the Division was conducted by increments from April 15 through August 19, 1978. The same Posts were again utilized as in the previous year, although the various commands were rotated. This rotation of Divisional commands from one annual training site to another from year to year has been done intentionally so that no one Post is "memorized" by the people of the Division and the efficacy of training reduced because topography has become familiar. Turnover of personnel, coupled with rotation of commands to the different sites, ensures that this does not happen, and enhances the training value of each annual training period. In 1978, however, two Divisional units were presented with some very special challenges and interesting and rewarding annual training experiences. Company C, 2-112 Infantry went to Sennybridge, South Wales, United Kingdom and Company F, 728 Maintenance Battalion went to Colemen Barracks, Germany.

The week-end of September 16 and 17 was the occasion of the 28th Infantry Division Military Skills Competition held at Fort Indiantown Gap. This first competition was so well received by the troops and so successful that another such competition was scheduled for September 1979. (The competition was subsequently cancelled due to severe fuel shortages which were prevalent throughout the United States.) The series of competitions was designed to promote interest among Division members in the skills in weaponry necessary to success in battle. The competition included individual rifle and pistol marksmanship, a machine gun team contest, as well as competitions between Anti-Tank Platoons, 81 mm Mortar Platoons, 107 mm Mortar Platoons, Rifle Squads, 105 mm Howitzer Sections, Tank crews, Scout Sections, Artillery Fire Direction Centers, and a timed four mile march. The latter was actually a run, since the individual crossing the finish line in the least elapsed time was the winner. Competitors in this event were required to complete the course in battle gear, excluding the individual weapon, but including the steel helmet.

In October, 1978, both the Finance and Adjutant General Companies were reorganized. Under the reorganization the Aviation Battalion, each artillery battalion and each infantry battalion of the Division was allocated a small number of personnel positions from the Adjutant General Company at the battalion headquarters. Further, the Aviation Battalion, each artillery battalion and Headquarters, 2d Brigade were allocated one or two positions from the Finance Company. This reorganization was made in conjunction with a reorganization of the State Staff Headquarters involving personnel, mess and other augmentations which had been authorized certain commands. The reorganization, which provided each command with small personnel-finance teams, was designed to enhance their ability to perform these tasks.

An incident at the Three Mile Nuclear Power Plant, located ten miles south of Harrisburg, on March 28, 1979, resulted in low levels of radiation being emitted into the air. Low level emissions continued for several more days. An additional emergency was created by the development of a large hydrogen-oxygen bubble within the reactor building. The need to cool down the reactor and to end other emergent problems was complicated by continuing intermittent emissions, a vapor bubble, equipment failures, personnel shortcomings and the fact that there was no precedent for such an emergency. If the nuclear reactor unit could not be cooled slowly in a controlled manner, it was probable that the reactor core would melt down, releasing toxic amounts of radioactivity throughout a six county area populated by half a million people.

The Governor, in consultation with his advisors, representatives of Metropolitan Edison Company (Power Plant operators) and the Federal Nuclear Regulatory Commission, decided on March 30 to declare a state of emergency. Colonel Robert A. Armstrong, Commander of the 28th Division Artillery, was named to lead a task force of Pennsylvania National Guardsmen in support of civil authorities, in the event that it became necessary to evacuate people from the area.

Colonel Armstrong's command was designated "Task Force 28th Division Artillery." It consisted, potentially, of the following commands:

HHB, 28 DivArty	1-110 Inf	103 Engr Bn
Btry F, 109 FA	1-108 FA	876 Engr Bn
1-109 Inf (Mech)	1-109 FA	228 S&T Bn (-)
2-109 Inf	2-112 Inf	103 Med Bn (-)
3-109 Inf	1-103 Armor	
1-111 Inf	1-104 Cav	
2-111 Inf		

The primary mission of the task force was to assist civil authorities in evacuation, traffic control and security of evacuated areas. Planning began at once. The Pennsylvania Emergency Management Agency (formerly named the State Council of Civil Defense) prepared evacuation plans, in coordination with the affected counties, involving five, ten and twenty mile radii from Three Mile Island. At the outset certain National Guard commands were assigned to work with civil authorities in the various affected counties. Assignments were:

Dauphin	-	3-109 Inf
York	-	2-112 Inf
Lebanon	-	Initially, 228 S&T Bn. (1-109 Inf slated later on)
Lancaster	-	Initially 103 Med Bn. 2-111 Inf assumed mission later
Perry and Cumberland	-	1-108 FA

Other commands of the Task Force were alerted

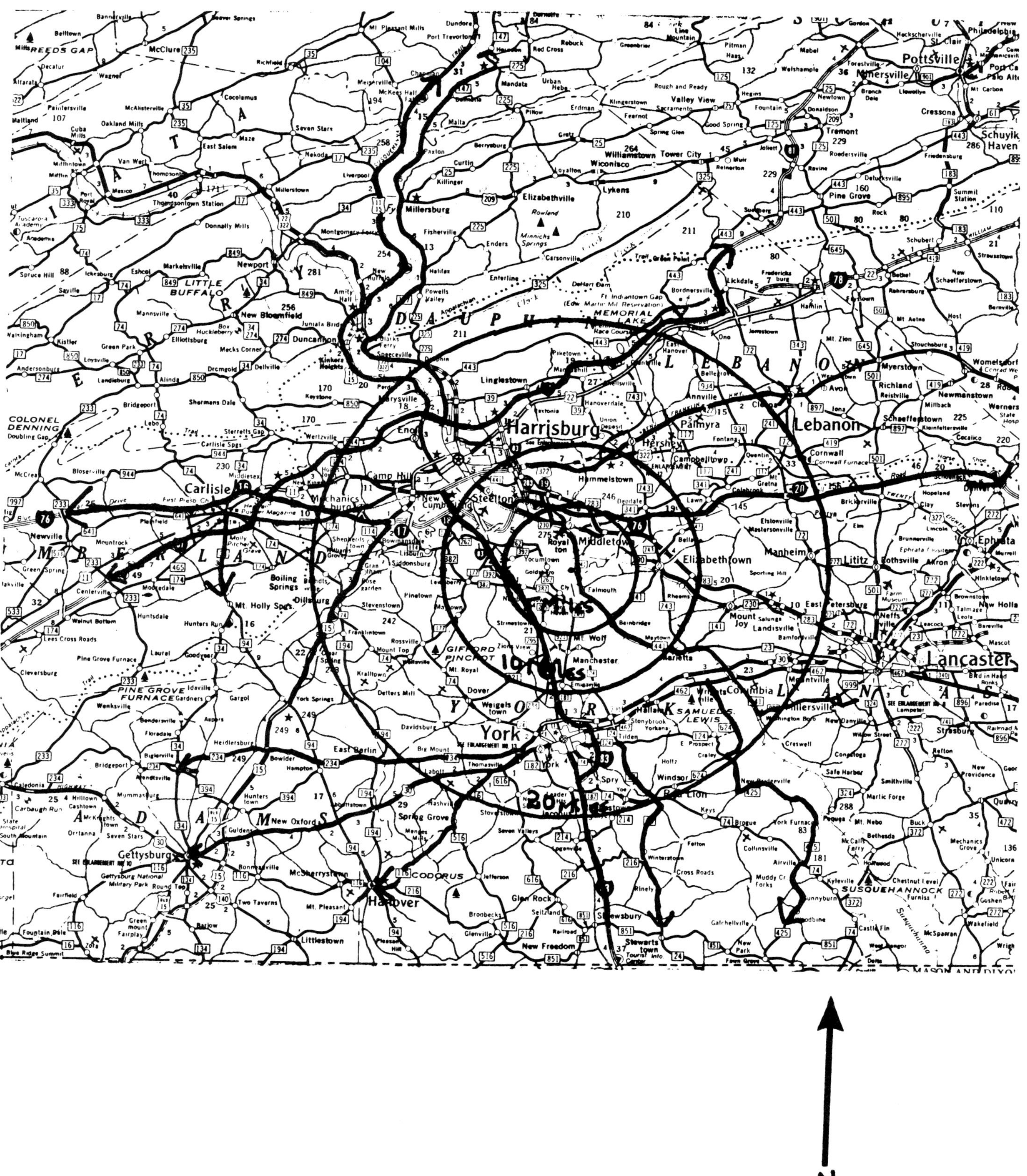

Evacuation routes and destinations, Three Mile Island incident.

for duty but not called. The Task Force Headquarters and commands having county assignments called only the people necessary to prepare plans and establish and maintain liaison with county officials. No troop call-up, as such, took place. Plans were completed and the leadership and staffs of the Adjutant General's headquarters, Division Headquarters and task force commands were completely oriented as to their provisions.

By April 4 the situation at Three Mile Island had stabilized enough for the termination of the National Guard alert. As a result, the people on duty were all released, except for one liaison officer with Dauphin County and six more from the task force headquarters. These seven were released from State Active Duty at midnight, April 5.

Although, fortunately, matters stabilized at Three Mile Island, and evacuation was not required, the National Guard, in its rapid and professional approach to this mission, instilled confidence in the minds of civil authorities that the National Guard is a ready and capable force.

Summary of Spanish-American War Service

Organization	Dates Muster-in Mt Gretna 1898	Muster-out Date	Muster-out Place	Str at Muster-in Off	Str at Muster-in EM	Str at Muster-out Off	Str at Muster-out EM	Losses: Off — Died, disease	Losses: Enlisted Men — KIA	Died, wounds	Died, disease	Died,	Drowned	Suicide	Deserted	Wounded Off	Wounded EM
Infantry																	
First	May 10-11	Oct 26, '98	Phila, Pa.	40	754	41	992	1			11	1			8		
Second	May 10-13	Nov 15, '98	Phila, Pa.	40	754	25	391				2				12		
Third	May 9-Jul 22	Oct 22, '98	Phila, Pa.	40	816	42	877				12	1			5		
Fourth	May 9-Jul 6	Nov 16, '98	Pennsylvania*	47	1,014	45	1,261	3			32						
Fifth	May 11-Jul 20	Nov 7-11, '98	Pennsylvania	41	1,022	48	1,231				16	1		1	1		
Sixth	May 10-13	Oct 17, '98	Pennsylvania	50	927	50	1,223				8				4		
Eighth	May 11-12	Mar 7, '99	Augusta, Ga.	41	774	41	949				9				4		
Ninth	May 11-Jul 12	Oct 29, '98	Pennsylvania	41	623	46	1,224	3			24						
Tenth	May 11-12	Aug 22, '99	San Francisco, Cal.	36	604	33	736	1	12	3	6					7	61
Twelfth	May 11-13	Oct 29, '98	Pennsylvania	36	604	35	829	1			20				3		
Thirteenth	May 12-13	Mar 11, '99	Augusta, Ga.	36	604	36	759				21				10		
Fourteenth	May 12	Feb 28, '99	Summerville, S.C.	36	604	35	687				2	1			8		
Fifteenth	May 10-11	Jan 31, '99	Athens, Ga.	36	604	33	675	1			4		2		4		
Sixteenth	May 10-Jul 18	Dec 22-29, '98	Pennsylvania	47	1,023	48	1,238			1	38				3		6
Eighteenth	May 11-13	Oct 22, '98	Pittsburgh, Pa.	34	604	34	837				1				2		
Cavalry																	
Phila City Trp	May 7	Nov 21, '98	Phila, Pa.	3	60	3	82				1						
Governor's Trp	May 13	Nov 21, '98	Harrisburg, Pa.	3	60	3	97										
Sheridan Trp	May 11	Nov 16, '98	Tyrone, Pa.	3	60	5	100										
Artillery																	
Light Btry A	May 6	Nov 19, '98	Phila, Pa.	3	60	4	160				4						
Light Btry B	May 8	Nov 27, '98	Pittsburgh, Pa.	3	60	4	170				1						
Light Btry C	May 6	Nov 28, '98	Phoenixville, Pa.	3	60	3	168				4						
			Totals:	619	11,696	614	14,636	10	12	4	216	4	2	1	64	7	67

SERVICE OUTSIDE THE UNITED STATES

Infantry

Fourth — Departed US for Puerto Rico on July 27, 1898 and arrived there on August 2. Left Puerto Rico on September 1 and arrived in the US on the 6th. (Aug 2 - Sep 1)

Tenth — Departed US for Philippine Islands on June 15, 1898 and arrived there July 17. Left Philippine Island July 1, 1899 and arrived in the US on August 1st. (Jul 17, 1898 - July 1, 1899)

Sixteenth — Departed US for Puerto Rico on July 22, 1898 and arrived there on the 28th. Left Puerto Rico on October 10 and arrived in the US on the 17th. (July 28 - Oct 10)

Cavalry

1st Trp — Departed US for Puerto Rico on July 28, 1898 and arrived there on August 2. Left Puerto Rico on September 3 and arrived in the US on the 10th. (Aug 2 - Sep 3)

Gov Trp — Departed US for Puerto Rico on August 5, 1898 and arrived there on the 10th. Left Puerto Rico on September 3 and arrived in the US on the 10th. (Aug 10 - Sep 3)

Sheridan Trp — Same itinerary as the Governor's Troop, above.

Artillery

Btry A — Same itinerary as the Governor's Troop, above.

Btry B — Departed US for Puerto Rico on July 28, 1898 and arrived there on August 2. Left Puerto Rico on September 8 and arrived in the US on the 15th (Aug 2 - Sep 8)

Btry C — Departed US for Puerto Rico on August 5, 1898 and arrived there on the 10th. Left Puerto Rico on September 8 and arrived in the US on the 15th. (Aug 10 - Sep 8)

* The notation "Pennsylvania" means that the command was mustered out at home station.

NOTE: The chart and information tabulated above was extracted from a chart appearing in *Pennsylvania Volunteers in the Spanish-American War, 1898*, compiled under the supervision of Thomas J. Stewart, Adjutant General (William Stanley Ray, State Printer, 1901)

APPENDIX **B**

The Private Slovik Case

Uzal W. Ent, Colonel, Hq, 28th Inf Division

ONLY IN A technical sense was the individual discussed below a part of the 28th Division, and then, only for one day. It would follow that his story should not be part of the history of a Division that counted a wartime casualty list of 26,286 men, 2,146 of whom were killed in action or died of wounds suffered in combat.

On the other hand, Private Edward Slovik and the 28th Division unfortunately figure together in the total picture of the War in Europe. In the resume below Colonel Ent summarizes what he has been able to read about the Slovik story and what he has been able to learn in discussions with some of the men who encountered him.

Perhaps Edward Donald Slovik should never have been drafted, but he was. Better known as "Eddie," Slovik was born in 1920 in a poor neighborhood of Detroit and quit school at fifteen when he was in the ninth grade. Eddie had several brushes with the law. First, in 1932, at the age of twelve, when he and some friends broke into a foundry to steal brass. He was arrested several more times between 1932 and 1937 for such offenses as petty theft, breaking and entering and disturbing the peace. He was always with associates but was never the leader. Eddie first went to jail in October 1937 for stealing change, candy, chewing gum and cigarettes from the drug store where he worked. He was paroled in September 1938. In January 1939 he and two pals got drunk, stole a car and accidentally wrecked it. Eddie was sentenced to two and a half to seven years in prison but was paroled in April 1942. Because of his prison record, he was classified 4-F and, therefore, was not eligible to be drafted.

When Eddie left prison, two good things happened: he had a job in Dearborn, and he met and married Antoinette Wisniewski. Described as a personable, good-looking young man, but easily led, he needed someone stronger to help and guide him along. Antoinette was that person.

Eddie's draft classification was changed 1-A in November 1943. In January 1944 he was drafted into the Infantry.

At training camp, he earned the reputation of being a good-hearted buddy and learned to fire a rifle—which he detested—and other military weapons. Arriving in France August 20, 1944, five days later he was assigned to Company G, 109th Infantry.

On the way to the front, his group of replacements came under fire. They dug in. Some way, he and a friend became separated from the group, who moved on in the night. Eddie and his friend "joined" the Canadian 13th Provost Corps and stayed with them until October 5. Slovik finally joined Company G on October 8 but deserted about an hour later, ignoring his friend's pleas not to leave.

After one day Eddie voluntarily surrendered to an officer of the 28th Infantry Division. He handed the officer a signed confession, stating that he had deserted and would run away again if he had "to go out their" (sic). Eddie was warned by the officer that the written confession was damaging and advised him to take it back and destroy it. Eddie refused and was then confined to the division stockade. On October 26 Slovik was offered a deal by the Division Judge Advocate, Lieutenant Colonel Henry P. Sommer, whereby the court-martial would be dropped if Eddie would go back to his unit. Eddie refused. As a result, on November 11 he was tried and convicted of desertion, although he pleaded "Not Guilty" at the trial. Because of the seriousness of the case, the court voted by secret ballot three different times. Each time the sentence of death was voted unanimously. Members of the court could not have been influenced by Eddie's police record, because they did not have that information.

On December 9 Slovik wrote a letter to General Eisenhower, pleading for clemency. No basis for clemency was found. On December 23, during the height of the Battle of the Bulge, General Eisenhower confirmed the sentence and exactly one month later ordered him executed by a firing squad from the 109th Infantry. The sentence was carried out at 10:04 on the morning of January 31, 1945. At the end, Eddie was braver against the rifles of the firing squad than he was against those of the Germans.

Private Slovik believed that he would not be executed but would be imprisoned until sometime after the war, when he could return to his beloved Antoinette. A number of factors entered into the decision. During the review of Slovik's case, his police record was included in the deliberation for clemency and counted against him. Desertion in Europe had become a problem to our Army. General Eisenhower and others felt that something had to be done about it. Lastly, the Slovik case entered the critical period of review and final action by General Eisenhower's headquarters just when the United States Army was fighting its bitterest and bloodiest campaign of the war—The Battle of the Bulge.

What many front line soldiers thought about the execution of Private Slovik was summarized by two members of his firing squad. One is reported to have said: "I got no sympathy for the sonofabitch! He deserted us didn't he? He didn't give a damn how many of us got the hell shot out of us, why should we care about him?"

Another man remarked: "I personally figured that Slovik was a no-good, and that what he had done was as bad as murder."

APPENDIX C

The Centennial Celebration

Uzal W. Ent, Colonel, Hq, 28th Inf Division

MAY 19 AND 20, 1979, saw the fruition of some of the most complex plans ever devised by the 28th Infantry Division. On these two dates the Division celebrated its first century of service to the people of the Commonwealth and Nation. The 19th, termed a Day of Celebration, was highlighted by displays of American military equipment, Soviet military equipment and field military installations, as well as a review and air show, all at Indiantown Gap. This was followed by a banquet, held in the Farm Show Building that night and attended by over 1,900 members, families of members and friends of the 28th. At the conclusion of the banquet the people were treated to a military pageant featuring the history of the Division, presented by the Third Regiment (Old Guard) from Washington, D. C. They were accompanied musically by the Army Concert Band (Pershing's Own), also from Washington. The Day of Celebration was followed on May 20 by a memorial service at the Division Shrine in Boalsburg, Pennsylvania—the Day of Commemoration. Announcement of plans for publishing a history of the 28th Infantry Division later in the year was made.

The 28th Infantry Division has never undertaken any more complex a task than that of celebrating its one hundredth birthday while concurrently the staff was preparing the usual annual training and CPX plans. Planning commenced in early 1977. A Memorandum for Record, dated November 1977, outlined the general concept which was carried through on May 19-20. Subcommittees were established in this document as follows:

Banquet - 28th DISCOM; Boalsburg - 28th DivArty; *Combined Arms* (Air) *Show* - 28th Avn Bn; *Finance* - 28th Finance Co; *Public Affairs* - Public Affairs Section, Div Hq; *Invitations* - Div Hq; *Parade* - Div Hq.

The original concept included a company combined arms show. A battalion sized assault was then considered; however this was eventually reduced to a company sized operation, reinforced by artillery and helicopters. Although not all Divisional troops were to be in the review, every command was to be represented by at least the commander, staff and color guards. Some fifteen static displays were to be positioned around the periphery of the parade field. As planning proceeded, all military installations in the Harrisburg area were informed of the plans. Because the celebration was to take place during the annual Armed Forces Day Celebration (May 19), the New

Cumberland Army Depot and Ships Parts Control Center at Mechanicsburg both had displays of their own at Indiantown Gap. Recruiters from the various military services were also present. The authorities at New Cumberland, especially their Public Affairs Office, provided the Division with considerable assistance. Through their efforts, several thousand posters were made for the Division and TV slides for advertising, all at no expense to the command.

As planning progressed, the Division Commander Major General Fletcher C. Booker, Jr. saw that, since there were several major tasks, the effort had to be divided among a few leaders. Accordingly, he decreed that all activities scheduled for Fort Indiantown Gap would be under the direction of Brigadier General Gerald T. Sajer, senior Assistant Division Commander. Activities planned for Harrisburg, which included a reception for selected guests, the banquet and pageant, were all under the direction of Brigadier General Harold J. Lavell, the other Assistant Division Commander. Colonel Uzal W. Ent was assigned the function of fund raising, the preparation of a centennial medallion and plate, as well as the publication of a Division history. Publicity was initiated by Captain Gary J. Vezza, of the Division Public Affairs Section. The 28th Infantry Division Centennial Committee was formed and incorporated as a non-profit organization in order to transact business attending the completion of centennial plans. Major Allen L. Kifer, Secretary of the General Staff, became Treasurer of the Corporation. Many more officers and enlisted personnel became deeply involved in this complex undertaking. Unfortunately, not all of them can be mentioned in this history, but as this narrative unfolds, those who contributed most will be recognized.

In order to commence officially the observance of the centennial, March 12, the actual date of birth for the Division, was selected for a special ceremony. The Division Commander and his principal staff, together with the commanders of major subordinate commands, the massed colors of the Division, carried by the Command Sergeant Major of each organization, and a mounted escort from the First Troop, Philadelphia City Cavalry, formed a procession which marched from behind the Main Capitol Building in Harrisburg along Walnut Street to Third, north on Third to a position in front of the Capitol steps. At this point the flag bearers marched up the steps to form a double line of colors. When the colors were in position, the formation of commanders marched up the steps and halted just short of the entryway. The color bearers then marched into the Capitol and formed on either side of the rotunda entry. The commanders and staff followed the colors into the rotunda and stood in a mass formation between the lines of colors. This placed the Division Commander centered about twenty feet from the foot of the stairs leading from the rotunda balcony. At the appointed time Governor Richard Thornburgh and the Chairmen of both the Senate and House Military Affairs Committees, escorted by Captain James L. Walsh, of the Division Headquarters, came down the stairs and took their places at a table provided for them. Major General Richard M. Scott, the Adjutant General, Major General Nicholas P. Kafkalas, former Division Commander and Adjutant General, Major Generals Henry K. Fluck and Daniel B. Strickler, former Division Commanders, and Brigadier General Clarence D. Bell, former Deputy Adjutant General, all observed the ceremony from the rotunda steps. The Governor presented a Proclamation commemorating the Division's one hundredth birthday. Resolutions from both houses of the Legislature were also read and presented. An artillery salute to the Governor, fired by a battery of artillery located on the plaza in front of the building, unfortunately broke several windows in a vacant building across Third Street from the Capitol.

At the conclusion of ceremonies in the rotunda the officers and colors reformed outside and marched to the statue of General John F. Hartranft, the first Division Commander, where a wreath was placed. At the conclusion of the ceremony, the troops returned to the foot of the steps.

The Division was presented a flag which flew over the Nation's Capitol on March 12, 1979. It is planned to raise the flag at the Division Headquarters annually hereafter on this anniversary date.

From the outset every effort was inhibited by the lack of funds. The Department of Military Affairs early in the planning promised monetary support for rental of the Farm Show Building and associated expenses, but estimates of the total cost

Led by a color guard of troopers from the First Troop, Philadelphia City Cavalry, the Division Commander, Major General Fletcher C. Booker Jr., and (Left to right) Brigadier General Gerald T. Sajer; Brigadier General Harold J. Lavell and Colonel Uzal W. Ent head the commanders, massed colors and command sergeant majors of the Division into the rotunda of the State Capitol Building, where Governor Richard Thornburgh presented a proclamation commemorating the birth of the 28th Infantry Division. The artillery, meanwhile, prepare to fire the salute.

Sergeant Majors Kenneth C. Kelsey (left) and Michael J. Wabby (right) with the wreath which was placed at the statue of General John F. Hartranft.

came to somewhat over $175,000. Clearly, fund raising was immediately important. It was decided to strike 5,000 centennial medallions, manufacture 2,000 commemorative pewter plates and sell these. But rather large amounts of money was found to be needed to pay for this merchandise when it was delivered. Three plans for raising money were put into effect. The first of these, the Sponsorship program, provided the participant with two tickets to the banquet, a medallion, a plate and a copy of the history book, all for the donation of $100 dollars. Next was the Friendship Program. For a donation of $50 each participant in this program received a plate, a medallion and a history book. The third program was solicitation of funds from corporations. The Harrisburg office of ITT Terryphone not only made a considerable monetary gift to the Division but also spearheaded a drive which netted a donation of one half of the paper for the publication of this book from the P. H. Glatfelter Company of Spring Grove, Pennsylvania, in addition to $4,400 from ten corporations. Some $47,000 was raised by the Sponsor - Friendship programs. Plates, medallions and histories were offered for sale, bringing in some more revenue. While some cost estimates went down, others went up.

Although both the Commonwealth and Federal Governments recognized the 28th Infantry Division Centennial Committee as a non-profit entity, the Commonwealth decreed that the Division would have to charge and collect Pennsylvania Sales Tax on the sale of medallions, plates and books. However, the Department of Military Affairs has offered to purchase $10,000 worth of the division history books, which will be distributed to Pennsylvania Army National Guard units and to members of the Pennsylvania General Assembly.

The Department of Military Affairs graciously provided escort officers for distinguished guests of the Division, and Reproduction Section of the Adjutant General's office published a twenty-four page program for the Division. Non-divisional commands, such as the 165th Military Police Battalion and 154th Transportation Battalion supported the Division's centennial in a superb fashion. General Lavell organized his committee into a number of sub-committees:

Invitations:	**Captain Jonathan Myers**
Decorations:	**Captain John W. Wetterau**
Reception:	**CW4 Richard J. Spohn and CW3 Frederick B. Dayhoff**
Dinner:	**Major Thomas G. Bertz**
Entertainment:	**Captain James L. Walsh**
General Support:	**Lieutenant Colonel Henry P. Brown**
Finance:	**Major Allen L. Kifer**

Invitations were sent to the President of the United States, the Governor of Pennsylvania and General of the Army Omar N. Bradley (once a commander of the 28th). Neither the President nor Governor attended any of the functions of celebration. Advanced age and ill health prevented General Bradley's attendance.

General Sajer's task force included Major Albert B. Rutherford, Review; Major Terry L. Eckert, Static Displays; Major John L. Martin, Combined Arms Show; Major Robert E. Altier, Protocol; Major Edward A. Salter, Captain Edwin L. Rice and Captain J. Craig Nannos assisted Major Eckert. Major James Kroh was Major Altier's principal assistant. The bulk of the logistical support for the Centennial was arranged through the efforts of Major Philip Jackson and SFC Darryl Hamm. Major Robert M. Fisher and the Division Public Affairs Section provided publicity for the events.

Lieutenant Colonel Henry J. Palmieri coordinated the activities at Fort Indiantown Gap. Major Altier, assisted by Major Kroh and Captains Edward G. Gensel and James F. Crawford prepared a program for publication, assisted in raising funds and helped the effort in many other ways.

The static displays were disposed along a roadnet bordering Muir Field at Indiantown Gap on the appointed day, which dawned overcast, cool and raining. Wind, low lying clouds and intermittent rain marked the entire day until the conclusion of the review. The road next to which the displays were set up was some three miles long. These displays were working displays of equipment and troops, almost like life-like models. The men at these displays were not "standing around." Each display contained weapons and equipment with which the troops were working. The displays included:

Division Tactical Operations Center
Signal Communications Center
World War II Military Equipment
Soviet Military Equipment
Medical Clearing Station
Maintenance Company Field Site
Engineer Platoon & Equipment
Synthetic Flight Facility
Flight Facility
FA Battery & Howitzer Display
Armored Cav Platoon
Tank Platoon
Mech Infantry Platoon
Company field Kitchen
Infantry Battalion Weapons

On Friday afternoon, May 18, Major Rutherford and General Sajer, assisted by a crew of officers and men, put the finishing touches to review plans. In former years the Division had formed in a concealed position behind a hill mask across the parade field from the reviewing stand. At the appointed time they would march over the hill in mass formation, halting in front of the reviewing stand on the Ready Line. For the one hundredth anniversary, the Division put something extra into the maneuver. Some 5,000 troops of the Division, but with all the colors of each subordinate command, as well as the commanders and staffs of all battalion and higher headquarters, were massed behind a slight hill adjacent to the National Guard Flight Facility and to the right front of the reviewing stand and across the parade field from it. At the appointed time, the Division marched over this hill, approaching the reviewing stand at an acute angle. At a predetermined point, the massed troops wheeled to the left, bringing them to the front and centered on the reviewing stand. As the wheel was completed, the entire formation moved forward and halted on the Ready Line. The ceremony proceeded from that point without a flaw.

As the troops passed in review to the music of the 28th Division Band, they marched to the end of the field, turned and returned to a position on the opposite side of the parade ground from the reviewing stand and were seated in mass formation.

Within minutes after the last soldier was seated, the Combined Arms Show began, narrated by Major Luke Shade, Executive officer of the Aviation Battalion. As in an actual operation, Pathfinders, rappeling from a hovering helicopter, came into the landing zone ahead of the Infantry. They were soon followed by helicopters which delivered elements of a Rifle Company into the landing zone. Sling-loaded artillery pieces with their accompanying crews were then brought in to support the infantry's attack. Soon these guns were firing as the Infantry drove in on the enemy. As the demonstration proceeded the hitherto low clouds lifted somewhat, and a weak sun peaked tentatively through them. The break provided the aviators and airborne Pathfinders the excuse they needed to complete the show with a parachute jump by seven Pathfinders.

Following the demonstration, several CH-47, Chinook helicopters, came over the field with huge clamshell devices suspended by cables from

Some of the displays and visitors, Fort Indiantown Gap, May 19, 1979.

Troops rappel into the landing zone, followed by the Infantry in helicopters. More helicopters bring in artillery and other supporting forces.

The troops execute "Eyes Right," as the Reviewing Party troops the line, Fort Indiantown Gap, May 19, 1979.

their bellies. The clamshells contained water. These aircraft demonstrated how these devices are used in suppressing forest fires. As a finale to the aerial demonstration, some twenty-six helicopters, including every type flown by the 28th Infantry Division and attached 228th Aviation Company performed a fly-over. Over 5,000 spectators braved rainy weather to attend the celebration at the Gap. The banquet and pageant continued the celebration that night.

Just prior to the banquet a reception was held at the Harrisburg Military Post for distinguished guests of the Division. The guests were then escorted by a mounted contingent of the First Troop, Philadelphia City Cavalry, in their traditional uniforms, to the Farm Show building. Once the guests had been announced and seated the Division Colors were posted, followed by the organizational colors, carried by the Command Sergeant Majors. Music was provided by both the 28th Division Band and the 276th Army Band, a National Guard band from Philadelphia. The caterer, D. F. McCallister and Sons, of Philadelphia, served an excellent meal to the gathering. After dinner, remarks were made by General Frederick J. Kroesen, Vice Chief of Staff, United States Army.

At the conclusion of the banquet the people adjourned to the large arena where they witnessed a show and pageant presented by the Third Regiment (Old Guard) accompanied by the Army Concert Band. The show included precision drill and a pageant depicting the history of the 28th Infantry Division. On this high point of observance, the Day of Celebration ended.

May 20 dawned with a weak sun and little promise of much improvement. However, as the day progressed, no rain marred the Day of Commemoration at Boalsburg. Each year, the Society of the 28th Division, in cooperation with the 28th Infantry Division, holds memorial services at the Division Shrine in Boalsburg, normally on the third Sunday in May. From that standpoint, 1979 was no different. What was different, however, was the significance of the year—the one hundredth anniversary of America's oldest military division. A band concert by the 28th Division Band preceded the ceremony. This was followed by the posting of the colors and guidons of all Divisional units, the invocation and remarks by the Commander of the Society of the 28th Division, George Iyoob, Jr., and Major William O. Hickok V. A very moving and impressive address was given by Lieutenant General Daniel B. Strickler, a former Division Commander. He was followed by the placing of wreaths at the monuments representing the commands of the 28th, in honor of the Division's dead. The playing of "Taps" benediction and the National Anthem completed the services. In every previous ceremony the salute has been fired by a squad of riflemen. This year the salute was fired by a battery of artillery. This added touch by Colonel Robert Armstrong, in charge of the Division's arrangements, made the entire ceremony more impressive than ordinary. So ended the week-end of ceremonies observing the centennial of the 28th Infantry Division. May it continue in proud and honorable service for another century!—U.W.E.

Division Patch, March and Prayer

THE OFFICIAL DIMENSIONS OF THE KEYSTONE INSIGNIA, 28TH INFANTRY DIVISION

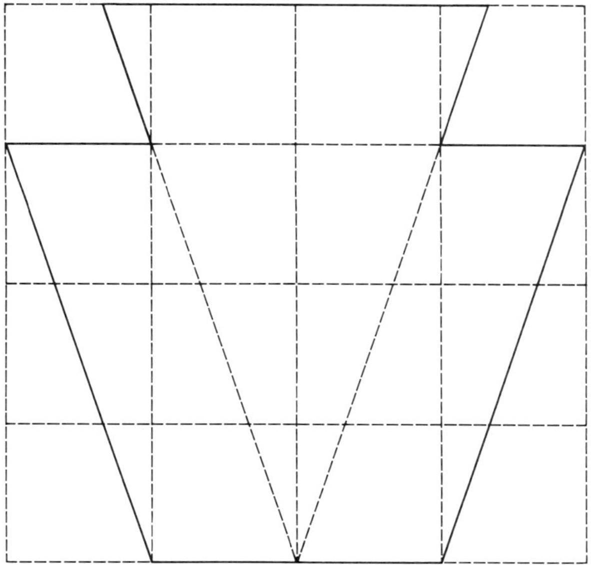

ROLL ON, 28TH

BY
SGT Emil Raab

We're the 28th Men,
And we're out to fight again
For the Good Old U.S.A.
We're the Guys who know
Where to strike the blow
And you'll know just why
After we say:

Roll on, 28th
Roll on, set the pace,
Hold the banners high
And raise the cry,
"We're off to Victory!"
Let the Keystone shine
Right down the line
For all the world to see.
When we meet the foe
We'll let them know
We're Iron Infantry,
So, Roll On, 28th,
Roll On!

DIVISION PRAYER

I

Our Father, Creator of beauty and life,
Spare this earth from wanton strife.
28th men, who through history long,
To thy Kingdom now belong.
Bless and abide with them there.

II

God in Heaven, Eternal Love,
Protect their homes with Grace from above;
Their loved ones, who Thy will endure,
Comfort their hearts, keep them secure,
Bless and abide with them here.

III

We, on earth, who await Thy call,
Help us to keep the Faith with all,
With banners high and hearts in line,
In Thy holy light may the Keystone shine,
Bless and abide with us all.

AMEN

APPENDIX E

Tables of Casualties and Awards

CASUALTIES

	WORLD WAR I	WORLD WAR II	
		OFFICERS	EM
Killed in Action	2,133	110	1,731
Died of Wounds	704	18	287
Wounded In Action	11,120	448	7,339
Injured In Action	****	68	2,038
Missing In Action	****	396*	4,937*
POW	****	7**	14**
Died, Other Causes	399		
Total Battle Casualties:	13,957	1,047	16,346
Non-battle Casualties:	399***	244	8,649
Grand Total:	14,356	1,291	24,995

* As of June 15, 1945
** Verified
*** Figures incomplete
**** No figures available

References: Page 462, Volume V, *The Twenty-Eighth Division, Pennsylvania's Guard in the World War*, COL Edward Martin, 1924;
Pages 35-36; 71; 85, *28th Division Summary of Operations in the World War*, American Battle Monuments Commission, U.S. Government Printing Office, 1944; Table of Casualties - 28th Division World War II, *Historical and Pictorial Review of the 28th Infantry Division in World War II*, Albert Love Enterprises, Atlanta, Georgia, 1945.

AWARDS

	WORLD WAR I	WORLD WAR II	TOTALS
Congressional Medal of Honor	2	1	3
Distinguished Service Cross	176	18	194
Distinguished Service Medal	20	1	21
Legion of Merit	*	8	8
Silver Star	*	359	359
Soldier's Medal	*	15	15
Bronze Star	*	2,627	2,627
Air Medal	*	101	101

FOREIGN DECORATIONS

	WORLD WAR I	WORLD WAR II	TOTALS
French			
Ordre de l'Etoile Noir	1		1
Legion of Honor	10	7	17
Croix de Guerre	105	135**	240
Military Medal	3		3
Belgium			
Order of Leopold	2		2
Order of Leopold II	3	12	15
Military Decoration 2d Class		8	8
Croix de Guerre	23	21	44
Order of the King	1		1
Military Medal	1		1
Luxembourg			
Croix de Guerre		13	13
Ordre Nationale De La Couronne De Cherre		7	7
British			
Distinguished Service Order		1	1
Distinguished Conduct Medal		1	1
Military Cross		1	1
Order of St Michael and St George	1		1
Order of the Bath	1		1
Italian			
Order of the Crown	1		1
Cross of War	1		1

* Not awarded in World War I.
** 109th Infantry awarded as a unit.

References: Table of Awards, *Historical and Pictorial Review of the 28th Infantry Division in World War II*, Albert Love Enterprises, Atlanta, Georgia, 1945; Citations, pp. 465-483, Vol V, *The Twenty-Eighth Division, Pennsylvania's Guard in the World War*, COL Edward Martin, 1924.

Pictures of Division Commanders

Major General John F. Hartranft
1879-1889

Major General George R. Snowden
1889-1900

Major General Charles Miller
1900-1906

Major General J.P.S. Gobin
1906-1907

Major General John W. Schall
1907

Major General John A. Wiley
1907-1909

Major General Wendell P. Bowman
1909-1910

Major General
Charles B. Dougherty
1910-1915

Major General Charles M. Clement
1915-1917

Major General Charles H. Muir
1917-1918

Major General William H. Hay
1918-1920

Major General
William G. Price, Jr.
May 1920 - Mar 1933

Major General Edward C. Shannon
Mar 1933- Jun 1939

Major General Edward Martin
Jun 1939 - Jan 1942

Major General J. Garesch Ord
Feb 1942 - Jun 1942

Major General Omar N. Bradley
Jun 1942 - Feb 1943

Major General Lloyd D. Brown
Feb 1943 - Aug 1944

Brigadier General
James E. Wharton*
13 Aug 1944

Major General Norman D. Cota
Aug 1944 - Dec 1945

Major General
Edward J. Stackpole
Mar 1946 - Jun 1947

Major General
Daniel B. Strickler
Jun 1947 - Nov 1952

Major General
Cortland Van R. Schuyler
Nov 1952 - Jul 1953

Major General Donald Booth
Jul 1953 - Apr 1954

Major General
Charles C. Curtis (NGUS)
Jun 1953 - Oct 1953

Major General Henry K. Fluck
Oct 1953 - Apr 1967

Major General
Nicholas P. Kafkalas
Apr 1967 - May 1977

Major General
Fletcher C. Booker, Jr.
May 1977 - Present

*Killed in Action

THE DIVISION STAFF AND PRINCIPAL COMMANDERS
28TH INFANTRY DIVISION
JUNE 1979

CG: Major General Fletcher C. Booker, Jr.
ADC: Brigadier General Gerald T. Sajer
ADC: Brigadier General Harold J. Lavell

GENERAL STAFF

Chief of Staff:	COL Uzal W. Ent
ACofS, G1:	MAJ Robert E. Altier
AcofS, G2:	LTC John F. Stever
AcofS, G3:	LTC Henry J. Palmieri
AcofS, G4:	LTC Thomas I. Brunton
AcofS, G5:	LTC Stephen J. Zubach

SPECIAL STAFF

CmlO:	LTC Santo Dettore	PM:	LTC Howard K. White
Chap:	MAJ Earl M. Brooks	AG:	LTC James B. Samuels
IG:	LTC John J. Costello	Fin O:	LTC Gethin J. Kurtz
SJA:	MAJ Paul L. Ziegler	Wea O:	LTC Elvin E. Stambaugh (PA Air NG)
PAO:	MAJ Robert M. Fisher		
Trans O:	MAJ John E. Biddle, Jr.	Surgeon:	LTC Ralph A. Hendershott
Hq Comdt:	MAJ James J. DiBella	CSM:	CSM Benjamin H. Foltz, Jr.

COMMANDERS

Hq Co, 28th Inf Div:	1LT Lawrence E. Smarr	28 MMC:	MAJ Ernest M. Shott
28 M.P. Co:	CPT Merrill D. Brant	103 Med Bn:	LTC Ralph A. Hendershott
103 Engr Bn:	LTC Robert C. McQue	228 S&T Bn:	LTC Henry P. Brown
28 Sig Bn:	LTC Gilbert R. Steele	728 Maint Bn:	LTC James B. Stodart
1-104 Cav:	LTC William C. Troxell	1-104 Cav:	LTC William C. Troxell
1-103 Armor:	LTC Bruce D. Boring	1-103 Armor:	LTC Bruce D. Boring
28 Avn Bn:	LTC George E. Breslin	28 Avn Bn:	LTC George E. Breslin
228 Avn Co (ASH):	MAJ Tommy Klutts, Jr.	228 Avn Co (ASH):	MAJ Tommy Klutts, Jr.
2d Bde:	COL Kenneth R. Craig	55 Bde:	COL Thomas A. Sharpe
1-110 Inf:	LTC John V. Aleandri	1-109 Inf (Mech):	LTC Peter Krenitsky
2-110 Inf:	LTC Peter Uram	2-109 Inf:	LTC John T. Loftus
1-112 Inf:	LTC Thomas A. Knobloch	3-109 Inf:	LTC James W. MacVay
56 Bde:	COL Vernon E. James	28 Div Arty:	COL Robert D. Armstrong
1-111 Inf:	LTC George W. Lynch	1-107 FA:	LTC Edward A. Reynolds
2-111 Inf:	LTC William D. Carter	1-108 FA:	LTC James C. Dunkelberger
2-112 Inf:	LTC Herbert F. Werner	1-109 FA:	LTC Stanley E. Smith, Jr.
		1-229 FA:	LTC John Q. Filegar
28 DISCOM:	COL Albert G. Kuhn	Btry F, 109 FA:	CPT Normand E. Thomas
28 AG Co:	CPT William R. Sanner	28th Div. Band:	WO1 Richard W. Potter
28 Fin Co:	CPT John R. Wise	276th Army Band:	WO1 Jeffry S. Twiford

Index

This is an index of selected entries, concentrating on Guard units, Guard members and place names. References to certain persons were eliminated if their name was included in listings of commanders and staffs. Non-divisional generals are not included unless they are mentioned several times in the text. National Guard units not listed separately in the index may be found in the text dealing with the various reorganizations of the Division. Liberal use is made of abbreviations in the index. Most abbreviations are well known. Others are listed herewith: AA—Anti-aircraft artillery. Acq—Acquisition. Adm—Administration. AG—Adjutant General. Ar—Armor(ed). Arty—Artillery. Aslt—Assault. BG—Brigadier General. COL—Colonel. CPL—Corporal. CPT—Captain. CW—Civil War. Div—Division. GEN—General. Ger—German. Gov—Governor. LT—Lieutenant. LTC—Lieutenant Colonel. MAJ—Major. MG—Major General. MGn—Machine gun. MMC—Material Management Center. MSG—Master Sergeant. NGP—National Guard of Pennsylvania. PI—Phillipine Islands. PNG—Pennsylvania National Guard. PR—Puerto Rico. PVT—Private. Pz—Panzer. S&S—Supply and Service. S&T—Supply and Transport. SFC—Sergeant First Class. SGM—Sergeant Major. SGT—Sergeant. SP5—Specialist Fifth Class. SSG—Staff Sergeant. Spt—Support. TD—Tank Destroyer. Tgt—Target. Tk—Tank. VG—Volks Grenadier.